the
SCIENTIFIC AMERICAN
book of the

COSMOS

the
SCIENTIFIC AMERICAN

book of the

COSMOS

DAVID H. LEVY, EDITOR

A BYRON PREISS BOOK

ST. MARTIN'S PRESS
NEW YORK

Dedication:

To a certain star I named for you, Wendee:

One hundred and twenty-six light years away,

It's called Beta Ursae Minoris, or Wendee's star.

From there to here, you brighten my life.

DL

The Scientific American Book of the Cosmos

A Byron Preiss Book
Contributing Editors: Howard Zimmerman and Dinah Dunn
Associate Editor: Dwight Jon Zimmerman
Design by Tom Draper Design
Layout by Erin Bosworth

ISBN 0-312-25453-9

First Edition: November 2000

10 9 8 7 6 5 4 3 2 1

Books are available in quantity for promotional and premium use.
Write Director of Special Sales, St. Martin's Press,
175 Fifth Avenue, New York, NY 10010, for information on discounts and terms,
or call toll-free 1-(800)-221-7945. In New York, call (212)-674-5151 (ext. 645).

TABLE OF CONTENTS

CHAPTER NINE: SPECULATION ON ENDINGS AND BEGINNINGS

PREFACE

David H. Levy

Imagine a world with continents and oceans, cumulus clouds and a blue sky, snow-capped mountains and deserts drenched in sunlight, a place where the stars come out at night—in short, a place just like Earth. Only this majestic world is not the Earth. It is a planet the same size as Earth, but instead of orbiting its sun directly, it is a moon of a much larger world—a world much larger than Jupiter that orbits a distant star near the Big Dipper called 47 Ursae Majoris.

Is there really such a world? With humanity's present technology, we cannot tell. But thanks to the work of a team led by Geoffrey Marcy, we do know that 47 Ursae Majoris, a star very much like our Sun, has a planet larger than Jupiter that orbits at approximately the same distance that the Earth orbits the Sun. It is possible that this planet has an Earth-sized moon; or maybe there is another Earth-sized world orbiting that star at the right distance. If either world is large enough, it might be capable of holding a large supply of liquid water. And if it has water, it might have life. And if it has life, could it have intelligent life?

The Scientific American Book of the Cosmos is meant for those who look up at the night sky and wonder. How does gravity work, and how does it make the universe? What needs to happen to make a star? What can we learn from the stars we see in the sky? What do they tell us about the history of our own world and of our solar system? Are the comets we occasionally see reminiscent of other comets that brought the seeds of life to Earth billions of years ago? Will one of the new comets destined to be discovered in the next few years pose a threat to the Earth?

Throughout its long history, *Scientific American* has followed the road of astronomical progress, presenting both the research and the astronomical experience through first-person articles. Many scientists in today's forefront of astronomical discovery published their first articles in *Scientific American* and still use the magazine to bone up on subjects other than their specialties. The magazine's cover story for its November 1925 issue, for example, was written by Alfred Ingalls, an editor at the time, and titled "The Heavens Declare the Glory of God." That title is taken from Psalm 19. The words were also emblazoned on the front of a then-new clubhouse called Stellar Fane, located near Springfield in southern Vermont. At the time, small telescopes were not easily affordable. What Ingalls' article did was describe and publicize Stellar Fane's goal, which was to enable thousands of people to look at the stars through telescopes they could make by themselves. One of the readers of that issue was a young farm boy named Clyde Tombaugh. Years later, Clyde told me how he followed the directions in that article and built his own 9-inch reflecting telescope to observe the heavens. He then submitted his drawings of Jupiter and Saturn to the Lowell Observatory in Flagstaff, Arizona. The senior staff there was so impressed with the young skywatcher's ingenuity and work that they hired him to conduct their search for new planets. On February 18, 1930, Clyde discovered the solar system's ninth major planet. It is fair to say that if Ingalls' 1925 article did not appear, Pluto might not have been discovered for decades.

The Scientific American Book of the Cosmos tells the story of the universe we live in through articles commissioned by and written for the magazine. Each chapter begins with a brief discussion that sets the scene and plots the map for the road ahead.

"Cosmos" is a small word with an enormous meaning. It comes from the Greek *kosmos*, which means "order," specifically order as applied to the world or the universe. Today we define the cosmos as the universe—the ordered system that encompasses all we know in science, from the subatomic quark to the giant superclusters of galaxies, as they exist now and as they evolved throughout time.

Our understanding of the cosmos has also evolved through time, and through the work of dedicated men and women who have devoted their lives to science and to telling the world of their research and discoveries. Accordingly, the essays and articles in this book that were first published in the pages of *Scientific American* have been left as they originally appeared. Newly commissioned essays in each chapter update this previously published work. As you read a chapter from beginning to end, you will gain a historical perspective on the gathering and expansion of our knowledge on a particular subject through the years, with the final essay representing the current state of our understanding.

This includes the article I wrote for *Scientific American* along with codiscoverers Gene and Carolyn Shoemaker, called "Comet Shoemaker-Levy 9 Meets Jupiter." The comet and its fate utterly dominated my life from the moment we began writing the article up until the Comet Shoemaker-Levy 9 meeting in Paris, where, atop the Eiffel Tower one evening, I proposed to my wife, Wendee. In fact, although I didn't know it at the time, the comet entered my life on a summer night in 1960, when I was looking through my first telescope for the first time. That evening the night sky was an open book, its pages filled with wondrous worlds and images that I could not begin to understand. The explanations were absent, but the feeling and the excitement of discovery were there as I looked around me, without a guide, without a star map to tell me where to look. I inserted the eyepiece and turned the telescope to the brightest point of light in the sky that night. A little focusing resulted in a ball of light with two dark bands crossing it and nearby stars accompanying it. I knew enough to exclaim to my parents that this world was Jupiter and that the nearby stars were its moons. In the starry symmetry of a distant world, I had made my first discovery.

I could not know that a future turn in my own astronomical road was hidden in the field of view that August night in 1960. Comet Shoemaker-Levy 9 was indeed there, having been orbiting Jupiter for some time, but it was far too faint for my telescope or any other telescope that night. That discovery would have to wait for 33 years, when the comet, unable to keep a safe distance from Jupiter, passed too close to the planet's cloud tops and broke up into pieces, releasing a tremendous amount of dust. Eight months later, the Shoemakers and I, as part of our photographic search for comets and asteroids, turned our telescope to an area that happened to include Jupiter, took two eight-minute-long exposures, and discovered the comet.

The finding of a comet is not enough to warrant a specific research paper. But here was a real unicorn in the cosmic zoo, a comet so badly disrupted that it looked, as one astronomer wrote, like a string of pearls in the night. First, we learned that the comet was in orbit about Jupiter, and then, on May 22, 1993, we read that astronomers had calculated that its current orbit would be its last. Fourteen months later, Comet Shoemaker-Levy 9 collided with Jupiter in the largest series of explosions ever witnessed in our solar system.

Our discovery team was asked to write a review article about the impacts several months after Shoemaker-Levy 9 collided with Jupiter. We felt that the article we had been invited to write gave us the chance to reflect on the significance of the collision before a readership consisting mostly of intelligent, nonprofessional scientists. We realized that the collision of our comet with Jupiter was epochal in the history of science. Not only did it mark the first time in the history of astronomy that a planet was subjected to the stress of collision, but this was an event eagerly witnessed by chemists, physicists, and even those interested in the origin and evolution of life. The collisions were also noted by a good part of the world's population, who could follow up-to-the-hour developments through the press or on the Internet, which in 1994 was emerging as a new way of bringing large numbers of people into the process of science.

All this excitement needed to calm somewhat before Gene and Carolyn Shoemaker and I could look at the event as the scientific bonanza it turned out to be and focus on what kind of article we wanted to compose for *Scientific American*. It would have to begin with a little history, to put the event into context. But as we got into the science, we understood that we would be writing not for the scientific community but for a variety of readers, from science students in high school and college to people with other interests who might be inspired by the science-adventure story provided by the comet and its impacts. Working with the magazine's editors, we produced the article that appeared in the magazine's August 1995 issue, about a year after the impacts ended.

The impacts reminded us that two things happen in our solar system. One, the planets orbit the Sun, and two, comets strike planets. We can see the first process every night in the sky—in fact, the ancient Greeks considered the planets (Greek for "wanderers") part of the order that defined the cosmos they understood. The second operation is something humanity had never witnessed before. During the third week of July 1994, we got a lesson in how the solar system works that would have astounded the ancient Greeks as much as it amazed us. In the early years of our solar system, cometlike bodies called planetesimals hit each other to create and build up our system of planets. After the planets were formed, comets continued to pound them, bringing with them their precious organic materials and eventually setting off the origin of life on at least one planet (ours). Other comets have changed the course of terrestrial life through the heroic forces of their collisions: 65 million years ago, a cometary collision ended the era of the dinosaurs (although some think it was an asteroid), opening the way for the evolution of mammals and, ultimately, humans. And in 1994, the breakup and collision of tiny comet Shoemaker-Levy 9 was showing us how the process worked.

The cosmos is truly a wonder to behold, whether we concentrate on its very first subatomic particle, the formation of its oldest galaxies, the panorama of the Milky Way, or a small comet that taught us so much about our solar system. Each essay in this book seeks to explain how a specific part of the cosmos was understood at the time of its writing. Supplement your reading by going outdoors to look up at the sky. One of the multitude of stars you might see, if the sky is dark enough, is 47 Ursae Majoris, that sunlike star that might conceal a world like ours. Does that world have life like ours? Have its inhabitants looked up at the stars and then written articles and books that try to make sense of it all?

On our world, *The Scientific American Book of the Cosmos* can help explain some of the details of the wondrous thing we call the cosmos. But to appreciate that wonder as a whole, close this book, go outdoors on a clear night, and look up at the stars.

Chapter One:
Some Pieces from History

Introduction

In the life of anyone interested in science, there are bound to appear certain articles that capture the heart—seminal pieces that point the way. *Scientific American* has published a number of such articles over the years, but the importance of some of them was not evident at the time of publication. In this chapter we reprint three of these articles. "The Heavens Declare the Glory of God" was the magazine's cover story for November 1925, and it started a revolution. Before that article appeared, anyone who wanted to use a telescope had to visit an observatory or have the financial wherewithal to buy a telescope. This article showed how amateur astronomers who wanted to see the universe for themselves could build a telescope by grinding their own mirrors. The article, written by *Scientific American* editor Albert Ingalls, also described a new outdoor conference getting started atop Breezy Hill, near Springfield, Vermont. Reading this, a young Kansan named Clyde Tombaugh learned about Russell Porter, an amateur astronomer who was showing the world how to make inexpensive telescopes. Thanks to Porter's enthusiasm, Tombaugh was inspired to build the telescope that led to his hiring by Lowell Observatory and his subsequent discovery of Pluto.

The essays that follow Ingalls', by Albert Einstein and Erwin Schrödinger, address two of the most important scientific achievements of the century, the evolution of our understanding of gravitation and of matter. Put together, this trio offers an introduction to the sky from the most unusual perspective of history. From Einstein we learned so much of gravity, and from Schrödinger we learned of the nature of matter. And from the hand of a *Scientific American* editor and amateur astronomer, we learned how to look at the amazing result.

The Heavens Declare the Glory of God

Albert G. Ingalls

"Why not make your own telescope?" said Mr. R. W. Porter, the telescope maker, as the waiter in a famous Broadway chop house started for the kitchen with our order. "Astronomy would mean a lot more to you if you did."

We had met to talk about Porter's hobby, astronomy. I had already heard quite a lot about this versatile man whose whole life had centered about the study of the stars. In his earlier years he had spent a dozen winters in the Arctic as astronomer, topographer and artist. Three years he had been with Peary, three more with Fiala in Franz Josef Land, and two years with Cook, who Porter says certainly did not climb Mt. McKinley. Other seasons he spent in northwestern Canada and in unknown Labrador. During all these years in the Far North where the Arctic stars fairly snap in the cold, clear air, he was studying astronomy.

Now he had settled down in the picturesque manufacturing village of Springfield, Vermont, tucked away in a deep valley in the foothills of the Green Mountains, where, as everyone in the mechanical industry knows, a famous type of flat turret lathe is made. Here he had fired a score of men with his own keen enthusiasm for the stars and had organized them into a group which is perhaps unique—machinists by day, amateur astronomers by night.

"You'd have no trouble in making a good telescope," he assured me.

"I could make the mounting all right," I replied, "but when it came to making the optical parts I'd be out of it. Only a handful of men in the world are skilled enough to do that fine work."

"You come up to Springfield, where I live," he laughed, "and I'll show you a good many home-made telescopes, made in spare time by men who knew nothing about it when they began. They'll tell you how any amateur—even an editor—can make his own telescope for less than fifty dollars, providing he's reasonably handy and will take pains. And it will be a real telescope fit for serious work, not just a toy or a makeshift."

THE "POOR MAN'S TELESCOPE"

He went on to tell me how in the Vermont village a group of men, most of them mechanics in the local machine shops, had banded together to study the stars; how each one had made and mounted his concave mirror; how they had later pooled their efforts and built a sort of combined clubhouse, lodge and observatory on the top of a mountain near their homes. Here they gathered when the week's work was done, to study the stars. "The Telescope Makers of Springfield," they call their club, and none may join who has not made his own telescope.

When summer rolled around, I went to Springfield, as Porter had suggested, and there the amateur astronomers told me how they had learned their new avocation.

There are two common types of telescopes, the refractor and the reflector. The refractor is the ordinary type that everyone knows. It is like a big spyglass; you look *through* it, the light actually passing through its lenses. For serious amateur work such a telescope, having an objective lens four inches in diameter, is very valuable, but it costs several hundred dollars to buy, while the ordinary amateur cannot hope to make it himself.

But the reflector works on a different principle. It is a shorter, thicker instrument having a large, round, concave mirror in its lower end. The light coming from a star strikes this concave mirror and is reflected upward in a converging cone. Near the upper end of the big tube, which is open at the top, a small diagonal mirror or sometimes a three-sided prism of glass is mounted in such a position that the cone of light reflected by the large mirror is intercepted and is turned at right angles toward the eyepiece in the side of the telescope. Owing to the fact that the light does not pass through the glass as in the other type of telescope, the mirror does not have to be made of optical glass—simply ordinary thick plate glass; and since the mounting of the telescope does not have to be very accurately constructed, this type of telescope may be made for fifty dollars or even less. Therefore it is called "the poor man's telescope." One having a six-inch mirror will magnify from 100 to 200 diameters, and more in transparent atmosphere, and will do really effective astronomical work.

The Springfield amateurs set to work enthusiastically, and before many weeks most of them had surprised themselves by making the most difficult part, the mirror. The best work was done by the elderly men of the group, for they proved to be most patient and painstaking and did not try to rush the job through. The only feminine member turned out an excellent mirror, without a scratch on its polished surface.

When the telescopes were completed the back yards of Springfield bristled like Mt. Wilson, the California mecca of astronomers, with heaven-pointed instruments. This was great fun, but the observers soon discovered that they were missing a lot because they and their telescopes were scattered. They were not within talking distance of one another.

Several expeditions to neighboring peaks resulted, the would-be astronomers and their wives, telescopes, coffee pots, frying pans and bean kettles, all partaking together. But shivering, shelterless nights on windy mountain tops set the telescope makers planning further. Why not buy one of these peaks, they asked themselves, and build a shelter on it, with a warm fireplace, cots and a kitchen, as well as a place to store the heavy telescopes when not in use? Thus resulted Stellar Fane, "The Temple of the Stars."

The Saturday afternoon of my visit we climbed the mountain in cars piled high with provisions, for at least half the fun in one of these astronomical jaunts to Stellar Fane is the gathering of the observers about the long board and the stowing away of acres of johnny cake and other good things prepared by one of the members, Mr. Redfield, the duly appointed "cook-laureate" of the club. His double title is due to the fact that with his edibles he also serves up poetry.

From the highway we passed through an ancient, rustic gate and churned our way spasmodically up across a boulder-strewn slope, and then up a steeper pitch. The radiator boiled furiously. The upper half of the peak was clad with virgin forest of birch and beech and black spruce, so that I got no glimpse of Stellar Fane until at last we came out on a level clearing at the summit. There, enclosed in a semicircle of trees was the Fane, a bizarre little house with steeply sloping roof anchored to the solid rock at all four corners by means of steel cables in order to keep it from blowing entirely off the mountain.

A CLOSE-UP OF THE SUN

All around the north and west horizon stood a ring of wooded mountain peaks, thin blue in the distance and as untouched as the day before man was man. Not a sound came up from the world below to annoy the star lovers in their lofty retreat.

"That peak over there is Ascutney," Porter explained, "and just behind that ridge is the place where President Coolidge grew up. But let's go inside and look around—we've got some things in there that may interest you."

In the front of the building there was a long room, finished in gray-stained pine, timbers naked. On the walls were a few pictures of the moon and other celestial bodies. There were several astronomical drawings, and a small blackboard was built into one corner for use in demonstrating disputed points raised by the amateurs. Sundry books on astronomy were tucked into odd niches in the walls. A folding staircase led aloft somewhere. A massive, homemade table was decorated with sawed-out signs of the zodiac. One end of the room was crowded with reflecting telescopes of various shapes and sizes, waiting to be dragged out by their owners and set up on selected spots nearby for the night's observation.

In the rear was a complete kitchen, with a workbench at one end for quick repairs to damaged telescopes. Upstairs were two rooms, one packed with cots, the other used for the solar telescope.

"While the sun is still up," said Porter, "let's set up the solar telescope." He drew out a big flat round mirror and attached it to a heavy bracket just outside the window opening. This mirror reflects the sun's light to another mirror on the ground seventy–five feet distant. There is no telescope tube in this type of instrument, for none is necessary.

With a wheel and worm gear the flat mirror was moved into proper relation with the sun and the concave mirror back of the Fane, when suddenly a powerful shaft of sunlight bored into the darkened room and the beautiful, silvery image of the sun appeared on a perpendicular screen. Each separate sunspot and every prominent detail showed sharply and clearly.

"We gather around this screen," said Porter, "while one of us keeps the sun's image centered on it with this wheel. We can study old Sol's face here in comfort and with precision, and at night we can see the moon, too, but not so vividly."

By the time we got downstairs again the telescope makers had set up their instruments for the night's vigil.

"This fat, wooden one is Mr. Fullam's," said John Pierce, the vocational teacher at the Springfield School. "Fullam is a pattern maker, so naturally he used wood for his mounting, and it has proved very satisfactory."

"Here's Marshall," said the cook-laureate, "let's have a look at his telescope. He's a foreman in the shops."

Marshall's telescope is unique. Its main feature is its ever-upright eyepiece. With ordinary telescopes one often has to take up very awkward and iresome positions to see stars directly overhead, but with a telescope like Marshall's you always look down into the eyepiece much as if it were a microscope.

The light reflected by the eight-inch main mirror is intercepted several inches short of its focus by means of a prism of glass which turns it through an angle of ninety degrees to a second prism, and this in turn turns it another ninety degrees into the eyepiece. Powers up to 560 diameters are available for exceptionally clear nights, though the 140 power is usually used. Provision is made for a driving clock in the turret, which is mounted on seventy-two steel balls. Marshall's telescope was a thoroughly workmanlike job which took him two weeks to finish.

By the time I had inspected everything in sight and taken some photographs, it was dark. We all sat down at the long table inside and "stoked up" for the night with the cook-laureate's excellent provisions.

"There'll be another feed or two during the night," said one of the men, "for when we're not star gazing we're always eating."

One by one the telescope makers drifted away from the table, as I sat talking with a professor from a New England university who had motored over to visit Stellar Fane. Someone touched my arm.

The original caption, from the 1925 issue of *Scientific American*: On the spruce-clad summit of Breezy Mountain the Vermont astronomers have built their stellar fane. A seventy-five foot solar telescope projects the Sun's image on a screen indoors. In this type of telescope, no tube is required. The light from the Sun is reflected by the sixteen-inch, flat, pivoted mirror (B), to an equally large concave, paraboloidad mirror mounted on a stone pier at (A). Thence the light converges through a circular opening just above the first mirror, and focusses on the screen, (C), where the amateurs study the Sun's image.

"Come out and have a peep at Saturn," said Marshall, disappearing into the night. I followed him. There was Saturn, looking just like the Saturn of the pictures, but far more beautiful. Even the narrow Cassini division between the two pearly rings was clearly visible. Pretty good for an amateur's first telescope, is it not?

YOU CAN MAKE A TELESCOPE

Marshall now turned his telescope on Jupiter, revealing four of its satellites, tiny yellow balls whirling around the parent planet. Then we hunted up a spiral nebula, setting the two graduated circles on the telescope for the exact number of degrees and minutes called for in the ephemeris, the book which is used for locating the stars. With its spiral structure looking like a whirling pinwheel the nebula stood out sharp and clear, a whole universe of suns, distant so far that the light had required a million years to travel the 6,000,000,000,000,000,000 miles from it to our eye.

The night grew chill. Inside, a fire burned cheerfully on the hearth and someone had found a big, flat pan of johnny cake—"about a square mile of it," one remarked. I thought they ought to measure across Redfield's enormous pans in astronomer's terms, by light years.

I was getting dozy but someone brewed some heavy, black coffee as strong as dynamite, and guaranteed it to break up all desire for sleep. Pretty soon—for it was June with its short nights—the birds were chirping in the trees. Faint dawn.

"You haven't seen the moon yet," someone put in, "and it's just rising above the trees." Damon's six-inch telescope brought the moon's yellow face, now turning in the dawn to gray, right up into Vermont.

The lunar landscape forms a striking telescopic study. Men become so interested in its infinitude of minute detail that they spend years, nightly inspecting its volcanoes, craterlets, clefts, ridges, ramparts, rills, terraces, cracks, fault lines and cliffs, all of which look different from night to night as the sunlight strikes the moon at different angles.

So ended the night at Stellar Fane, and breakfast came on with the sun. A Vermont breakfast since time immemorial is traditionally incomplete unless topped off with pie. We didn't have pie, we had strawberry shortcake! Some folks say that men, when they are alone, will not bother to fix up fancy things to eat. They ought to put in a night at Stellar Fane.

The next day we visited the workroom in Springfield, where some of the telescope mirrors were made. We silvered a telescope mirror, ten inches in diameter—at least I watched Porter do it. This job, often said to be tedious, took only twenty-five minutes. It requires a few chemicals, not many, and a willingness to take pains and to follow directions minutely. In fact, from those with whom I talked and from work which I have subsequently done I have gathered that the whole art of making telescopes is pretty much a matter of taking pains. You must be handy, of course, but you do not have to be a genius. Patience is necessary, but no knowledge of mathematics, abstruse science or astronomy itself is required for telescope making.

The tools are simply a barrel to work on, two inexpensive plate glass disks, a bit of common pitch, half a dollar's worth of optical rouge, a very few household tools, about four dollars' worth of abrasive, and your two hands to keep the upper disk moving back and forth over the lower one.

On the Generalized Theory of Gravitation

Albert Einstein

The editors of *Scientific American* have asked me to write about my recent work which has just been published. It is a mathematical investigation concerning the foundations of field physics.

Some readers may be puzzled: Didn't we learn all about the foundations of physics when we were still at school? The answer is "yes" or "no," depending on the interpretation. We have become acquainted with concepts and general relations that enable us to comprehend an immense range of experiences and make them accessible to mathematical treatment. In a certain sense these concepts and relations are probably even final. This situation is true, for example, of the laws of light refraction, of the relations of classical thermodynamics as far as it is based on the concepts of pressure, volume, temperature, heat and work, and of the hypothesis of the nonexistence of a perpetual motion machine.

What, then, impels us to devise theory after theory? Why do we devise theories at all? The answer to the latter question is simply because we enjoy "comprehending," that is, reducing phenomena by the process of logic to something already known or (apparently) evident. New theories are first of all necessary when we encounter new facts that cannot be "explained" by existing theories. But this motivation for setting up new theories is, so to speak, trivial, imposed from without. There is another, more subtle motive of no less importance. This is the striving toward unification and simplification of the premises of the theory as a whole (that is, Mach's principle of economy, interpreted as a logical principle).

There exists a passion for comprehension, just as there exists a passion for music. That passion is rather common in children but gets lost in most people later on. Without this passion, there would be neither mathematics nor natural science. Time and again the passion for understanding has led to the illusion that man is able to comprehend the objective world rationally, by pure thought, without any empirical foundations—in short, by metaphysics. I believe that every true theorist is a kind of tamed metaphysicist, no matter how pure a "positivist" he may fancy himself. The metaphysicist believes that the logically simple is also the real. The tamed metaphysicist believes that not all that is logically simple is embodied in experienced reality but that the totality of all sensory experience can be "comprehended" on the basis of a conceptual system built on

13

premises of great simplicity. The skeptic will say that this is a "miracle creed." Admittedly so, but it is a miracle creed which has been borne out to an amazing extent by the development of science.

The rise of atomism is a good example. How may Leucippus have conceived this bold idea? When water freezes and becomes ice—apparently something entirely different from water—why is it that the thawing of the ice forms something that seems indistinguishable from the original water? Leucippus is puzzled and looks for an "explanation." He is driven to the conclusion that in these transitions the "essence" of the thing has not changed at all. Maybe the thing consists of immutable particles, and the change is only a change in their spatial arrangement. Could it not be that the same is true of all material objects which emerge again and again with nearly identical qualities?

This idea is not entirely lost during the long hibernation of occidental thought. Two thousand years after Leucippus, Bernoulli wonders why gas exerts pressure on the walls of a container. Should this phenomenon be "explained" by mutual repulsion of the parts of the gas, in the sense of Newtonian mechanics? This hypothesis appears absurd, for the gas pressure depends on the temperature, all other things being equal. To assume that the Newtonian forces of interaction depend on temperature is contrary to the spirit of Newtonian mechanics. Since Bernoulli is aware of the concept of atomism, he is bound to conclude that the atoms (or molecules) collide with the walls of the container and in doing so exert pressure. After all, one has to assume that atoms are in motion; how else can one account for the varying temperature of gases?

A simple mechanical consideration shows that this pressure depends only on the kinetic energy of the particles and on their density in space. This should have led the physicists of that age to the conclusion that heat consists in random motion of the atoms. Had they taken this consideration as seriously as it deserved to be taken, the development of the theory of heat—in particular the discovery of the equivalence of heat and mechanical energy—would have been considerably facilitated.

This example is meant to illustrate two things. The theoretical idea (atomism in this case) does not arise apart from and independent of experience; nor can it be derived from experience by a purely logical procedure. It is produced by a creative act. Once a theoretical idea has been acquired, one does well to hold fast to it until it leads to an untenable conclusion.

As for my latest theoretical work, I do not feel justified in giving a detailed account of it before a wide group of readers interested in science. That should be done only with theories which have been adequately confirmed by experience. So far it is primarily the simplicity of its premises and its intimate connection with what is already known (namely, the laws of the pure gravitational field) that speak in favor of the theory to be

discussed here. It may, however, be of interest to a wide group of readers interested in science to become acquainted with the train of thought that can lead to endeavors of such an extremely speculative nature. Moreover, it will be shown what kinds of difficulties are encountered and in what sense they have been overcome.

In Newtonian physics the elementary theoretical concept on which the theoretical description of material bodies is based is the material point, or particle. Thus, matter is considered a priori to be discontinuous. This assumption makes it necessary to consider the action of material points on one another as "action at a distance." Because the latter concept seems quite contrary to everyday experience, it is only natural that the contemporaries of Newton—and indeed Newton himself—found it difficult to accept. Because of the almost miraculous success of the Newtonian system, however, the succeeding generations of physicists became accustomed to the idea of action at a distance. Any doubt was buried for a long time to come.

But when, in the second half of the 19th century, the laws of electrodynamics became known, it turned out that these laws could not be satisfactorily incorporated into the Newtonian system. It is fascinating to muse: Would Faraday have discovered the law of electromagnetic induction if he had received a regular college education?

Unencumbered by the traditional way of thinking, he felt that the introduction of the "field" as an independent element of reality helped him to coordinate the experimental facts. It was Maxwell who fully comprehended the significance of the field concept; he made the fundamental discovery that the laws of electrodynamics found their natural expression in the differential equations for the electric and magnetic fields. These equations implied the existence of waves, whose properties corresponded to those of light as far as they were known at that time. This incorporation of optics into the theory of electromagnetism represents one of the greatest triumphs in the striving toward unification of the foundations of physics. Maxwell achieved this unification by purely theoretical arguments, long before it was corroborated by Hertz's experimental work. The new insight made it possible to dispense with the hypothesis of action at a distance, at least in the realm of electromagnetic phenomena; the intermediary field now appeared as the only carrier of electromagnetic interaction between bodies, and the field's behavior was completely determined by contiguous processes, expressed by differential equations.

Now a question arose: Since the field exists even in a vacuum should one conceive of the field as a state of a "carrier," or should it rather be endowed with an independent existence not reducible to anything else? In other words, is there an "ether" which carries the field; the ether being considered in the undulatory state, for example, when it carries light waves?

The question has a natural answer: because one cannot dispense with the field concept, it is preferable not to introduce in addition a carrier with hypothetical properties. However, the pathfinders who first recognized the indispensability of the field concept were still too strongly imbued with the mechanistic tradition of thought to accept unhesitatingly this simple point of view. But during the course of the following decades this view imperceptibly took hold.

The introduction of the field as an elementary concept gave rise to an inconsistency of the theory as a whole. Maxwell's theory, although adequately describing the behavior of electrically charged particles in their interaction with one another, does not explain the behavior of electrical densities, that is, it does not provide a theory of the particles themselves. They must therefore be treated as mass points on the basis of the old theory. The combination of the idea of a continuous field with the conception of material points discontinuous in space appears inconsistent. A consistent field theory requires continuity of all elements of the theory, not only in time but also in space and in all points of space. Hence, the material particle has no place as a fundamental concept in a field theory. Thus, even apart from the fact that gravitation is not included, Maxwell's electrodynamics cannot be considered a complete theory.

Maxwell's equations for empty space remain unchanged if the spatial coordinates and the time are subjected to a particular kind of linear transformations—the Lorentz transformations ("covariance" with respect to Lorentz transformations). Covariance also holds, of course, for a transformation composed of two or more such transformations; this is called the "group" property of Lorentz transformations.

Maxwell's equations imply the "Lorentz group," but the Lorentz group does not imply Maxwell's equations. Indeed, it is possible to redefine the Lorentz group independently of Maxwell's equations as a group of linear transformations that leave a particular value of the velocity—the velocity of light—invariant. These transformations hold for the transition from one "inertial system" to another which is in uniform motion relative to the first. The most conspicuous novel property of this transformation group is that it does away with the absolute character of the concept of simultaneity of events distant from one another in space. On this account it is to be expected that all equations of physics are covariant with respect to Lorentz transformations (special theory of relativity). Thus it came about that Maxwell's equations led to a heuristic principle valid far beyond the range of the applicability or even validity of the equations themselves.

Special relativity has this in common with Newtonian mechanics: the laws of both theories are supposed to hold only with respect to certain coordinate systems—those known as inertial systems. An inertial system is a system in a state of motion such that "force-free" material points within it are not accelerated with respect to the coordinate system. Yet this definition is empty if there is no independent means for recognizing the absence of forces. But such a means of recognition does not exist if gravitation is considered as a "field."

Let A be a system uniformly accelerated with respect to an inertial system I. Material points, not accelerated with respect to I, are accelerated with respect to A, the acceleration of all the points being equal in magnitude and direction. They behave as if a gravitational field exists with respect to A, for it is a characteristic property of the gravitational field that the acceleration is independent of the particular nature of the body.

There is no reason to exclude the possibility of interpreting this behavior as the effect of a "true" gravitational field (*principle of equivalence*). This interpretation implies that A is an inertial system, even though it is accelerated with respect to another inertial system. (It is essential for this argument that the introduction of independent gravitational fields is considered justified even though no masses generating the field are defined. Therefore, to Newton such an argument would not have appeared convincing.)

Thus, the concepts of inertial system, the law of inertia and the law of motion are deprived of their concrete meaning—not only in classical mechanics but also in special relativity. Moreover, following up this train of thought, it turns out that with respect to A time cannot be measured by identical clocks; indeed, even the immediate physical significance of coordinate differences is generally lost. In view of all these difficulties, should one not try, after all, to hold on to the concept of the inertial system, relinquishing the attempt to explain the fundamental character of the gravitational phenomena that manifest themselves in the Newtonian system as the equivalence of inert and gravitational mass? Those who trust in the comprehensibility of nature must answer: No.

This is the gist of the principle of equivalence: in order to account for the equality of inert and gravitational mass within the theory, it is necessary to admit nonlinear transformations of the four coordinates. That is, the group of Lorentz transformations, and hence the set of the "permissible" coordinate systems, has to be extended.

What group of coordinate transformations can then be substituted for the group of Lorentz transformations? Mathematics suggests an answer which is based on the fundamental investigations of Gauss and Riemann: namely, that the appropriate substitute is the group of all continuous (analytical) transformations of the coordinates. Under these transformations the only thing that remains invariant is the fact that neighboring points have nearly the same coordinates; the coordinate system expresses only the topological order of the points in space (including its four-dimensional character). The equations expressing the laws of nature must be covariant with respect to all continuous transformations of the coordinates. This is the principle of general relativity.

The procedure just described overcomes a deficiency in the foundations of mechanics that had already been noticed by Newton and was criticized by Leibnitz and, two centuries later, by Mach; inertia resists acceleration, but acceleration relative to what? Within the frame of classical mechanics the only answer is: inertia resists acceleration relative to space. This is a physical property of space—space acts on objects, but objects do not act on space. Such is probably the deeper meaning of Newton's assertion *spatium est absolutum* (space is absolute). But the idea disturbed some, in particular Leibnitz, who did not ascribe an independent existence to space but considered it merely a property of "things" (contiguity of physical objects). Had his justified doubts won out at that time, it hardly would have been a boon to physics, inasmuch as the empirical and theoretical foundations necessary to follow up his idea were not available in the 17th century.

According to general relativity, the concept of space detached from any physical content does not exist. The physical reality of space is represented by a field whose components are continuous functions of four independent variables—the coordinates of space and time. It is just this particular kind of dependence that expresses the spatial character of physical reality.

Since the theory of general relativity implies the representation of physical reality by a *continuous field*, the concept of particles or material points cannot play a fundamental part, and neither can the concept of motion. The particle can only appear as a limited region in space in which the field strength or the energy density is particularly high.

A relativistic theory has to answer two questions: namely, What is the mathematical character of the field? and What equations hold for this field?

Concerning the first question: From the mathematical point of view, the field is essentially characterized by the way its components transform if a coordinate transformation is applied. Concerning the second question: The equations must determine the field *to a sufficient extent* while satisfying the postulates of general relativity. Whether or not this requirement can be satisfied depends on the choice of the field type.

The attempt to comprehend the correlations among the empirical data on the basis of such a highly abstract program may at first appear almost hopeless. The procedure amounts, in fact, to putting the question: What most simple property can be required from what most simple object (field) while preserving the principle of general relativity? Viewed from the standpoint of formal logic, the dual character of the question appears calamitous, quite apart from the vagueness of the concept "simple." Moreover, from the standpoint of physics, there is nothing to warrant the assumption that a theory that is "logically simple" should also be "true."

Yet every theory is speculative. When the basic concepts of a theory are comparatively "close to experience" (for example, the concepts of force, pressure, mass), its speculative character is not so easily discernible. If, however, a theory is such as to require the application of complicated logical processes in order to reach conclusions from the premises that can be confronted with observation, everybody becomes conscious of the speculative nature of the theory. In such a case an almost irresistible feeling of aversion arises in people who are inexperienced in epistemological analysis and who are unaware of the precarious nature of theoretical thinking in those fields with which they are familiar.

On the other hand, it must be conceded that a theory has an important advantage if its basic concepts and fundamental hypotheses are "close to experience," and greater confidence in such a theory is certainly justified. There is less danger of going completely astray, particularly since it takes so much less time and effort to disprove such theories by experience. Yet more and more, as the depth of our knowledge increases, we must give up this advantage in our quest for logical simplicity and uniformity in the foundations of physical theory. It has to be admitted that general relativity has gone further than previous physical theories in relinquishing "closeness to experience" of fundamental concepts in order to attain logical simplicity. This holds already for the theory of gravitation, and it is even more true of the new generalization, which is an attempt to comprise the properties of the total field. In the generalized theory the procedure of deriving from the premises of the theory conclusions that can be confronted with empirical data is so difficult that so far no such result has been obtained. In favor of this theory are, at this point, its logical simplicity and its "rigidity." Rigidity means here that the theory is either true or false, but not modifiable.

The greatest inner difficulty impeding the development of the theory of relativity is the dual nature of the problem, indicated by the two questions we have asked. This duality is the reason why the development of the theory has taken place in two steps so widely separated in time. The first of these steps, the theory of gravitation, is based on the principle of equivalence discussed above and rests on the following consideration: According to the theory of special relativity, light has a constant velocity of propagation. If a light ray in a vacuum starts from a point, designated by the coordinates x_1, x_2 and x_3 in a three-dimensional coordinate system, at the time x_4, it spreads as a spherical wave and reaches a neighboring point $(x_1 + dx_1, x_2 + dx_2, x_3 + dx_3)$ at the time $x_4 + dx_4$. Introducing the velocity of light, c, we write the expression:

$$\sqrt{dx_1{}^2 + dx_2{}^2 + dx_3{}^2} = c\,dx_4$$

This can also be written in the form:

$$dx_1{}^2 + dx_2{}^2 + dx_3{}^2 - c^2 dx_4{}^2 = 0$$

This expression represents an objective relation between neighboring space-time points in four dimensions, and it holds for all inertial systems, provided the coordinate transformations are restricted to those of special relativity. The relation loses this form, however, if arbitrary continuous transformations of the coordinates are admitted in accordance with the principle of general relativity. The relation then assumes the more general form:

$$\sum_{ik} g_{ik}\, dx_i\, dx_k = 0$$

The g_{ik} are certain functions of the coordinates which transform in a definite way if a continuous coordinate transformation is applied. According to the principle of equivalence, these g_{ik} functions describe a particular kind of gravitational field: a field that can be obtained by transformation of "field-free" space. The g_{ik} satisfy a particular law of transformation. Mathematically speaking, they are the components of a "tensor" with a property of symmetry which is preserved in all transformations; the symmetrical property is expressed as follows:

$$g_{ik} = g_{ki}$$

The idea suggests itself: May we not ascribe objective meaning to such a symmetrical tensor, even though the field *cannot* be obtained from the empty space of special relativity by a mere coordinate transformation? Although we cannot expect that such a symmetrical tensor will describe the most general field, it may well describe the particular case of the "pure gravitational field." Thus it is evident what kind of field, at least for a special case, general relativity has to postulate: a symmetrical tensor field.

Hence, only the second question is left: What kind of general covariant field law can be postulated for a symmetrical tensor field?

This question has not been difficult to answer in our time, since the necessary mathematical conceptions were already at hand in the form of the metric theory of surfaces, created a century ago by Gauss and extended by Riemann to manifolds of an arbitrary number of dimensions. The result of this purely formal investigation has been amazing in many respects. The differential equations that can be postulated as field law for g_{ik} cannot be of lower than second order, that is, they must at least contain the second derivatives of the g_{ik} with respect to the coordinates. Assuming that no higher than second derivatives appear in the field law, *it is mathematically determined by the principle of general relativity.* The system of equations can be written in the form:

$R_{ik} = 0$

The R_{ik} transform in the same manner as the g_{ik}, that is, they too form a symmetrical tensor.

These differential equations completely replace the Newtonian theory of the motion of celestial bodies, provided the masses are represented as singularities of the field. In other words, they contain the law of force as well as the law of motion while eliminating inertial systems.

The fact that the masses appear as singularities indicates that these masses themselves cannot be explained by symmetrical g_{ik} fields, or "gravitational fields." Not even the fact that only *positive* gravitating masses exist can be deduced from this theory. Evidently a complete relativistic field theory must be based on a field of more complex nature, that is, a generalization of the symmetrical tensor field.

Before considering such a generalization, two remarks pertaining to gravitational theory are essential for the explanation to follow.

The first observation is that the principle of general relativity imposes exceedingly strong restrictions on the theoretical possibilities. Without this restrictive principle it would be practically impossible for anybody to hit on the gravitational equations, not even by using the principle of special relativity, even though one knows that the field has to be described by a symmetrical tensor. No amount of collection of facts could lead to these equations unless the principle of general relativity were used.

This is the reason why all attempts to obtain a deeper knowledge of the foundations of physics seem doomed to me unless the basic concepts are in accordance with general relativity from the beginning. This situation makes it difficult to use our empirical

knowledge, however comprehensive, in looking for the fundamental concepts and relations of physics, and it forces us to apply free speculation to a much greater extent than
is currently assumed by most physicists.

I do not see any reason to assume that the heuristic significance of the principle of general relativity is restricted to gravitation and that the rest of physics can be dealt with
separately on the basis of special relativity, with the hope that later on the whole may
be fitted consistently into a general relativistic scheme. I do not think that such an atti-
tude, although historically understandable, can be objectively justified. The comparative smallness of what we know today as gravitational effects is not a conclusive reason
for ignoring the principle of general relativity in theoretical investigations of a fundamental character. In other words, I do not believe that it is justifiable to ask: What
would physics look like without gravitation?

The second point we must note is that the equations of gravitation are 10 differential
equations for the 10 components of the symmetrical tensor g_{ik}. In the case of a non-
general relativistic theory, a system is ordinarily not overdetermined if the number of
equations is equal to the number of unknown functions. The manifold of solutions is
such that within the general solution a certain number of functions of three variables
can be chosen arbitrarily. For a general relativistic theory, this cannot be expected as a
matter of course. Free choice with respect to the coordinate system implies that out of
the 10 functions of a solution, or components of the field, four can be made to assume
prescribed values by a suitable choice of the coordinate system. In other words, the
principle of general relativity implies that the number of functions to be determined by
differential equations is not 10 but 10 - 4 = 6. For these six functions only six independent differential equations may be postulated. Only six out of the 10 differential
equations of the gravitational field ought to be independent of one another, while the
remaining four must be connected to those six by means of four relations (identities).
And indeed there exist among the left-hand sides, R_{ik}, of the 10 gravitational equations
four identities—known as "Bianchi's identities"—which assure their "compatibility."

In a case like this—when the number of field variables is equal to the number of differential equations—compatibility is always assured if the equations can be obtained
from a variational principle. This is indeed the case for the gravitational equations.

However, the 10 differential equations cannot be entirely replaced by six. The system
of equations is indeed "overdetermined," but because of the existence of the identities
it is overdetermined in such a way that its compatibility is not lost, that is, the manifold
of solutions is not critically restricted. The fact that the equations of gravitation imply
the law of motion for the masses is intimately connected with this (permissible) overde-
termination.

After this preparation it is now easy to understand the nature of the present investigation without entering into the details of its mathematics. The problem is to set up a relativistic theory for the total field. The most important clue to its solution is that there exists already the solution for the special case of the pure gravitational field. The theory we are looking for must therefore be a generalization of the theory of the gravitational field. The first question is: What is the natural generalization of the symmetrical tensor field?

This question cannot be answered by itself, but only in connection with the other question: What generalization of the field is going to provide the most natural theoretical system? The answer on which the theory under discussion is based is that the symmetrical tensor field must be replaced by a nonsymmetrical one. This change means that the condition $g_{ik} = g_{ki}$ for the field components must be dropped. In that case the field has 16 instead of 10 independent components.

There remains the task of setting up the relativistic differential equations for a nonsymmetrical tensor field. In the attempt to solve this problem, one meets with a difficulty that does not arise in the case of the symmetrical field. The principle of general relativity does not suffice to determine completely the field equations, mainly because the transformation law of the symmetrical part of the field alone does not involve the components of the antisymmetrical part or vice versa. Probably this is the reason why this kind of generalization of the field has hardly ever been tried before. The combination of the two parts of the field can only be shown to be a natural procedure if in the formalism of the theory only the total field plays a role, and not the symmetrical and antisymmetrical parts separately.

It turned out that this requirement can indeed be satisfied in a natural way. But even this requirement, together with the principle of general relativity, is still not sufficient to determine uniquely the field equations. Let us remember that the system of equations must satisfy a further condition: the equations must be compatible. It has been mentioned above that this condition is satisfied if the equations can be derived from a variational principle.

This has indeed been achieved, although not in so natural a way as in the case of the symmetrical field. It has been disturbing to find that it can be achieved in two different ways. These variational principles furnished two systems of equations—let us denote them by E_1 and E_2—which were different from each other (although only slightly so), each of them exhibiting specific imperfections. Consequently, even the condition of compatibility was insufficient to determine the system of equations uniquely.

It was, in fact, the formal defects of the systems E_1 and E_2 that indicated a possible way out. There exists a third system of equations, E_3, which is free of the formal defects of the systems E_1 and E_2 and represents a combination of them in the sense that every solution of E_3 is a solution of E_1 as well as of E_2. This suggests that E_3 may be the system we have been looking for. Why not postulate E_3, then, as the system of equations? Such a procedure is not justified without further analysis, since the compatibility of E_1 and that of E_2 do not imply compatibility of the stronger system E_3, where the number of equations exceeds the number of field components by four.

An independent consideration shows that irrespective of the question of compatibility the stronger system, E_3, is the only really natural generalization of the equations of gravitation.

But E_3 is not a compatible system in the same sense as are the systems E_1 and E_2, whose compatibility is assured by a sufficient number of identities, which means that every field that satisfies the equations for a definite value of the time has a continuous extension representing a solution in four-dimensional space. The system E_3, however, is not extensible in the same way. Using the language of classical mechanics, we might say of this situation: In the case of the system E_3 the "initial condition" cannot be freely chosen. What really matters is the answer to the question: Is the manifold of solutions for the system E_3 as extensive as must be required for a physical theory? This purely mathematical problem is as yet unsolved.

The skeptic will say: "It may well be true that this system of equations is reasonable from a logical standpoint. But this does not prove that it corresponds to nature." You are right, dear skeptic. Experience alone can decide on truth. Yet we have achieved something if we have succeeded in formulating a meaningful and precise question. Affirmation or refutation will not be easy, in spite of an abundance of known empirical facts. The derivation, from the equations, of conclusions which can be confronted with experience will require painstaking efforts and probably new mathematical methods.

What Is Matter?

Erwin Schrödinger

Fifty years ago science seemed on the road to a clear-cut answer to the ancient question which is the title of this article. It looked as if matter would be reduced at last to its ultimate building blocks—to certain submicroscopic but nevertheless tangible and measurable particles. But it proved to be less simple than that. Today a physicist no longer can distinguish significantly between matter and something else. We no longer contrast matter with forces or fields of force as different entities; we know now that these concepts must be merged. It is true that we speak of "empty" space (that is, space free of matter), but space is never really empty, because even in the remotest voids of the universe there is always starlight—and that is matter. Besides, space is filled with gravitational fields, and according to Einstein gravity and inertia cannot very well be separated.

Thus, the subject of this article is in fact the total picture of space-time reality as envisaged by physics. We have to admit that our conception of material reality today is more wavering and uncertain than it has been for a long time. We know a great many interesting details, learn new ones every week. But to construct a clear, easily comprehensible picture on which all physicists would agree—that is simply impossible. Physics stands at a grave crisis of ideas. In the face of this crisis, many maintain that no objective picture of reality is possible. However, the optimists among us (of whom I consider myself) look upon this view as a philosophical extravagance born of despair. We hope that the present fluctuations of thinking are only indications of an upheaval of old beliefs which in the end will lead to something better than the mess of formulas that today surrounds our subject.

Since the picture of matter that I am supposed to draw does not yet exist, since only fragments of it are visible, some parts of this narrative may be inconsistent with others. Like Cervantes's tale of Sancho Panza, who loses his donkey in one chapter [of *Don Quixote*] but a few chapters later, thanks to the forgetfulness of the author, is riding the dear little animal again, our story has contradictions. We must start with the well-established concept that matter is composed of corpuscles or atoms, whose existence has been quite "tangibly" demonstrated by many beautiful experiments, and with Max Planck's discovery that energy also comes in indivisible units, called quanta, which are supposed to be transferred abruptly from one carrier to another.

But then Sancho Panza's donkey will return. For I shall have to ask you to believe neither in corpuscles as permanent individuals nor in the suddenness of the transfer of an energy quantum. Discreteness is present, but not in the traditional sense of discrete single particles, let alone in the sense of abrupt processes. Discreteness arises merely as a structure from the laws governing the phenomena. These laws are by no means fully understood; a probably correct analogue from the physics of palpable bodies is the way various partial tones of a bell derive from its shape and from the laws of elasticity to which, of themselves, nothing discontinuous adheres.

The idea that matter is made up of ultimate particles was advanced as early as the fifth century B.C. by Leucippus and Democritus, who called these particles atoms. The corpuscular theory of matter was lifted to physical reality in the theory of gases developed during the 19th century by James Clerk Maxwell and Ludwig Boltzmann. The concept of atoms and molecules in violent motion, colliding and rebounding again and again, led to full comprehension of all the properties of gases: their elastic and thermal properties, their viscosity, heat conductivity and diffusion. At the same time, it led to a firm foundation of the mechanical theory of heat, namely, that heat is the motion of these ultimate particles, which becomes increasingly violent with rising temperature.

Within one tremendously fertile decade at the turn of the century came the discoveries of x rays, of electrons, of the emission of streams of particles and other forms of energy from the atomic nucleus by radioactive decay, of the electric charges on the various particles. The masses of these particles, and of the atoms themselves, were later measured very precisely, and from this was discovered the mass defect of the atomic nucleus as a whole. The mass of a nucleus is less than the sum of the masses of its component particles; the lost mass becomes the binding energy holding the nucleus firmly together. This is called the packing effect. The nuclear forces of course are not electrical forces—those are repellent—but are much stronger and act only within very short distances, about 10^{-13}.

Here I am already caught in a contradiction. Didn't I say at the beginning that we no longer assume the existence of force fields apart from matter? I could easily talk myself out of it by saying: "Well, the force field of a particle is simply considered a part of it." But that is not the fact. The established view today is rather that everything is at the same time both particle and field. Everything has the continuous structure with which we are familiar in fields, as well as the discrete structure with which we are equally familiar in particles. This concept is supported by innumerable experimental facts and is accepted in general, although opinions differ on details, as we shall see.

In the particular case of the field of nuclear forces, the particle structure is more or less known. Most likely, the continuous force field is represented by the so-called pi mesons. On the other hand, the protons and neutrons, which we think of as discrete particles, also have a continuous wave structure, as is shown by the interference patterns they form when diffracted by a crystal. The difficulty of combining these two so very different character traits in one mental picture is the main stumbling block that causes our conception of matter to be so uncertain.

Neither the particle concept nor the wave concept is hypothetical. The tracks in a photographic emulsion or in a Wilson cloud chamber leave no doubt of the behavior of particles as discrete units. The artificial production of nuclear particles is being attempted right now with terrific expenditure, defrayed in the main by the various state ministries of defense. It is true that one cannot kill anybody with one such racing particle, or else we should all be dead by now. But their study promises, indirectly, a hastened realization of the plan for the annihilation of mankind which is so close to all our hearts.

You can easily observe particles yourself by looking at a luminous numeral of your wrist watch in the dark with a magnifying glass. The luminosity surges and undulates, just as a lake sometimes twinkles in the sun. The light consists of sparklets, each produced by a so-called alpha particle (helium nucleus) expelled by a radioactive atom which in this process is transformed into a different atom. A specific device for detecting and recording single particles is the Geiger-Müller counter. In this short résumé I cannot possibly exhaust the many ways in which we can observe single particles.

Now to the continuous field or wave character of matter. Wave structure is studied mainly by means of diffraction and interference—phenomena that occur when wave trains cross each other. For the analysis and measurement of light waves the principal device is the ruled grating, which consists of a great many fine, parallel, equidistant lines, closely engraved on a specular metallic surface. Light impinging from one direction is scattered by them and collected in different directions depending on its wavelength. But even the finest ruled gratings we can produce are too coarse to scatter the very much shorter waves associated with matter. The fine lattices of crystals, however, which Max von Laue first used as gratings to analyze the very short x rays, will do the same for "matter waves." Directed at the surface of a crystal, high-velocity streams of particles manifest their wave nature. With crystal gratings, physicists have diffracted and measured the wavelengths of electrons, neutrons and protons.

What does Planck's quantum theory have to do with all this? Planck told us in 1900 that he could comprehend the radiation from red-hot iron, or from an incandescent star such as the sun, only if this radiation was produced in discrete portions and transferred in such discrete quantities from one carrier to another (for example, from atom to atom). This was extremely startling, because up to that time energy had been a highly abstract concept. Five years later Einstein told us that energy has mass and mass is energy; in other words, that they are one and the same. Now the scales begin to fall from our eyes: our dear old atoms, corpuscles, particles are Planck's energy quanta. *The car-*

riers of those quanta are themselves quanta. One gets dizzy. Something quite fundamental must lie at the bottom of this, but it is not surprising that the secret is not yet understood. After all, the scales did not fall suddenly. It took 20 or 30 years. And perhaps they still have not fallen completely.

The next step was not quite so far-reaching, but important enough. By an ingenious and appropriate generalization of Planck's hypothesis, Niels Bohr taught us to understand the line spectra of atoms and molecules and how atoms were composed of heavy, positively charged nuclei with light, negatively charged electrons revolving around them. Each small system—atom or molecule—can harbor only definite discrete energy quantities, corresponding to its nature. In transition from a higher to a lower "energy level," it emits the excess energy as a radiation quantum of definite wavelength, inversely proportional to the quantum given off. This means that a quantum of given magnitude manifests itself in a periodic process of definite frequency that is directly proportional to the quantum; the frequency equals the energy quantum divided by the famous Planck's constant, h.

According to Einstein, a particle has the energy mc^2, m being the mass of the particle and c the velocity of light. In 1925 Louis de Broglie drew the inference, which rather suggests itself, that a particle might have associated with it a wave process of frequency mc^2 divided by h. The particle for which he postulated such a wave was the electron. Within two years the "electron waves" required by his theory were demonstrated by the famous electron diffraction experiment of C. J. Davisson and L. H. Germer. This was the starting point for the cognition that everything—anything at all—is simultaneously particle and wave field. Thus, de Broglie's dissertation initiated our uncertainty about the nature of matter. Both the particle picture and the wave picture have truth value, and we cannot give up either one or the other. But we do not know how to combine them.

That the two pictures are connected is known in full generality with great precision and down to amazing details. But concerning the unification to a single, concrete, palpable picture, opinions are so strongly divided that a great many deem it altogether impossible. I shall briefly sketch the connection. But do not expect that a uniform, concrete picture will emerge before you, and do not blame the lack of success either on my ineptness in exposition or your own denseness—nobody has yet succeeded.

One distinguishes two things in a wave. First, a wave has a front, and a succession of wave fronts forms a system of surfaces like the layers of an onion. A two-dimensional analogue is the beautiful wave circles that form on the smooth surface of a pond when a stone is thrown in. The second characteristic of a wave, less intuitive, is the path along which it travels—a system of imagined lines perpendicular to the wave fronts. These lines are known as the wave "normals" or "rays."

We can make the provisional assertion that these rays correspond to the trajectories of particles. Indeed, if you cut a small piece out of a wave, approximately 10 or 20 wavelengths along the direction of propagation and about as much across, such a "wave packet" would actually move along a ray with exactly the same velocity and change of velocity as we might expect from a particle of this particular kind at this particular place, taking into account any force fields acting on the particle.

Here I falter. For what I must say now, though correct, almost contradicts this provisional assertion. Although the behavior of the wave packet gives us a more or less intuitive picture of a particle, which can be worked out in detail (for example, the momentum of a particle increases as the wavelength decreases; the two are inversely proportional), yet for many reasons we cannot take this intuitive picture quite seriously. For one thing, it is, after all, somewhat vague, the more so the greater the wavelength. For another, quite often we are dealing not with a small packet but with an extended wave. For still another, we must also deal with the important special case of very small "packelets" which form a kind of "standing wave" that can have no wave fronts or wave normals.

One interpretation of wave phenomena extensively supported by experiments is this: at each position of a uniformly propagating wave train, there is a twofold structural connection of interactions, which may be distinguished as "longitudinal" and "transversal." The transversal structure is that of the wave fronts and manifests itself in diffraction and interference experiments; the longitudinal structure is that of the wave normals and manifests itself in the observation of single particles. However, these concepts of longitudinal and transversal structures are not sharply defined and absolute, since the concepts of wave front and wave normal are not, either.

The interpretation breaks down completely in the special case of the standing waves mentioned above. Here the whole wave phenomenon is reduced to a small region of the dimensions of a single or very few wavelengths. You can produce standing water waves of a similar nature in a small basin if you dabble with your finger rather uniformly in its center, or else just give it a little push so that the water surface undulates. In this situation we are not dealing with uniform wave propagation; what catches the interest are the normal frequencies of these standing waves. The water waves in the basin are an analogue of a wave phenomenon associated with electrons, which occurs in a region just about the size of the atom. The normal frequencies of the wave group washing around the atomic nucleus are universally found to be exactly equal to Bohr's atomic "energy levels" divided by Planck's constant h. Thus, the ingenious yet somewhat artificial assumptions of Bohr's model of the atom, as well as of the older quantum theory in general, are superseded by the far more natural idea of de Broglie's wave phenomenon. The wave phenomenon forms the "body" proper of the atom. It takes the place of the individual pointlike electrons, which in Bohr's model are supposed to swarm around the nucleus. Such pointlike single particles are completely out of the question within the atom, and if one still thinks of the nucleus itself in this way, one does so quite consciously for reasons of expediency.

What seems to me particularly important about the discovery that "energy levels" are virtually nothing but the frequencies of normal modes of vibration is that as a result one can do without the assumption of sudden transitions, or quantum jumps, since two or more normal modes may very well be excited simultaneously. The discreteness of the normal frequencies fully suffices—so I believe—to support the considerations from which Planck started and many similar and just as important ones—I mean, in short, to support all of quantum thermodynamics.

The theory of quantum jumps is becoming more and more unacceptable, at least to me personally, as the years go on. Its abandonment has, however, far-reaching consequences. It means that one must give up entirely the idea of the exchange of energy in well-defined quanta and replace it with the concept of resonance between vibrational frequencies. Yet we have seen that because of the identity of mass and energy, we must consider the particles themselves as Planck's energy quanta. This is at first frightening. For the substituted theory implies that we can no longer consider the individual particle as a well-defined permanent entity.

That it is, in fact, no such thing can be reasoned in other ways. For one thing, there is Werner Heisenberg's famous uncertainty principle, according to which a particle cannot simultaneously have a well-defined position and a sharply defined velocity. This uncertainly implies that we cannot be sure that the same particle could ever be observed twice. Another conclusive reason for not attributing identifiable sameness to individual particles is that we must obliterate their individualities whenever we consider two or more interacting particles of the same kind, for example, the two electrons of a helium atom. Two situations that are distinguished only by the interchange of the two electrons must be counted as one and the same; if they are counted as two equal situations, nonsense obtains. This circumstance holds for any kind of particle in arbitrary numbers without exception.

Most theoreticians will probably accept the foregoing reasoning and admit that the individual particle is not a well-defined permanent entity of detectable identity or sameness. Nevertheless, this inadmissible concept of the individual particle continues to play a large role in their ideas and discussions. Even deeper rooted is the belief in "quantum jumps," which is now surrounded with a highly abstruse terminology whose common-sense meaning is often difficult to grasp. For instance, an important word in the standing vocabulary of quantum theory is "probability," referring to transition from one level to another. But, after all, one can speak of the probability of an event only assuming that, occasionally, it actually occurs. If it does occur, the transition must be sudden, since intermediate stages are disclaimed. Moreover, if it takes time, it might be interrupted halfway by an unforeseen disturbance. This possibility leaves one at sea.

The wave versus corpuscle dilemma is supposed to be resolved by asserting that the wave field merely serves for the computation of the probability of finding a particle of given properties at a given position if one looks for it there. But once one deprives the waves of reality and assigns them only a kind of informative role, it becomes very difficult to understand the phenomena of interference and diffraction on the basis of the combined action of discrete single particles. It seems easier to explain particle tracks in terms of waves than to explain the wave phenomenon in terms of corpuscles.

"Real existence" is, to be sure, an expression that has been virtually chased to death by many philosophical hounds. Its simple, naive meaning has almost become lost to us. Therefore, I want to recall something else. I spoke of a corpuscle's not being an individual. Properly speaking, one never observes the same particle a second time—very much as Heraclitus says of the river. You cannot mark an electron, you cannot paint it red. Indeed, you must not even *think* of it as marked; if you do, your "counting" will be false and you will get wrong results at every step—for the structure of line spectra, in thermodynamics and elsewhere. A wave, on the other hand, can easily be imprinted with an individual structure by which it can be recognized beyond doubt. Think of the beacon fires that guide ships at sea. The light shines according to a definite code; for example, three seconds light, five seconds dark, one second light, another pause of five seconds, and again light for three seconds—the skipper knows that is San Sebastian. Or you talk by wireless telephone with a friend across the Atlantic; as soon as he says, "Hello there, Edward Meier speaking," you know that his voice has imprinted on the radio wave a structure which can be distinguished from any other. But one does not have to go that far. If your wife calls, "Francis!" from the garden, it is exactly the same thing, except that the structure is printed on sound waves and the trip is shorter (though it takes somewhat longer than the journey of radio waves across the Atlantic). All our verbal communication is based on imprinted individual wave structures. And, according to the same principle, what a wealth of details is transmitted to us in rapid succession by the movie or the television picture!

This characteristic, the individuality of the wave phenomenon, has already been found to a remarkable extent in the very much finer waves of particles. One example must suffice. A limited volume of gas, say, helium, can be thought of either as a collection of many helium atoms or as a super-position of elementary wave trains of matter waves. Both views lead to the same theoretical results as to the behavior of the gas upon heating, compression and so on. But when you attempt to apply certain somewhat involved enumerations to the gas, you must carry them out in different ways according to the mental picture with which you approach it. If you treat the gas as consisting of particles, no individuality must be ascribed to them. If, however, you concentrate on the matter wave trains instead of on the particles, every one of the wave trains has a well-defined structure that is different from that of any other. It is true that there are many pairs of waves so similar to each other that they could change roles without any noticeable effect on the gas. But if you should count the very many similar states formed in this way as merely a single one, the result would be quite wrong.

In spite of everything, we cannot completely banish the concepts of quantum jump and individual corpuscle from the vocabulary of physics. We still require them to describe many details of the structure of matter. How can one ever determine the weight of a carbon nucleus and of a hydrogen nucleus, each to the precision of several decimals, and detect that the former is somewhat lighter than the 12 hydrogen nuclei combined in it, without accepting for the time being the view that these particles are something quite concrete and real? This view is so much more convenient than the roundabout consideration of wave trains that we cannot do without it, just as the chemist does not discard his valence-bond formulas, although he fully realizes that they represent a drastic simplification of a rather involved wave-mechanical situation.

If you finally ask me: "Well, what *are* these corpuscles, really?" I ought to confess honestly that I am almost as little prepared to answer that as to tell where Sancho Panza's second donkey came from. At the most, it may be permissible to say that one can think of particles as more or less temporary entities within the wave field whose form and general behavior are nevertheless so clearly and sharply determined by the laws of waves that many processes take place *as if* these temporary entities were substantial permanent beings. The mass and the charge of particles, defined with such precision, must then be counted among the structural elements determined by the wave laws. The conservation of charge and mass in the large must be considered as a statistical effect, based on the "law of large numbers."

Chapter Two:
The Birth of the Universe

Introduction

Not so long ago, humanity lived in caves, busily developing brawn and brain for the one purpose of survival. In 1610, Galileo first turned his telescope toward the Moon and Jupiter and dared to propose a model showing that the Earth, the cave of humanity, was not the center of the universe.

Then came 1929, the year of the great stock market crash and of Edwin Hubble's theory that the universe began with a colossal explosion. Today virtually everything we see in space supports the idea that the universe is expanding. Everything, that is, except several clusters of galaxies, each containing one member that apparently moves differently than the other. Today much of our understanding of the expanding universe is still based on the shift to the red end of the spectrum of all the distant galaxies; the more extreme the shift, the faster they are racing away from us. Although some astronomers doubt that the universe began with a single explosion, the vast majority accept the model. The following essays tell the story.

The Evolution of the Universe

P. James E. Peebles, David N. Schramm, Edwin L. Turner and Richard G. Kron

At a particular instant roughly 12 to 15 billion years ago, all the matter and energy we can observe, concentrated in a region smaller than a dime, began to expand and cool at an incredibly rapid rate. By the time the temperature had dropped to 100 million times that of the Sun's core, the forces of nature assumed their present properties, and the elementary particles known as quarks roamed freely in a sea of energy. When the universe had expanded an additional 1,000 times, all the matter we can measure filled a region the size of the solar system.

At that time, the free quarks became confined in neutrons and protons. After the universe had grown by another factor of 1,000, protons and neutrons combined to form atomic nuclei, including most of the helium and deuterium present today. All of this occurred within the first minute of the expansion. Conditions were still too hot, however, for atomic nuclei to capture electrons. Neutral atoms appeared in abundance only after the expansion had continued for 300,000 years and the universe was 1,000 times smaller than it is now. The neutral atoms then began to coalesce into gas clouds, which later evolved into stars. By the time the universe had expanded to one fifth its present size, the stars had formed groups recognizable as young galaxies.

When the universe was half its present size, nuclear reactions in stars had produced most of the heavy elements from which terrestrial planets were made. Our solar system is relatively young: it formed five billion years ago, when the universe was two thirds its present size. Over time the formation of stars has consumed the supply of gas in galaxies, and hence the population of stars is waning. Fifteen billion years from now stars like our Sun will be relatively rare, making the universe a far less hospitable place for observers like us.

Our understanding of the genesis and evolution of the universe is one of the great achievements of 20th-century science. This knowledge comes from decades of innovative experiments and theories.

Our best efforts to explain this wealth of data are embodied in a theory known as the standard cosmological model or the big bang cosmology. The major claim of the theory is that in the large-scale average, the universe is expanding in a nearly homogeneous way from a dense early state. At present, there are no fundamental challenges to the big bang theory, although there are certainly unresolved issues within the theory itself.

Yet the big bang model goes only so far, and many fundamental mysteries remain. What was the universe like before it was expanding? (No observation we have made allows us to look back beyond the moment at which the expansion began.) What will happen in the distant future, when the last of the stars exhaust the supply of nuclear fuel? No one knows the answers yet.

Our universe may be viewed in many lights—by mystics, theologians, philosophers or scientists. In science we adopt the plodding route: we accept only what is tested by experiment or observation. Albert Einstein gave us the now well-tested and accepted general theory of relativity, which establishes the relations between mass, energy, space and time. Einstein showed that a homogeneous distribution of matter in space fits nicely with his theory. He assumed without discussion that the universe is static, unchanging in the large-scale average.

In 1922 the Russian theorist Alexander A. Friedmann realized that Einstein's universe is unstable; the slightest perturbation would cause it to expand or contract. At that time, Vesto M. Slipher of Lowell Observatory was collecting the first evidence that galaxies are actually moving apart. Then, in 1929, the eminent astronomer Edwin P. Hubble showed that the rate a galaxy is moving away from us is roughly proportional to its distance from us.

The existence of an expanding universe implies that the cosmos has evolved from a dense concentration of matter into the present broadly spread distribution of galaxies. Fred Hoyle, an English cosmologist, was the first to call this process the big bang. Hoyle intended to disparage the theory, but the name was so catchy it gained popularity. It is somewhat misleading, however, to describe the expansion as some type of explosion of matter away from some particular point in space.

That is not the picture at all: in Einstein's universe the concept of space and the distribution of matter are intimately linked; the observed expansion of the system of galaxies reveals the unfolding of space itself. An essential feature of the theory is that the average density in space declines as the universe expands; the distribution of matter forms no observable edge. In an explosion the fastest particles move out into empty space, but in the big bang cosmology, particles uniformly fill all space. The expansion of the universe has had little influence on the size of galaxies or even clusters of galaxies that are bound by gravity; space is simply opening up between them. In this sense, the expansion is similar to a rising loaf of raisin bread. The dough is analogous to space, and the raisins, to clusters of galaxies. As the dough expands, the raisins move apart. Moreover, the speed with which any two raisins move apart is directly and positively related to the amount of dough separating them.

The evidence for the expansion of the universe has been accumulating for some 60 years. The first important clue is the redshift. A galaxy emits or absorbs some wavelengths of light more strongly than others. If the galaxy is moving away from us, these emission and absorption features are shifted to longer wavelengths—that is, they become redder as the recession velocity increases. This phenomenon is known as the redshift.

Hubble's measurements indicated that the redshift of a distant galaxy is greater than that of one closer to Earth. This relation, now known as Hubble's Law, is just what one would expect in a uniformly expanding universe. Hubble's Law says the recession velocity of a galaxy is equal to its distance multiplied by a quantity called Hubble's constant. The redshift effect in nearby galaxies is relatively subtle, requiring good instrumentation to detect it. In contrast, the redshift of very distant objects—radio galaxies and quasars—is an awesome phenomenon; some appear to be moving away at greater than 90 percent of the speed of light.

Hubble contributed to another crucial part of the picture. He counted the number of visible galaxies in different directions in the sky and found that they appear to be rather uniformly distributed. The value of Hubble's constant seemed to be the same in all directions, a necessary consequence of uniform expansion. Modern surveys confirm the fundamental tenet that the universe is homogeneous on large scales. Although maps of the distribution of the nearby galaxies display clumpiness, deeper surveys reveal considerable uniformity.

To test Hubble's Law, astronomers need to measure distances to galaxies. One method for gauging distance is to observe the apparent brightness of a galaxy. If one galaxy is four times fainter than an otherwise comparable galaxy, then it can be estimated to be twice as far away. This expectation has now been tested over the whole of the visible range of distances.

Some critics of the theory have pointed out that a galaxy that appears to be smaller and fainter might not actually be more distant. Fortunately, there is a direct indication that objects whose redshifts are larger really are more distant. The evidence comes from observations of an effect known as gravitational lensing. An object as massive and compact as a galaxy can act as a crude lens, producing a distorted, magnified image (or even many images) of any background radiation source that lies behind it. Such an object does so by bending the paths of light rays and other electromagnetic radiation. So if a galaxy sits in the line of sight between Earth and some distant object, it will bend the light rays from the object so that they are observable. The object behind the lens is always found to have a higher redshift than the lens itself, confirming the qualitative prediction of Hubble's Law.

Hubble's Law has great significance not only because it describes the expansion of the universe but also because it can be used to calculate the age of the cosmos. To be precise, the time elapsed since the big bang is a function of the present value of Hubble's constant and its rate of change. Astronomers have determined the approximate rate of the expansion, but no one has yet been able to measure the second value precisely.

Still, one can estimate this quantity from knowledge of the universe's average density. One expects that because gravity exerts a force that opposes expansion, galaxies would tend to move apart more slowly now than they did in the past. The rate of change in expansion is thus related to the gravitational pull of the universe set by its average density. If the density is that of just the visible material in and around galaxies, the age of the universe probably lies between 10 and 15 billion years. (The range allows for the uncertainty in the rate of expansion.)

Yet many researchers believe the density is greater than this minimum value. So-called dark matter would make up the difference. A strongly defended argument holds that the universe is just dense enough that in the remote future the expansion will slow almost to zero. Under this assumption, the age of the universe decreases to the range of seven to 13 billion years.

Estimates of the expansion time provide an important test for the big bang model of the universe. If the theory is correct, everything in the visible universe should be younger than the expansion time computed from Hubble's law.

These two time scales do appear to be in at least rough concordance. For example, the oldest stars in the disk of the Milky Way galaxy are about nine billion years old—an estimate derived from the rate of cooling of white dwarf stars. The stars in the halo of the Milky Way are somewhat older, about 12 billion years—a value derived from the rate of nuclear fuel consumption in the cores of these stars. The ages of the oldest known chemical elements are also approximately 12 billion years—a number that comes from radioactive dating techniques. Workers in laboratories have derived these age estimates from atomic and nuclear physics. It is noteworthy that their results agree, at least approximately, with the age that astronomers have derived by measuring cosmic expansion.

Another theory, the steady-state theory, also succeeds in accounting for the expansion and homogeneity of the universe. In 1946 three physicists in England—Hoyle, Hermann Bondi and Thomas Gold—proposed such a cosmology. In their theory the universe is forever expanding, and matter is created spontaneously to fill the voids. As this material accumulates, they suggested, it forms new stars to replace the old. This steady-state hypothesis predicts that ensembles of galaxies close to us should look statistically the same as those far away. The big bang cosmology makes a different prediction: if galaxies were all formed long ago, distant galaxies should look younger than those nearby because light from them requires a longer time to reach us. Such galaxies

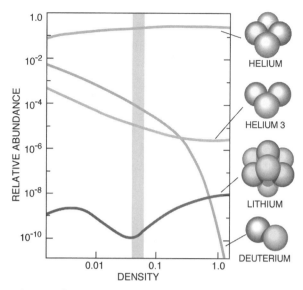

The density of neutrons and protons in the universe determined the abundances of certain elements. For a higher-density universe, the computed helium abundance is little different, and the computed abundance of deuterium is considerably lower. The shaded region is consistent with the observations, ranging from an abundance of 24 percent for helium to one part in 10^{10} for the lithium isotope. This quantitative agreement of theory and observation is a prime success of the big bang cosmology.

should contain more short-lived stars and more gas out of which future generations of stars will form.

The test is simple conceptually, but it took decades for astronomers to develop detectors sensitive enough to study distant galaxies in detail. When astronomers examine nearby galaxies that are powerful emitters of radio wavelengths, they see, at optical wavelengths, relatively round systems of stars. Distant radio galaxies, on the other hand, appear to have elongated and sometimes irregular structures. Moreover, in most distant radio galaxies, unlike the ones nearby, the distribution of light tends to be aligned with the pattern of the radio emission.

Likewise, when astronomers study the population of massive, dense clusters of galaxies, they find differences between those that are close and those far away. Distant clusters contain bluish galaxies that show evidence of ongoing star formation. Similar clusters that are nearby contain reddish galaxies in which active star formation ceased long ago. Observations made with the Hubble Space Telescope confirm that at least some of the enhanced star formation in these younger clusters may be the result of collisions between their member galaxies, a process that is much rarer in the present epoch.

So if galaxies are all moving away from one another and are evolving from earlier forms, it seems logical that they were once crowded together in some dense sea of matter and energy.

When the universe was very young and hot, radiation could not travel very far without being absorbed and emitted by some particle. This continuous exchange of energy maintained a state of thermal equilibrium; any particular region was unlikely to be much hotter or cooler than the average. When matter and energy settle to such a state, the result is a so-called thermal spectrum, where the intensity of radiation at each wavelength is a definite function of the temperature. Hence, radiation originating in the hot big bang is recognizable by its spectrum.

In fact, this thermal cosmic background radiation has been detected.

Astronomers have studied this radiation in great detail using the Cosmic Background Explorer (COBE) satellite and a number of rocket-launched, balloon-borne and ground-based experiments. The cosmic background radiation has two distinctive properties. First, it is nearly the same in all directions. Second, the spectrum is very close to that of an object in thermal equilibrium at 2.726 kelvins above absolute zero. To be sure, the cosmic background radiation was produced when the universe was far hotter than 2.726 kelvins, yet researchers anticipated correctly that the apparent temperature of the radiation would be low. In the 1930s Richard C. Tolman of the California Institute of Technology showed that the temperature of the cosmic background would diminish because of the universe's expansion.

The cosmic background radiation provides direct evidence that the universe did expand from a dense, hot state, for this is the condition needed to produce the radiation. In the dense, hot early universe thermonuclear reactions produced elements heavier than hydrogen, including deuterium, helium and lithium. It is striking that the computed mix of the light elements agrees with the observed abundances. That is, all evidence indicates that the light elements were produced in the hot young universe, whereas the heavier elements appeared later, as products of the thermonuclear reactions that power stars.

The theory for the origin of the light elements emerged from the burst of research that followed the end of World War II. George Gamow and graduate student Ralph A. Alpher of George Washington University and Robert Herman of the Johns Hopkins University Applied Physics Laboratory and others used nuclear physics data from the war effort to predict what kind of nuclear processes might have occurred in the early universe and what elements might have been produced. Alpher and Herman also realized that a remnant of the original expansion would still be detectable in the existing universe.

Despite the fact that significant details of this pioneering work were in error, it forged a link between nuclear physics and cosmology. The workers demonstrated that the early universe could be viewed as a type of thermonuclear reactor. As a result, physicists have now precisely calculated the abundances of light elements produced in the big bang and how those quantities have changed because of subsequent events in the interstellar medium and nuclear processes in stars.

Our grasp of the conditions that prevailed in the early universe does not translate into a full understanding of how galaxies formed. Nevertheless, we do have quite a few pieces of the puzzle. Gravity causes the growth of density fluctuations in the distribution of matter, because it more strongly slows the expansion of denser regions, making them grow still denser. This process is observed in the growth of nearby clusters of galaxies, and the galaxies themselves were probably assembled by the same process on a smaller scale.

The growth of structure in the early universe was prevented by radiation pressure, but that changed when the universe had expanded to about 0.1 percent of its present size. At that point, the temperature was about 3,000 kelvins, cool enough to allow the ions and electrons to combine to form neutral hydrogen and helium. The neutral matter was able to slip through the radiation and to form gas clouds that could collapse into star clusters. Observations show that by the time the universe was one fifth its present size, matter had gathered into gas clouds large enough to be called young galaxies.

A pressing challenge now is to reconcile the apparent uniformity of the early universe with the lumpy distribution of galaxies in the present universe. Astronomers know that the density of the early universe did not vary by much, because they observe only slight irregularities in the cosmic background radiation. So far it has been easy to develop theories that are consistent with the available measurements, but more critical tests are in progress. In particular, different theories for galaxy formation predict quite different fluctuations in the cosmic background radiation on angular scales less than about one degree. Measurements of such tiny fluctuations have not yet been done, but they might be accomplished in the generation of experiments now under way. It will be exciting to learn whether any of the theories of galaxy formation now under consideration survive these tests.

The universe may expand forever, in which case all the galaxies and stars will eventually grow dark and cold. The alternative to this big chill is a big crunch. If the mass of the universe is large enough, gravity will eventually reverse the expansion, and all matter and energy will be reunited. During the next decade, as researchers improve techniques for measuring the mass of the universe, we may learn whether the present expansion is headed toward a big chill or a big crunch.

In following the debate on such matters of cosmology, one should bear in mind that all physical theories are approximations of reality that can fail if pushed too far. Physical science advances by incorporating earlier theories that are experimentally supported into larger, more encompassing frameworks. The big bang theory is supported by a wealth of evidence: it explains the cosmic background radiation, the abundances of light elements and the Hubble expansion. Thus, any new cosmology surely will include the big bang picture. Whatever developments the coming decades may bring, cosmology has moved from a branch of philosophy to a physical science where hypotheses meet the test of observation and experiment.

The Evolution of Galaxy Clusters

J. Patrick Henry, Ulrich G. Briel and Hans Böhringer

In the night of April 15, 1779, Charles Messier watched from his Paris observatory as the Comet of 1779 slowly passed between the Virgo and Coma Berenices constellations on its long journey through the solar system. He noticed three fuzzy patches that looked like comets yet they did not move from night to night; he added them to his list of such impostors so as not to be misled by them during his real work, the search for comets. Later he commented that a small region on the Virgo-Coma border contained 13 of the 109 stationary splotches that he, with the aid of Pierre Mechain, eventually identified—the Messier objects well known to amateur and professional astronomers today.

As so often happens in astronomy, Messier found something completely different from what he was seeking. He had discovered the first example of the most massive things in the universe held together by their own gravity: clusters of galaxies. Clusters are assemblages of galaxies in roughly the same way that galaxies are assemblages of stars. In fact, they are more massive relative to a human being than a human being is relative to a subatomic particle.

In many ways, clusters are the closest that astronomers can get to studying the universe from the outside. Because a cluster contains stars and galaxies of every age and type, it represents an average sample of cosmic material—including the dark matter that choreographs the movements of celestial objects yet cannot be seen by human eyes. And because a cluster is the result of gravity acting on immense scales, its structure and evolution are tied to the structure and evolution of the universe itself. Thus, the study of clusters offers clues to three of the most fundamental issues in cosmology: the composition, organization and ultimate fate of the universe.

A few years after Messier's observations in Paris, William Herschel and his sister, Caroline, began to examine the Messier objects from their garden in England. Intrigued, they decided to search for others. Using substantially better telescopes than their French predecessor had, they found more than 2,000 fuzzy spots—including 300 in the Virgo cluster alone. Both William and his son, John, noticed the lumpy arrangement of these objects on the sky. What organized these objects (which we now know to be galaxies) into the patterns they saw?

A second question emerged in the mid-1930s, when astronomers Fritz Zwicky and Sinclair Smith measured the speeds of galaxies in the Virgo cluster and in a slightly more distant cluster in Coma. Just as the planets orbit about the center of mass of the solar system, galaxies orbit about the center of mass of their cluster. But the galaxies were orbiting so fast that their collective mass could not provide enough gravity to hold them all together. The clusters had to be nearly 100 times as heavy as the visible galaxies, or else the galaxies would have torn out of the clusters long ago. The inescapable conclusion was that the clusters were mostly made of unseen, or "dark," matter. But what was this matter?

LIGHT FROM DARK MATTER

Impelled by these mysteries, the pace of discovery in the study of clusters has accelerated over the past 40 years. Astronomers now know of some 10,000 of them. American astronomer George Abell compiled the first large list in the early 1950s, based on photographs of the entire northern sky taken at Palomar Observatory in California. By the 1970s astronomers felt they at least understood the basic properties of clusters: They consisted of speeding galaxies bound together by huge amounts of dark matter. They were stable and immutable objects.

Then came 1970. In that year a new satellite, named Uhuru ("freedom" in Swahili) in honor of its launch from Kenya, began observing a form of radiation hitherto nearly inaccessible to astronomers: x rays. Edwin M. Kellogg, Herbert Gursky and their colleagues at American Science and Engineering, a small company in Massachusetts, pointed Uhuru at the Virgo and Coma clusters. They found that the clusters consist not only of galaxies but also of huge amounts of gas threading the space between the galaxies. The gas is too tenuous to be seen in visible light, but it is so hot—more than 25 million degrees Celsius—that it pours out x rays.

In short, astronomers had found some of the dark matter—20 percent of it by mass. Although the gas is not enough to solve the dark matter mystery completely, it does account for more mass than all the galaxies put together. In a way, the term "clusters of galaxies" is inaccurate. These objects are balls of gas in which galaxies are embedded like seeds in a watermelon.

Since the early 1970s, the x-ray emission has been scrutinized by other satellites, such as the Einstein X-Ray Observatory, the Roentgen Satellite (ROSAT) and the Advanced Satellite for Cosmology and Astrophysics (ASCA). The first x-ray telescope to record images of the entire sky, ROSAT is well suited for observations of large diffuse objects such as clusters and is now engaged in making detailed images of these regions. With this new technology, astronomers have extended the discoveries of Messier, Zwicky and the other pioneers.

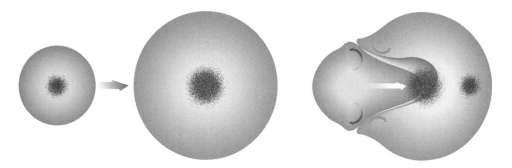

Absorption of a galaxy group allows a cluster to grow to colossal size. Pulled in by gravity, the group slams into the cluster, pushing gas out the sides. The galaxies themselves pass through the cluster, their progress unimpeded by the tenuous gas. Eventually the galaxies and gas mix together, forming a unified cluster that continues to draw in other groups until no more are to be found.

When viewed in x rays, the Coma cluster has a mostly regular shape with a few lumps. These lumps appear to be groups of galaxies—that is, miniature clusters. One lump to the southwest is moving into the main body of the cluster, where other lumps already reside. Virgo, by comparison, has an amorphous shape. Although it has regions of extra x-ray emission, these bright spots are coming from some of the Messier galaxies rather than from clumps of gas. Only the core region in the northern part of Virgo has a nearly symmetrical structure.

Such x–ray images have led astronomers to conclude that clusters form from the merger of groups. This dynamic view of clusters gobbling up and digesting nearby matter is in stark contrast to the static view that astronomers held just a few years ago.

TAKING THEIR TEMPERATURE

Ever since astronomers obtained the first good x-ray images in the early 1980s, they have wanted to measure the variation of gas temperature across clusters. But making these measurements is substantially more difficult than making images, because it requires an analysis of the x-ray spectrum for each point in the cluster. Only in 1994 did the first temperature maps appear.

The maps have proved that the formation of clusters is a violent process. Images of the cluster Abell 2256, for example, show that x-ray emission has not one but rather two peaks. The western peak is slightly flattened, suggesting that a group slamming into the main cluster has swept up material just as a snowplow does. A temperature map supports this interpretation. The western peak, it turns out, is comparatively cool; its temperature is characteristic of the gas in a group of galaxies. Because groups are smaller than clusters, the gravitational forces within them are weaker; therefore, the speed of the gas molecules within them—that is, their temperature—is lower. A typical group is 50 trillion times as massive as the Sun and has a temperature of 10 million degrees C. By comparison, a typical cluster weighs 1,000 trillion Suns and registers a temperature of 75 million degrees C; the heaviest known cluster is five times as massive and nearly three times as hot.

Two hot regions in Abell 2256 appear along a line perpendicular to the presumed motion of the group. The heat seems to be generated as snowplowed material squirts out the sides and smashes into the gas of the main cluster. In fact, these observations match computer simulations of merging groups. The group should penetrate to the center of the cluster in several hundred million years. Thus, Abell 2256 is still in the early stages of the merger.

The late stages of a merger are apparent in another cluster, Abell 754. This cluster has two distinguishing features. First, optical photographs show that its galaxies reside in two clumps. Second, x-ray observations reveal a bar-shaped feature from which the hot cluster gas fans out. One of the galaxy clumps is in the bar region, and the other is at the edge of the high-temperature region to the west.

A third cluster, Abell 1795, shows what a cluster looks like billions of years after a merger. The outline of this cluster is perfectly smooth, and its temperature is nearly uniform, indicating that the cluster has assimilated all its groups and settled into equilibrium. The exception is the cool region at the very center. The lower temperatures occur because gas at the center is dense, and dense gas emits x rays more efficiently than tenuous gas. If left undisturbed for two or three billion years, dense gas can radiate away much of its original energy, thereby cooling down.

As the gas cools, substantial amounts of lukewarm material build up—enough for a whole new galaxy. So where has all this material gone? Despite exhaustive searches, astronomers have yet to locate conclusively any pockets of tepid gas. That the cluster gas is now losing heat is obvious from the temperature maps. Perhaps the heat loss started only fairly recently, or perhaps the collision of galaxy groups prevents cool gas from collecting in one spot. These so-called cooling flows remain yet another unsolved mystery.

BOTTOMS UP

The sequence represented by these three Abell clusters is probably undergone by every cluster as it grows. Galaxy groups occasionally join the cluster; with each, the cluster gains hot gas, bright galaxies and dark matter. The extra mass creates stronger gravitational forces, which heat the gas and accelerate the galaxies. Most astronomers believe that almost all cosmic structures agglomerated in this bottom-up way. Star clusters merged to form galaxies, which in turn merged to form groups of galaxies, which are now merging to form clusters of galaxies. In the future it will be the clusters' turn to

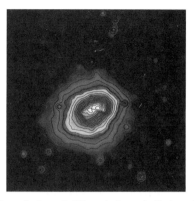
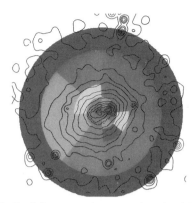

Three galaxy clusters at different stages in their evolution. On the left are x-ray images and on the right are temperature maps. The first cluster, Abell 2256 (above), is busily swallowing a small group of galaxies, which is identified by its relatively low temperature. Darker shades reflect comparatively cool temperatures, while lighter shades indicate intermediate and hot.

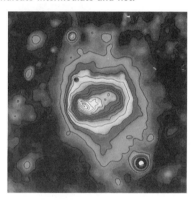
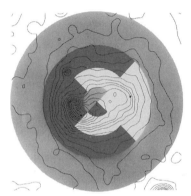

The second cluster, Abell 754 (above), is several hundred million years further along in its digestion of a galaxy group. The hapless group probably entered from the southeast, because the cluster is elongated in that direction. The galaxies of the group have separated from their gas and passed through the cluster.

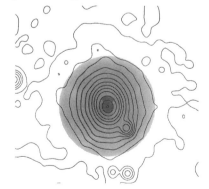

The third cluster, Abell 1795 (above), has gone several billion years since its last meal. Both its x-ray brightness and gas temperature are symmetrical. At the core of the cluster is a cool spot, a region of dense gas that has radiated away much of its heat.

merge to form still larger structures. There is, however, a limit set by the expansion of the universe. Eventually, clusters will be too far apart to merge. Indeed, the cosmos may be approaching this point already.

By cosmological standards, all the above-mentioned clusters (Coma, Virgo, and Abell 2256, 754 and 1795) are nearby objects.

Similarly, astronomers can observe clusters at ever increasing distances, which correspond to ever earlier times. On average, the clusters in a more distant sample are younger than those in a nearby one. The advantage of this approach is that it lets astronomers work with a whole sample of clusters, rather than just a few individual clusters. The disadvantage is that the younger objects are too far away to study in detail; only their average properties can be discerned.

One of us (Henry) applied this method to observations from the ASCA x-ray satellite. He found that distant, younger clusters are cooler than nearby, older ones. Such a temperature change shows that clusters become hotter and hence more massive over time—further proof of the bottom-up model. From these observations researchers have estimated the average rate of cluster evolution in the universe. The rate, which is related to the overall evolution of the universe and to the nature of the dark matter, implies that the universe will expand forever.

Those of us involved in this work feel a special bond with Charles Messier as he strained to glimpse those faint patches of light in Virgo, not knowing their true significance. As advanced as our technology has become, we still strain to understand these clusters. We have been helped by those who preceded us; we share our new understanding with those who follow.

The Lives of Quasars

Gerald Cecil, Jonathan Bland-Hawthorn, and Sylvain Veilleux

Quasars are the lighthouses of the cosmos. Their beacons reach us from baryon islands in a vast dark-matter ocean, and whisper of how things were as galaxies formed in a hotter, more compact universe. Their tremendous luminosity across the electromagnetic spectrum makes them prime targets for study by ground and space-based observatories. When we look back to the far reaches of the universe, we find an era like the Golden Age of ancient Greece, a brief episode when quasars produced a tremendous burst of activity. How did this come about and why did it stop? Almost 40 years after the discovery of quasars, astrophysicists have developed theories of how they formed, generate their tremendous energy and evolve. A consistent picture is now emerging that makes key predictions about the high-redshift universe. Remarkably, many quasars seem to have been overlooked because of heavy dust shrouds.

Quasar light appears to have had a profound influence on the formation of galaxies, and the thermal history of the universe. In this chapter we review recent developments in our understanding of quasars.

QUASARS TODAY

The distinguishing characteristic of quasars is that they emit photons over a much greater energy range than that generated by a galaxy of stars of different temperatures. Their spectrum is characteristic of gas that has disintegrated into electrons, protons and neutrons, and of emission from hot dust.

Abrupt x-ray brightening shows that the core traced by the highest-energy particles is comparable to the size of our solar system. Yet this tiny volume outshines the light of the rest of the galaxy. Based on this enormous luminosity yet compact volume, what compresses and accelerates gas is likely the gravitational field near a heavy black hole. Computer models show that gas near a black hole will first settle into an accretion disk, then be dragged inward by viscosity. Near the inner edge, gas shocks to high temperature and cools by emitting gamma-ray photons and by forming particle-antiparticle pairs. Some of these are absorbed by the black hole, adding to its mass and spin. The quasar core shines brightly by converting several solar masses of material into energy each year. At this rate, it would consume much of its host galaxy if it operated over much of the age of the universe. But quasars are rare today, so the energy consumption has somehow slowed.

There are many assumptions. First, what evidence exists for these accretion disks? In the summer of 1999, Japanese astronomers using an x-ray satellite saw for the first time gas entering a supermassive black hole. The observations confirmed a prediction of general relativity that the disappearing gas would emit x-ray spectral lines that are highly Doppler shifted to red wavelengths.

That we see molecules in the cores of certain active galaxies shows that accretion disks can be dense enough to shield gas from the tremendous heat of the black hole. Remarkably, water absorbs radiation from near the black hole and can beam this toward us in natural microwave lasers, or masers. Enough masers beam light to us to delineate accretion disks in two active galaxies—Messier 106 and NGC 1068. Although these nuclei are far less luminous than that of a typical quasar, gas motions are consistent with those expected in an accretion disk in orbit around a black hole tens of millions of times as massive as the Sun.

Another characteristic of quasar spectra is broad emission lines. Broadening seems to arise as gas particles whirl around the black hole. Indeed the Hubble Space Telescope (HST) and ground-based telescopes have found strong evidence for a dark, billion solar-mass concentration at the center of several galaxies. Theorists have been hard pressed to come up with alternatives to a black hole.

The great luminosity of quasars allows us to study their evolution. All we need do is study progressively fainter examples to see them on average farther back into the past. Of course, our expanding universe was more compact then, the dilation factor being $(1 + z)$ since redshift z. A galaxy at $z = 5$ emitted its light when the universe was less than $1/6$ of its age today.

It is therefore fortuitous that two luminous examples are comparatively close to us. These two galaxies—with catalog names 3C 273 and Arp 220—represent dust-clear and dust-enshrouded quasars. Clearly both are not starlike. In fact a "fuzz" surrounds the bright core of many quasars and contributes on average about 20 percent of the total light from the quasar. Pioneering spectra taken by astronomers Todd Boroson and Bev Oke nearly 20 years ago showed that this is starlight from the galaxy that hosts the quasar core. Boroson and Oke could make these measurements because their charge-coupled device allowed accurate subtraction of scattered core light from the much fainter galaxy starlight.

Such measurements fight two fundamental limits: first, astrophysicists have no clear expectation of how the mass of the quasar galaxy is distributed in space. The greatest uncertainty relates to the distribution of "dark matter." Was it clumped on the scale of the galaxy, as it seems to be today, or was it still distributed diffusely? Only by measuring the motions of stars in the quasar host galaxy can we see if mass was distributed

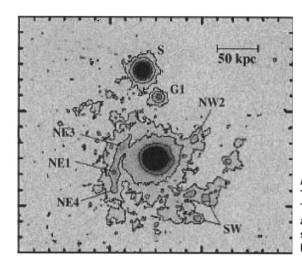

A nearby (z = 0.06) quasar, imaged with the Taurus Tunable Filter at the 39m Anglo-Australian Telescope. A gas envelope 700,000 light years across, companion galaxy, and spiral arms are shown with unusual clarity in the light of ionized hydrogen gas.

into patterns that resemble those seen in nearby galaxies today. But the second problem is that starlight fades as $(1 + z)^{-4}$, so rapidly that by $z = 2$ quasars show only the brightest clumps outside the nucleus. These features are unlikely to track the distribution of mass in the host galaxy.

Instead we would like to see how the older stars, which carry most of the galaxy mass, move. The quasar redshift displaces this starlight into the infrared. At these wavelengths the HST produces fuzzier images than large, ground-based telescopes because of its larger diffraction. Optical systems on 8–10m aperture telescopes will be activated early in the 21st century to deblur the effects of our atmosphere. Such so-called adaptive optics will also improve the contrast of faint material against the bright infrared sky, enabling study of distortions at higher redshifts. These studies will accurately assess the effects of distortions at different places and times in the past.

In fact, Bland-Hawthorn has developed a technique to improve contrast that works on even a small, 4m telescope without adaptive optics. The Taurus tunable filter uses extremely narrow wavelength intervals tuned to study gas around quasars, among other targets. The narrow band greatly reduces natural light pollution from molecules in the Earth's atmosphere. While not the dominant mass component, gas can be an excellent tracer of agitated motions. Such techniques, used on larger telescopes with adaptive optics, will clarify our views of distant quasars.

BACK INTO THE PAST

Even these techniques simply delay the time into the past when gas streams or starlight become too redshifted and, hence, too dim to see. Fortunately, quasars reveal their influence on nearby material in ways that do not depend on distance. As light from the quasar travels to us, it encounters tenuous gas clouds. The gas may absorb or scatter the quasar light anywhere along the vast distance to the quasar, attenuating it in narrow spectral lines. However, if the lines have redshifts close to the quasar, we can be confident that they are also close in space. When spectral absorption lines close to quasars are compared to those formed by more distant gas, astronomers infer that the quasar's strong radiation field has ionized the gas, disintegrating low-mass structures within a sphere millions of light-years across. Beyond, the reduced intensity of ionizing radiation from the quasar allows light-absorbing gas clumps to remain. This "proximity effect" allows us to measure the influence of the population of quasars near a certain redshift on the ionization of intergalactic gas. The degree to which these ionized bubbles overlap is very hard to pin down. Nonetheless, astronomers estimate that the number of dust-cleared quasars is insufficient to explain the inferred average ionization in intergalactic space.

Dust shrouds do not impede x-ray or submillimeter photons, and x rays ionize gas. Could dusty objects also have contributed to the ionization? Space observatories have separated the x-ray sky into numerous point sources, which are generally invisible to optical or infrared telescopes. Curiously, the sky we see today has proportionally more x-ray photons than seen in quasars. A natural explanation is that before the apparent peak in quasar activity near $z = 2$, most massive black holes were enshrouded in dust and gas. Only the most energetic x rays leaked out.

The redshifts of some of the x-ray background objects can be inferred from their overall spectrum when measurements from submillimeter telescopes are included. This combined technique suggests that 80 to 90 percent of the activity in galaxies at $z < 4$ is hidden by dust. When this recently discovered population is included, the x-ray background can be accounted for. Depending on how efficiently ionizing ultraviolet photons escape as the dust shroud clears, the ionization of the intergalactic medium may also be explained.

HOW WAS THE DUST VEIL LIFTED?

Gas in orbits like those found in galaxies today does not fall into a central accretion disk and thence a black hole. But, if agitated, the gas would shock and dissipate energy to flow inward. Exploding a gas clump certainly agitates it and also makes it bright enough to see at great distances. Massive stars explode as supernovae within tens to hundreds of millions of years after their birth. It turns out that in such "star bursting" galaxies much gas is shock heated to such high temperatures that it boils off the galaxy as a wind (see Chapter 3, "Colossal Galactic Explosions"), so it cannot fuel the black hole. In dust-enshrouded quasars, the wind will be channeled by dense gas to break out perpendicular to the galaxy disk.

A toolkit with which to study quasars. Above: Concept for the ALMA sub-millimeter array. Up to 96 antennas will be deployed at 18,000' altitude in the Chilean altiplano. Left: The 8m Next Generation Space Telescope and its large sunshade. Below: A 25m-aperture extremely large telescope (ELT) optimized for infrared.

While supernovae from young stars agitate the gas to keep the black hole obscured, the black hole continues to feed and grow on this gas. Heating and magnetic fields cause electrons to scatter against one another and high-energy photons in a rarefied plasma shell above the inner accretion disk similar to the corona of our Sun. Like the solar wind, the corona is hot enough to expand. This superwind is impeded in some directions by dense gas and dust and the equatorial accretion disk but can emerge near either pole. The age of quasars may be the epoch in which quasar winds blow away intervening dust to reveal the quasar nucleus. The bright clumps seen in quasar galaxies may be regions of intense star formation in leftover gas.

Why are black holes today so quiescent and have not grown larger? Few stars that form in a burst are massive enough to explode as supernovae. The rest form a stellar sphere and dribble gas from stellar winds into the accretion disk. Once most gas converts into stars, the black hole and stellar sphere cease growing and quasar activity fades.

What evidence supports this view? Pioneering work by Frank Low and George Rieke at the University of Arizona in the early 1970s unveiled a new class of galaxies whose "ultrahigh," far-infrared luminosities rivaled the total energy output by quasars. Their work paved the way for the Infrared Astronomy Satellite, which in 1983 discovered across the sky more "ultraluminous infrared galaxies" (ULIGs) than optical quasars, the only objects with comparable luminosities. Studies since have shown that most ULIGs are undergoing spectacular collisions and show signs of activity in their nuclei, including narrow emission lines from very young stars and broad lines that suggest an accretion disk. From these results, David Sanders and his collaborators at the California Institute of Technology suggested in the late 1980s that ULIGs were young quasars in formation, still surrounded by their dusty cocoons. These astronomers were motivated in part by computer simulations of colliding galaxies. In slow collisions, two galaxies coalesce to form what resembles a normal elliptical galaxy. Much of the gas in the model galaxy flows to the center of the merger, possibly to form stars. What happens next is speculative because it happens behind an opaque dust veil. Perhaps leftover gas is fed to a preexisting supermassive black hole. In its early evolution, the starburst coexists with the active nucleus. Much of the energy produced by the starburst and active nucleus heats surrounding dust, which then cools by emitting infrared radiation. The dust shroud is gradually swept by the combined wind action of the starburst and active nucleus. This housecleaning leaves little gas, so the starburst fades rapidly until only the more efficient black-hole engine shines as an optical quasar. Once two spiral galaxies have merged, the result is a normal-looking elliptical galaxy that hosts a quasar in its core.

Astronomers have invested much effort to test this evolutionary picture. The powerhouse of a ULIG is expected to switch from starburst to quasar as the merger progresses. Recently, M. Brotherton and colleagues obtained the spectrum of a quasar that shows the clear signature of a starburst. Optical and infrared studies by several groups, including one led by Veilleux, reveal quasars among the more luminous galaxies with "warm" quasarlike infrared spectra, as expected in this model. The host galaxies of quasars and quasarlike infrared galaxies also appear to be farther along the merger sequence.

HST images reveal that nearly all quasars are in elliptical galaxies, as expected in the merger model. The abundant molecular gas in ULIGs and many quasars, more than ten times that in our Milky Way Galaxy, also supports the idea that these two types of object are related. Tides can also warp the accretion disk, opening a more direct route for gas to flow to the black hole. Indeed, pictures of nearby active galaxies from the ground and the HST often show clear distortions by gravitational tides, and the maser measurements discussed before show that both of those accretion disks are warped.

If ULIGs give birth to quasars, their density in space should evolve like that of quasars. Only recently have sensitive detectors been developed to seek dusty objects at high redshifts to test this idea. Since 1997, the Submillimeter Common User Bolometer Array (SCUBA) camera on the James Clerk Maxwell Telescope at the Mauna Kea Observatories in Hawaii has carried out deep surveys of the distant infrared universe. Several groups recently announced the detection of ULIGs at redshifts inferred to be in the range 1 to 3. These results imply that ULIGs evolve strongly, like quasars. The importance of these findings may extend beyond that of the origin of quasars. The SCUBA sources may emit most of the luminosity of galaxies over cosmic time. They may, therefore, account for most of the far-infrared background light outside our Galaxy, and even possibly the x-ray background, if many of these distant objects prove to have buried quasars.

The evidence suggests that most quasars have faded to invisibility but that their number in the past equals that of galaxies somewhat more massive than our own Milky Way. In other words, many larger galaxies may each have a dead quasar in its core.

ANTICIPATING THE FUTURE

Technological advances like SCUBA stretch our knowledge of quasars. The Cosmic Origins Spectrograph to be installed by astronauts in 2003 will increase the efficiency of the HST to obtain ultraviolet spectra of nearby quasars more than tenfold and allow astronomers to determine the present-day proximity effect, and hence, the ionizing influence of nearby quasars. To study the high-redshift universe, astronomers in Europe, the United States, and Japan are preparing to construct a large array of submillimeter radio telescopes that will be located in Chile, 2 kilometers higher than the summit of Mauna Kea.

Tremendous gains in information will come from 25-100 m aperture optical telescopes currently on drawing boards. Costing $400 million to 2.5 billion each, this possibly final generation of ground-based facilities will work at 3.5–10 micron wavelengths to obtain spectra of objects at redshifts of up to z = 10, when most dust-enshrouded quasars are thought to have formed. One goal will be to learn how heavy elements were produced in the first generation of stars. Some can be used as cosmological clocks, to accurately time the formation of the quasar. Perhaps we will even learn what seeded the first black holes. Did they arise from the fabric of space itself?

By the year 2015, the L2 point between Earth and the Moon may seem as crowded as the summit of Mauna Kea is today. The 8m Next Generation Space Telescope (NGST) may hover there, along with an array of x-ray telescopes called Constellation-X. With both planets distant disks in the sky, the low background light will allow deep surveys. The NGST will be optimized for imaging and mid- and near-infrared spectroscopy of bright regions within the host galaxies of quasars to study their masses. These missions will firm up our understanding of quasar birth and no doubt uncover new puzzles in the life of quasars. Their enormous luminosities ensure that quasars will remain our best time machines.

Edwin Hubble and the Expanding Universe

Donald E. Osterbrock, Joel A. Gwinn and Ronald S. Brashear

During the 1920s and 1930s, Edwin Powell Hubble changed the scientific understanding of the universe more profoundly than had any astronomer since Galileo. Much as Galileo banished the Earth from the center of the solar system, Hubble proved that, rather than being unique, the Milky Way is but one of untold millions of galaxies, or "island universes." Hubble's work also helped to replace the notion of a static cosmos with the startling view that the entire universe is expanding, the ultimate extension of Galileo's defiant (if apocryphal) assertion, "Yet still it moves." Although many researchers contributed to those revolutionary discoveries, Hubble's energetic drive, keen intellect and supple communication skills enabled him to seize the problem of the construction of the universe and make it peculiarly his own.

Hubble's early years have become enmeshed in myth, in part because of his desire to play the hero and in part because of the romanticized image of him recorded by his wife, Grace, in her journals. Many accounts of Hubble consequently bear little relation to the real man. For example, his biographers often state that he was a professional-caliber boxer, yet there is no evidence to support that claim. Likewise, the oft-repeated story that he was wounded during service in World War I seems to have no basis in fact.

Even without such embellishments, Hubble's life has the ring of an all-American success story. The Hubble family established itself in Missouri in 1833, when Edwin's great-grandfather rode horseback from Virginia to settle in what is now Boone County. With characteristic overexuberance, Grace Hubble described her husband's ancestors thus: "Tall, well-made, strong, their bodily inheritance had come down to him, even to the clear, smooth skin that tanned in the Sun, and the brown hair with a glint of reddish gold. They had handed down their traditions as well, integrity, loyalty as citizens, loyalty to their families . . . and a sturdy reliance on their own efforts."

Edwin was born in Marshfield, Missouri, on November 20, 1889, the third of seven children who survived. In 1898 the Hubbles moved to Evanston, Illinois, and, two years later, to Wheaton, where Edwin showed a robust aptitude for both academics and athletics. He became a star athlete at Wheaton High School, especially on the track team. At the age of 16, he entered the University of Chicago, where he continued to display his twin talents. He earned high grades in mathematics, chemistry, physics, astronomy and languages. Despite being two years younger than most of his classmates, he won letters in track and basketball.

After graduation, Hubble received a Rhodes scholarship. Even then, it was far from clear what direction his wide-ranging abilities would finally take. When he "went up" to Queen's College at the University of Oxford in October 1910, Hubble "read" (studied) jurisprudence; both his father and grandfather hoped he would become a lawyer.

Edwin remained a Rhodes scholar for three years and then rejoined his family, who had since moved to Kentucky. He claimed to have passed the bar and practiced law in Louisville. According to Hubble's friend Nicholas U. Mayall, he "chucked the law for astronomy, and I knew that even if I were second-rate or third-rate, it was astronomy that mattered." No record of Hubble's bar exam or of his law practice exists, however.

Uninterested in the law and unsatisfied by his life as a teacher, Hubble decided to return to his true passion, astronomy. In the spring of 1914 Hubble wrote Forest Ray Moulton, his astronomy professor at Chicago, to ask about the possibility of returning as a graduate student. Moulton enthusiastically recommended him to Edwin B. Frost, the director of Yerkes Observatory in Williams Bay, Wisconsin, where the graduate program was given.

Frost gladly accepted Hubble as a student and gave him a scholarship covering tuition and living expenses. The American Astronomical Society was to meet at Northwestern University in nearby Evanston in August 1914; Frost suggested that Hubble come north in time for the event. Thus it happened that Hubble was present at the meeting when Vesto M. Slipher, a quiet, modest astronomer at Lowell Observatory, fueled a controversy as a result of his latest studies of nebulae.

"Nebulae" was a blanket term used by astronomers for centuries to designate faint, cloudy objects that, unlike comets, do not change in position or appearance. The nature of these objects defied easy explanation. In 1755 Immanuel Kant had postulated that some nebulae might be so-called island universes—self-contained systems of stars like our own Milky Way.

During the 19th century, improved telescopic observations showed that many nebulae definitely consist of clouds of luminous gas. One noteworthy class of spiral-shaped nebulae looks distinctly unlike the others, however. By the beginning of the 20th century, many astronomers had come to believe that spiral nebulae were in fact distant galaxies composed of a multitude of stars; skeptics continued to argue that these objects were nearby structures, possibly infant stars in the process of formation.

At the 1914 Astronomical Society meeting, Slipher personally presented the first well-exposed, well-calibrated photographs of the spectra of spiral nebulae. The spectra of stars contain dark lines, called absorption lines, caused by radiation captured by atoms in the stellar atmospheres. Slipher showed that the spectra of spiral nebulae exhibit the kinds of absorption-line spectra characteristic of collections of stars.

Edwin Hubble in 1949, shown putting the 48-inch Schmidt Photographic Telescope at Mt. Palomar through its final rehearsal before beginning an ambitious four-year program to create the first photo atlas of the heavens.

Moreover, Slipher found that the wavelengths of the nebular absorption lines are significantly offset from where they appear in the laboratory. The Doppler shifts indicated that the spiral nebulae observed by Slipher are moving far faster than is typical for stars within the Milky Way and hence strengthened the argument that the spiral nebulae are not part of our galaxy.

When Hubble arrived at Yerkes Observatory, he found a near-moribund institution engaging in routine research. Possibly inspired by Slipher's presentation, he began an intriguing program of photographing nebulae using the observatory's underused 24-inch reflecting telescope. That work grew into his Ph.D. thesis, "Photographic Investigations of Faint Nebulae," which contains foreshadowings of his later research on galaxies and on cosmology.

Hubble described and classified the numerous small, faint nebulae that did not resemble diffuse clouds of gas and noted that most of those nebulae are not spiral but elliptical. He emphasized that the distribution of faint nebulae avoids the parts of the sky near the Milky Way and that many of them appear clustered together. Hubble left no doubt as to the direction of his thinking. "Suppose them to be extra-sidereal and perhaps we see clusters of galaxies; suppose them within our system, their nature becomes a mystery," he wrote.

In October 1916 Hubble corresponded with George Ellery Hale, the director of Mount Wilson Observatory in Pasadena, California. Hale was scouting for staff members to work with the 100-inch reflecting telescope nearing completion on Mount Wilson. Hale offered Hubble a job, conditional on the completion of his doctorate. Hubble hoped to finish the following June and then take the job at Mount Wilson.

Those plans were scuttled by the entry of the United States into World War I on April 6, 1917. Hubble, a former Rhodes scholar who had deep emotional ties to Great Britain, decided that patriotic duty took precedence over scientific pursuits. On May 15, just three days after receiving his degree, he reported for duty and began an officer training course.

Hubble thrived in the military setting. He became a battalion commander and next a major in the 86th "Black Hawk" Division. In September 1918 the division embarked for Europe, but, much to Hubble's regret, the Armistice arrived before he went into combat.

Hubble was discharged on August 20, 1919, whereupon he immediately went to Mount Wilson. He had arrived at an opportune time. The observatory now had two huge reflecting telescopes, a 60-inch and the new 100-inch, then the largest in the world. Hubble's studies at Yerkes had taught him how to make efficient use of such telescopes, and his Ph.D. research had imparted a firm direction to his investigations.

Milton L. Humason, who later collaborated with Hubble on his studies of the distant universe, vividly recalled Hubble's first observing session at Mount Wilson: " 'Seeing' that night was rated as extremely poor on our Mount Wilson scale, but when Hubble came back from developing his plate he was jubilant. 'If this is a sample of poor seeing conditions,' he said, 'I shall always be able to get usable photographs with the Mount Wilson instruments.' The confidence and enthusiasm which he showed on that night were typical of the way he approached all his problems. He was sure of himself—of what he wanted to do, and of how to do it."

Hubble promptly returned to the question of the nature of the "nongalactic nebulae," those whose "members tend to avoid the galactic plane and concentrate in high galactic latitudes." At the time, he had not yet fully accepted the notion that these objects are galaxies outside our own. He initiated his detailed study of nongalactic nebulae by concentrating on the irregular object NGC 6822 ("NGC" denotes J. L. E. Dreyer's *New General Catalog* of nebulae). By 1923, using the 100-inch telescope, he had found several small nebulae and 12 variable stars within NGC 6822.

This endeavor suffered a happy interruption in 1924, when Hubble married Grace Burke Leib, a widow whose first husband had died in an accident in a coal mine in 1921. Grace had met Edwin during a trip to Mount Wilson in 1920 and idolized him for the rest of her life. Years later she recalled her electrifying first impression of Edwin as "an Olympian, tall, strong and beautiful, with the shoulders of the Hermes of Praxiteles, and the benign serenity."

The newlyweds honeymooned in Carmel and in Europe. Hubble then returned to Pasadena, where he turned his attention to M31, the famous Andromeda Galaxy. Drawing on his well-honed observing skills and the unequaled capabilities of the giant 100-inch telescope, Hubble managed to resolve distinctly six variable stars in M31. To Hubble's mind, the achievement unequivocally confirmed that M31 is a remote system composed of stars and, by implication, that other, fainter spiral nebulae are even more distant galaxies.

Although the brightest star that Hubble detected was only about 18th magnitude at its maximum—60,000 times fainter than the dimmest stars visible to the naked eye—he managed to make 83 measurements of its brightness, an impressive feat at the time. From these data, he determined that the star's brightness rose and fell in the manner characteristic of a class of stars known as Cepheid variables. These stars were of tremendous interest because the period of variation of a Cepheid directly relates to the star's absolute luminosity. By comparing that luminosity with his observations of the star's apparent brightness, Hubble was able to deduce the distance to the star and its surrounding galaxy.

A demon of energy, Hubble made discoveries in rapid succession. By the end of 1924 he had identified several more Cepheid variables in M31 and in M33, a spiral galaxy in the constellation Triangulum. In contrast to the rapid progress of his research, Hubble published his results in a leisurely fashion, trying to be as certain as possible before committing them to print. The first paper containing Hubble's results from his studies of variable stars in spiral nebulae appeared later in 1925. Based on observations of Cepheids in NGC 6822, he declared it to be "the first object definitely assigned to a region outside the galactic system." Hubble's epochal paper, "A Spiral Nebula as a Stellar System," presenting his evidence that M33 also is a galaxy lying far outside our own, appeared in 1926.

In addition to determining the distance to the "extra-galactic nebulae," Hubble sought to classify them and to comprehend their diversity of structure. Starting in 1923, he ordered the nebulae according to a perceived pattern of evolution. As an elliptical nebula ages, Hubble argued, it forms arms that enlarge and open up. To explain the existence of barred spirals, in which the spiral arms trail from a barlike formation, Hubble suggested that the elliptical could evolve into either a normal spiral or the barred form. The resulting double-pronged classification diagram is still widely used, though stripped of its evolutionary implications.

By far Hubble's most remarkable bequest to modern science was his research showing that we live in an expanding universe. Hints of that phenomenon had begun to surface well before Hubble tackled the problem.

Further progress required a bigger telescope; the 100-inch telescope on Mount Wilson was ideal for the task. In the mid-1920s Hubble was busy measuring the brightnesses of stars in spiral galaxies, so he enlisted his colleague Humason to determine the galaxies' radial velocities—that is, their motion toward or away from the observer.

To search for signs of an expanding cosmos, Hubble needed to know not only the velocities at which galaxies are moving but also how far away those galaxies are. Hubble's familiarity with measuring cosmic distances greatly assisted this effort. In each of the few galaxies whose distances he had determined by observing Cepheids, Hubble derived the mean absolute magnitude of the brightest stars and of the galaxy as a whole. He then extrapolated from these results to determine the distances to galaxies far beyond M31 and M33.

Hubble's first paper on a velocity-distance relation, published in 1929, sent shock waves through the astronomical community. The findings he presented became part of the basis for the big bang theory underpinning modern cosmology. The paper combined his distance estimates with precision measurements of radial velocities collected by Slipher, Humason and others.

From these data, Hubble derived a linear relation between velocity and distance, written as $v = Hd$ in modern notation. The statement that the velocities of recession of distant galaxies are proportional to their distances is now known as the Hubble law, and the constant H is referred to as the Hubble constant. The Hubble law implies that the universe is expanding: velocities seem to increase as one looks progressively farther outward from any point within the universe.

Between 1931 and 1936 Hubble and Humason extended the law of redshifts to increasingly great distances. Hubble had immediately grasped that his findings supported the notion of an expanding universe, but he never completely accepted that the redshifts resulted only from the radial motions of galaxies. He consistently maintained that the motions inferred from the redshifts should be described as "apparent velocities."

Hubble estimated H—now understood as the expansion rate of the universe—to be about 500 kilometers per second per megaparsec (one megaparsec equals about three million light-years). In retrospect, Hubble's value of the constant was much too large. Astronomers today find values between 50 and 100 kilometers per second per megaparsec. Hubble's errors arose chiefly because he severely underestimated the absolute magnitudes of the brightest stars in the galaxies he observed. Nevertheless, it was a splendid first effort.

After his investigation of the law of redshifts, Hubble concentrated his efforts in observational cosmology. In particular, he sought to measure how the apparent density of extremely remote galaxies changes with distance, in order to determine the overall geometry of the universe.

Although his later undertakings never equaled his remarkable findings made during the 1920s and early 1930s, Hubble continued to sway strongly the direction of astronomical research and to make that research better known to the public. His books *The*

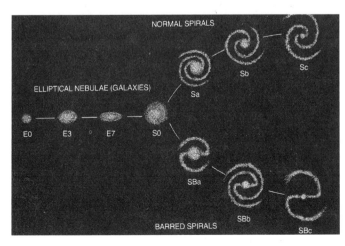

Galaxy classification scheme developed by Hubble in the 1920s assumed that elliptical galaxies develop arms and turn into normal or barred spiral galaxies. Astronomers have rejected Hubble's evolutionary theory but still use his diagram and nomenclature to categorize the wide variety of galaxy shapes.

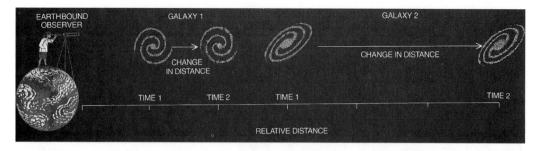

Hubble's Law states that galaxies appear to be receding at speeds in direct proportion to their distance from the observer. A graph of distance versus velocity, which Hubble published in 1929, plainly shows the linear relation. That relation is caused by the fact that the universe is expanding uniformly in all directions. The schematic illustration shows why that is so. Galaxy 2 is three times as far away as Galaxy 1, so it has moved three times as far during the duration from time 1 to time 2. Earthbound observers therefore see Galaxy 2 receding at three times the rate of Galaxy 1.

Realm of the Nebulae and *The Observational Approach to Cosmology* inspired a generation of young astronomers and physicists. Nobody who read Hubble's books or heard his lectures could ignore him; his compelling personality seemed less like those of other astronomers than like those of the movie stars and writers who became his friends in the later years of his life. Hubble often affected English clothes, expressions and mannerisms, habits that colleagues who knew of his Missouri background found unconvincing but that contributed to a dramatic public image.

In an effort to push the boundaries of observational cosmology, Hubble helped to organize the construction of a 200-inch Hale telescope on Mount Palomar in California. The advent of World War II delayed the construction of the telescope; soon after the Japanese attack on Pearl Harbor, Hubble joined the staff of the U.S. Army's Ballistics Research Laboratory at Aberdeen, Md., where he applied his early astronomical training to directing calculations of artillery-shell trajectories. The 200-inch telescope finally began operation in 1948. Hubble continued his research at Palomar until he died at the age of 63, felled by a stroke on September 28, 1953.

After Hubble's death, Humason and then Rudolph L. Minkowski used the 200-inch telescope to detect ever more remote, swiftly receding galaxies. Walter Baade recalibrated the period-luminosity relation for Cepheid variable stars and discovered that all Hubble's galactic distances, huge though they seemed, needed to be multiplied by a factor of two. The triumphant work of Hubble's successors stands as eloquent tribute to the innovative thinking and boundless energy that enabled Hubble to redirect the course of modern cosmology.

The Big Bang and Beyond—A Century of Modern Cosmology

Louis Bernstein

For ages, people have tried to comprehend the cosmos and its origins. Our earliest cosmologies were rooted in the myths and perceptions of the day. The Earth was thought to be flat simply because it looked flat; it floated in a vast "cosmic" sea (probably on the back of a giant turtle); and the stars were tiny pinholes in the celestial sphere, through which the fires of creation could be glimpsed shining beyond. Over time, as our knowledge of the world grew, so did our horizons: beyond the Earth-centered, Ptolemaic view, and past the Copernican, heliocentric order. Today, although our understanding of the universe is based on scientific observation, the impetus behind our age-old quest is ultimately the same: it is a quest to understand ourselves, our origins, and our place in the cosmos.

Nowadays we are used to the concept of living in a universe that contains billions of galaxies, receding from one another on an expanding tide of space and time. But this view of the cosmos is quite new: It arose about eight decades ago from the work of theorists who sought a more accurate description of the interaction between space, time, matter and energy. At first, this emerging new cosmology was generally regarded as speculative—interesting perhaps, but untestable because of technological limitations. As a result, many innovative ideas were largely ignored, or placed on the "back burner" at best. However, recent advances in technology and theory, have taken cosmology beyond the realm of speculation and made it a respectable and exciting science—one that has flourished over the past few decades, gaining substantial ground and insight. What is the scientific basis of this new cosmology? How certain are we of its findings? And how close can it take us to fulfilling our age-old quest of understanding the origin of the universe?

THE BIRTH OF MODERN COSMOLOGY

The theoretical foundations of modern cosmology were laid, almost inadvertently, during the first quarter of the 20th century. Among its principal architects were Albert Einstein, Hermann Minkowski, Willem de Sitter, Alexander Friedmann, and Georges Lemaître. What follows is a quick chronological tour.

In 1905 Einstein published his ground-breaking special theory of relativity, which showed that measurements of space and time vary according to the observer's motion and that mass and energy are equivalent, expressed as $E = mc^2$. In 1908 Minkowski, the noted Russian mathetician, formulated a unified, four-dimensional view of the universe, in which the standard three dimensions of space were combined with one dimension of time. Einstein later incorporated this important idea in his general theory of relativity, which was published in 1916. The general theory described gravity as a curvature in the geometry of space-time, produced by mass and energy.

Einstein published another classic paper, titled "Cosmological Considerations on the General Theory of Relativity," the following year. In it he pondered the effects of matter and energy on the geometry of the universe as a whole. This work became the cornerstone of modern cosmology. Einstein believed deeply in the aesthetic symmetry of nature, which led him to correctly assume that the universe is homogeneous (the same everywhere) and isotropic (the same in all directions). Similar reasoning also led him to believe that the universe is static. However, upon solving the equations of general relativity, Einstein was perplexed to find that they predicted a cosmic, gravitational collapse. In order to satisfy his esthetic sense, he added a cosmological term (())—a fictitious, universal repulsive force, designed to counter the predicted collapse. Although Einstein would later regret contriving a static solution to his equations, the idea of a universal repulsive force is currently being revived to explain recent observations that show that the expansion of the universe may be accelerating.

Einstein was also firmly convinced that his equations could have only one solution. He was, therefore, further perplexed to find that in 1917 the Dutch astronomer Willem de Sitter had formulated another static solution. De Sitter's model also included the cosmological term, but it was based on the idea that the mass-energy content of the cosmos was insignificantly small.

Between 1922 and 1924, the Russian meteorologist and mathematician Alexander Friedmann made a conceptual leap by considering the possibility of a nonstatic universe and discarding the cosmological constant. He discovered a number of expanding models, among which are the open, flat and closed geometries that form the basis of current cosmological theory. (The geometry and fate of the universe depend on the total density of cosmic mass and energy, compared with the rate of expansion. This density parameter is denoted (()0). An open Universe (()0 < 1) will simply expand forever. A flat universe (()0 = 1) will expand forever at an ever-decreasing rate that approaches, but never reaches, zero. A closed universe (()0 > 1) will eventually stop expanding and fall back upon itself.)

In 1922 Friedmann even suggested that the universe expanded from a denser, earlier state. In fact, by 1923, even the static, de Sitter model was shown to expand the moment mass and energy were considered significant. Finally, in 1927, George Lemaître, a Belgian astronomer and priest, formulated equations that suggested a cosmic beginning—"a day without a yesterday," as he phrased it. According to his calculations, the universe exploded from a tiny "primeval atom," an idea that would eventually evolve into the big bang theory. Lemaître also predicted that the residual radiation from the primordial explosion should permeate the cosmos, leaving behind a measurable trace.

SEEING IS BELIEVING

But the road to cosmological truth is not based on theory alone: theories must be validated by observation. This is, in essence, the scientific method. While many plausible theories can be proposed, only those confirmed by observation are destined to prevail. Through this process we move asymptotically closer to the truth, and while the road may be somewhat convoluted, in the end, the scientific method assures the validity of our findings.

For example, by the mid-1920s there were many cosmological theories but no way to determine which was correct—until the advent of the 100-inch Hooker telescope at Mount Wilson, California. The new instrument, then the largest in the world, provided Edwin Hubble, the noted American astronomer, with the optical power to study distant objects in detail. The resulting observations revolutionized cosmology: what were once thought of as "spiral nebulae" (cloudlike eddies in the Milky Way) were discovered to actually be distant galaxies; moreover, these galaxies were all observed to be receding from one another. There was only one possible conclusion—the universe is expanding! By 1929 Hubble had discovered that the recessional velocities of galaxies were directly proportional to their distances from Earth. This became known as Hubble's Law, expressed as $v = H_0 D$, where H_0 is the Hubble constant—about 65 km/sec/Mpc by today's best measurements. (One Megaparsec, mpc, equals 3.26 million light-years.)

Together, these preliminary theories and observations revealed a new universe. Cosmology was no longer rooted in earthly terms. Instead, it became rooted in the laws of physics and our belief in the symmetry of nature. However, the cycle of cosmological discovery progresses with the ebb and flow of theory and observation. In the 1930s, after the crest of Hubble's discoveries, cosmology went through a somewhat quiescent period. During this time, and over the next several decades, great strides were made in the field of particle physics and quantum mechanics (the study of the small-scale interactions of matter and energy), which would later prove invaluable to cosmology. By the late 1940s a new round of theories arose that divided theorists into two camps (on either side of the Atlantic). On one side of the cosmic debate, in the United States, there were the proponents of the big bang theory: George Gamow, Ralph Alpher, and Robert

Herman. On the other side, in England, there were the supporters of the steady state theory, proposed by Hermann Bondi, Thomas Gold and Fred Hoyle. In fact, Hoyle actually coined the term "big bang" in order to trivialize the rival theory. (Unfortunately, the name caught on.) According to the "Gamow school," the big bang represented the emergence and subsequent expansion of the universe from a very hot, dense primordial state. Remarkably, in 1948 Gamow, Alpher and Herman estimated that the background radiation left over by the cosmic fireball would have cooled to a present-day temperature of 5 Kelvins! (Today's best measurement is 2.73 Kelvins, that is, 2.73 degrees above absolute zero.)

However, an incomplete picture of nuclear processes led to the incorrect assumption that all the atoms in nature were synthesized from primeval neutrons during the big bang. While this process might have accounted for the production of hydrogen and helium, it could not explain the existence of heavier atoms. (In 1967 Fred Hoyle played an instrumental role in explaining the nucleosynthesis of heavier atoms in stellar cores.) Another hindrance for the big bang was that the best of galaxy distances and recessional velocities led to a cosmic age estimate of two billion years—much younger than the Earth. Alas, such is the effect of limited technology.

In response to these difficulties, Bondi, Gold and Hoyle conceived the steady state theory. Instead of a sudden cosmic creation, their theory proposed a universe that has always existed. Matter was envisaged to be continually created everywhere, at a rate sufficient to maintain a constant cosmic density. (As galaxies receded, new ones would appear, filling the expanding space in between.) However, the steady state theory had two substantial flaws: it predicted an unevolving universe that always looks the same, everywhere and at any time, and it also assumed that no cosmic background radiation (CBR) existed.

The demise of the steady state theory resulted from observations that clearly showed that the cosmos *is* evolving. The distant, younger universe was populated by quasars, and had many more active galaxies nearby than we observe today. In addition, the observed cosmic abundance of helium atoms proved that they had to have been "cooked" in vast numbers and at temperatures of over one billion degrees—far in excess of those predicted by the steady state.

It turns out that the observed 3:1 cosmic ratio of hydrogen to helium is the result of nuclear processes that we now understand quite well. These processes can only occur at extremely high temperatures and densities. This fact, coupled with the observation that the universe is expanding, points unequivocally to a time when the universe was much smaller, denser and hotter.

The final proof for the big bang came in 1965, when Arno Penzias and Robert Wilson, two researchers working at Bell Labs in Holmdel, New Jersey, detected the cosmic background radiation at a temperature of 2.73 K. This was the dying, microwave "glow" of the big bang itself. Subsequent observations confirmed that the CBR exhibits a perfect "blackbody" spectrum (the spectrum of a perfectly absorbing body whose temperature alone determines the wavelength of its radiation). This is exactly what one would expect if the CBR originated at a time when the universe was so hot and dense that it was opaque to radiation. The most notable of these CBR observations were made between 1990 and 1994 by the Cosmic Background Explorer (COBE) satellite. COBE not only confirmed the blackbody spectrum, but also detected subtle temperature variations in the CBR—an observation with deep cosmological implications (as we will soon see).

BACK TO THE FUTURE

Just as theories depend on observation, observations depend on technology. As technology improves the deeper we see, and the more accurate our theories become. When new observations impugn certain aspects of a theory, the theory is modified to conform to observed reality. This is how theories advance. In recent times, the big bang has undergone some conceptual modifications in order to conform to current particle physics. The most challenging of these modifications is the ongoing effort to harmonize the big bang (a gravity-based theory) with quantum theory. Ironically, in order to understand nature on the largest scale, it is necessary to understand nature on the smallest scale.

In the 1970s and 1980s, experiments conducted at the world's largest particle accelerators improved our understanding of the nuclear interactions that occur at very high energies—equivalent to those that existed during the early universe. The resulting picture evolved into what is now known as the standard model of elementary particle physics.

The standard model asserts that the entire universe is governed by four fundamental forces that orchestrate the interplay of space, time, matter and energy; and that all the matter we see consists of two fundamental particle groups: quarks and leptons. The strong nuclear force binds quarks together to form protons, neutrons and mesons, and it prevents the atomic nucleus from flying apart; the electromagnetic force binds electrons to atomic nuclei, forming atoms, and it determines how individual atoms interact; the weak force regulates the decay of elementary particles; and gravity, the weakest force, holds the large-scale aggregations of matter together. Furthermore, in quantum theory these forces are conceived as being mediated, or "carried" by exchange particles. The strong force is mediated by gluons, the electromagnetic force by photons; and the weak force by charged (W^+, W^-) and neutral (Z^o) bosons. Each of these exchange particles has been experimentally confirmed. However, gravitons, carriers of the gravitational force, remain undetected, simply because the force of gravity is extremely weak between elementary particles. (Although an understanding of quantum gravity is still a ways off, progress has recently been made based on superstring theory—the notion that particles behave like one-dimensional strings rather than points.)

Today we know that the big bang theory is essentially correct. Its validity rests on four pillars of solid observational evidence:

1. The observed expansion of space

2. The relative abundance of hydrogen and helium

3. The cosmic background radiation

4. The temperature pattern in the CBR

But, the term "big bang" is rather misleading because it was neither big nor loud, nor was it an explosion in the usual sense. In fact, the big bang did not occur anywhere in space, nor did it have an origin in time, because initially space and time did not exist. Instead, our current view of the big bang is that spacetime and energy were initially combined in an infinitely dense and infinitely hot state. Under these conditions everything was extremely simple. The four fundamental forces were unified; there were no particles because energy and mass were interchangeable; and there were no measurable events. Suddenly, 12 to 15 billion years ago, spacetime began expanding and as it did, mass-energy began cooling. Almost instantly, in a process known as symmetry breaking, the force of gravity separated from the grand unified force (the still unified "strong-electroweak force"). At this epoch, quarks and leptons, and their antiparticles were in equilibrium with energy (they materialized from energy and dematerialized back to energy in rapid succession). Another brief instant later, spacetime began inflating exponentially, at speeds faster than light, in a process called inflation. (According to Einstein's theories, spacetime can expand faster than light. Mass-energy cannot travel through spacetime faster than light.) Inflation had the effect of flattening the geometry of spacetime, and expanding spacetime beyond the "light horizon."

Yet the old dichotomy between the big bang and the steady state still points to the heart of a great cosmological dilemma—one that has yet to be resolved. The choice is simple: either the universe always existed or it did not. If the universe always existed (in any form at all), then we have to accept something that simply is—something infinite in time. Conversely, if the universe did not always exist, then we are forced to accept that existence arose from nonexistence—an equally daunting concept. Either way, both ideas stretch the boundaries of human experience, and both are equally difficult to accept.

Nowadays, orbiting observatories explore the universe over the entire range of the electromagnetic spectrum. Adaptive optics (correcting for atmospheric distortion) and interferometry (combining the optical power of two or more telescopes) have initiated a new era of ground-based, mega-observatories. (Instruments like the two 10-meter

Keck telescopes in Hawaii; the Gemini twin 8-meter telescopes, one in Hawaii and one in Chile; and the four 8-meter telescope array of the Very Large Telescope in Chile are providing unprecedented studies of processes that occurred deep in cosmological time.) In addition, computers now facilitate every aspect of cosmological research, from controlling telescopes and optics to data acquisition and analysis. Meanwhile, supercomputers are being used to simulate theoretical models and determine which ones best approximate reality. Finally, innovative new observatories such as LIGO (Laser Interferometry Gravatational-wave Observatory) and SNO (Sudbury Neutrino Observatory) promise to open new windows of observation for the next century. LIGO will search for gravity waves, which are ripples in spacetime caused by supernovae, merging black holes and perhaps even by the big bang itself. The SNO is poised to detect the tau neutrino for the first time and measure neutrinos as they oscillate (change from one "flavor" to another), thereby confirming the mass of these elusive particles. Such rapid advancements in technology have led to equally rapid developments in cosmological research and theory; and we are currently at the brink of a new age of discovery that is expected to surpass the Hubble era. Scheduled for launch in 2000, the Microwave Anisotropy Probe (MAP) will boast twenty times the resolution of COBE. Over the next few years, cosmologists expect its sharper view of the CBR to provide accurate measurements of the Hubble constant (H_0) and the cosmic mass density parameter ((0), and to determine if the cosmological constant (() exists. They should even be able to tell if the large-scale filamentary structure of the universe resulted from quantum fluctuations at the first instant of time.

As for fulfilling our age-old quest—a unified theory of everything—we will probably have to wait awhile longer, perhaps to the end of the 21st century. Until then, if we listen very carefully, we will be able to hear nature whispering its deepest secrets in our ears. They are the secrets of our existence.

Chapter Three:
Galaxy Formation

Introduction

Sometime during World War II, Edwin Hubble, the great cosmologist, and Clyde Tombaugh, the discoverer of Pluto who had earlier seen the sky through telescopes he had built on the basis of a *Scientific American* article, sat down in Hubble's office to talk about how the galaxies were distributed in space. Hubble was convinced that aside from the "zone of avoidance," where we see few galaxies because of the dust of our own galaxy, the galaxies, in their clusters, are evenly distributed through space. This "cosmological principle" was a cornerstone of the idea that the universe started with a single explosion.

Tombaugh, who had scanned almost two-thirds of the entire sky while searching for distant planets, didn't agree. That afternoon he reported to Hubble that the galaxies he photographed were clumped into clusters, and that the clusters were themselves clustered. He also said that his pictures showed unusual voids in the sky, large areas where few galaxies were present.

Hubble did not take the younger astronomer seriously, and the conversation ended in a stalemate. A decade later, the great Sky Survey at Palomar Observatory dramatically confirmed Tombaugh's view: the clusters of galaxies were gathered into superclusters of vast proportion. Voids, walls, and other structures have since been found. Now, at the end of this remarkable scientific century, we suspect that at the level of the superclusters, the universe still seems to be evenly distributed.

Within those galaxies, however, amazing things are happening. Galaxies are colliding with one another, and some galaxies have extremely energetic centers called quasars. Atop all that, we are probably unable to see most of the matter in and around these galaxies—the "dark matter" that might offer up most of the mass of the universe.

Considering that it all started with virtually nothing, it is amazing that we have a universe at all. But we do, and it is a magnificent place. In this chapter we explore how all this fits together. We continue our journey through time and space by visiting the universe's showpieces—the superclusters of galaxies, the mighty quasars, the supposed dark matter and the quasars. It was only in the second half of the twentieth century that we understood to some extent how the galaxies are distributed and structured.

Galaxies Behind the Milky Way

Renée C. Kraan-Korteweg and Ofer Lahav

In a dark night, far from city lights, we can clearly see the disk of our galaxy shimmering as a broad band across the sky. This diffuse glow is the direct light emitted by hundreds of billions of stars as well as the indirect starlight scattered by dust grains in interstellar space. We are located about 28,000 light-years from the center of the galaxy in the midst of this disk. But although the Milky Way may be a glorious sight, it is a constant source of frustration for astronomers who study the universe beyond our galaxy. The disk blocks light from a full 20 percent of the cosmos, and it seems to be a very exciting 20 percent.

Somewhere behind the disk, for example, are crucial parts of the two biggest structures in the nearby universe: the Perseus-Pisces supercluster of galaxies and the "Great Attractor," a gargantuan agglomeration of matter whose existence has been inferred from the motions of thousands of galaxies through space. Without knowing what lies in our blind spot, researchers cannot fully map the matter in our corner of the cosmos. This in turn prevents them from settling some of the most important questions in cosmology: How large are cosmic structures? How did they form? What is the total density of matter in the universe?

Only in recent years have astronomers developed the techniques to peer through the disk and to reconstruct the veiled universe from its effects on those parts that can be seen. Among other things, astronomers have found a new galaxy so close that it would dominate our skies were it not obscured by the disk. They have found colossal galaxy clusters never before seen and have even taken a first peek at the core of the elusive Great Attractor.

The obscuration of galaxies by the Milky Way was first perceived when astronomers began distinguishing external galaxies from internal nebulae, both seen simply as faint, extended objects. Because galaxies appeared everywhere except in the region of the Milky Way, this region was named the "zone of avoidance." Scientists now know that external galaxies consist of billions of stars as well as countless clouds of dust and gas. In the zone of avoidance the light of the galaxies is usually swamped by the huge number of foreground stars or is absorbed by the dust in our own galaxy.

Extragalactic astronomers have generally avoided this zone, too, concentrating instead on unobscured regions of the sky. But 20 years ago a crucial observation hinted at what

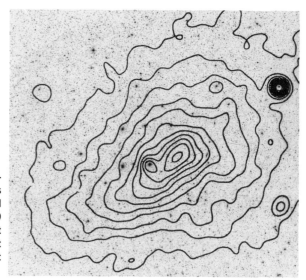

Core of the "Great Attractor" has been identified as galaxy cluster Abell 3627. It appears
both in a visible-light image (background) and
in x-ray observations (contours). Over 100
galaxies show up in this negative image; most
of the dots are stars in our own galaxy. The
tight concentric contours (top right) mark a
bright galaxy in the cluster.

they might be missing. Crude measurements of the cosmic microwave background radiation, a relic of the big bang, showed a 180-degree asymmetry, known as a dipole. It is
about 0.1 percent hotter than average at one location in the sky and equally colder in
the catercornered site. These measurements, confirmed by the Cosmic Background
Explorer satellite in 1989 and 1990, suggest that our galaxy and its neighbors, the so-
called Local Group, are moving at 600 kilometers per second (1.34 million miles per
hour) in the direction of the constellation Hydra. This vector is derived after correcting
for known motions, such as the revolution of the Sun around the galactic center and the
motion of our galaxy toward its neighbor spiral galaxy, Andromeda.

Where does this motion, which is a small deviation from the otherwise uniform expansion of the universe, come from? Galaxies are clumped into groups and clusters, and
these themselves agglomerate into superclusters, leaving other regions devoid of galaxies. The clumpy mass distribution surrounding the Local Group may exert an unbalanced gravitational attraction, pulling it in one direction.

The expected velocity of the Local Group can be calculated by adding up the gravitational forces caused by known galaxies. Although the resulting vector is within 20
degrees of the observed cosmic background dipole, the calculations remain highly
uncertain, partly because they do not take into account the galaxies behind the zone
of avoidance.

The lingering discrepancy between the dipole direction and the expected velocity vector
has led astronomers to postulate "attractors." One research group, later referred to as
the Seven Samurai, used the motions of hundreds of galaxies to deduce the existence of

the Great Attractor about 200 million light-years away. The Local Group seems to be caught in a cosmic tug of war between the Great Attractor and the equally distant Perseus-Pisces supercluster, which is on the opposite side of the sky.

Both are components of a long chain of galaxies known as the Supergalactic Plane. The formation of such a megastructure is thought to depend on the nature of the invisible dark matter that makes up the bulk of the universe.

Nearby galaxies are not to be ignored in the bulk motion of the Local Group. Because gravity is strongest at small distances, a significant force is generated by the nearest galaxies, even if they are not massive. And it is intriguing that five of the eight apparently brightest galaxies lie in the zone of avoidance; they are so close and bright that they shine through the murk. These galaxies belong to the galaxy groups Centaurus A and IC 342, close neighbors to our Local Group. For each member of these groups that astronomers manage to see, there are probably many others whose light is entirely blocked.

LIFTING THE FOG

Because the orbit of the Sun about the galactic center is inclined to the galactic plane, the solar system partakes in an epicyclic motion above and below the plane. Currently we are elevated only 40 light-years from the plane. If we had been born 15 million years from now, we would be located nearly 300 light-years above the plane—beyond the thickest layer of obscuration—and could view one side of the current zone of avoidance. It will take another 35 million years to cross the disk of the Milky Way to the other side.

Most astronomers do not want to wait that long to learn about the extragalactic sky behind the zone of avoidance. What can they do in the meantime? A first step is careful review of existing visible-light images. So various groups of astronomers have gone back to the old-fashioned way of examining images—by eye. Photographic plates from the Palomar Observatory sky survey and its Southern Hemisphere counterpart, conducted in the 1950s, have been painstakingly searched over the past 10 years. Researchers have covered a major fraction of the zone of avoidance, identifying 50,000 previously uncatalogued galaxies.

In areas where the extinction of light by dust is too severe, however, galaxies care fully obscured, and other methods are required. The leading option is to observe at longer wavelengths; the longer the wavelength, the less the radiation interacts with microscopic dust particles. The 21-centimeter spectral line emitted by electrically neutral hydrogen gas is ideal in this respect. It traces gas-rich spiral galaxies, intrinsically dim galaxies and dwarf galaxies—that is, most galaxies except gas-poor elliptical galaxies.

In 1987 a pioneering 21-centimeter project was launched by Patricia A. Henning of the

The Sagittarius dwarf spheroidal galaxy cannot be seen directly even with hindsight; its stars are jumbled with those of the Milky Way and must be identified one by one. (The Sagittarius image is a mosaic of pictures mainly from the southern sky, projected so that the Milky Way runs horizontally across the center.)

University of New Mexico and Frank J. Kerr of the University of Maryland. They pointed the 91-meter radio telescope at Green Bank, West Virginia, toward random spots in the zone of avoidance and detected 18 previously unknown galaxies. Unfortunately, the telescope collapsed spectacularly before they could finish their project. A more systematic survey was initiated by an international team that includes us. Conducted at the 25-meter Dwingeloo radio telescope in the Netherlands, this longer-term project is mapping all the spiral galaxies in the northern part of the zone of avoidance out to a distance of 175 million light-years. So far it has discovered 40 galaxies.

Last year [1997] another international collaboration, led by Lister Staveley-Smith of the Australia Telescope National Facility in Marsfield and one of us (Kraan-Korteweg), began an even more sensitive survey of the southern Milky Way. This survey, which maps galaxies out to 500 million light-years, uses a custom-built instrument at the 64-meter radio telescope at Parkes, Australia. More than 100 galaxies have already been detected, and thousands more are expected when the survey reaches its full depth.

The radio-wave bands are not the only possible peepholes through the zone of avoidance. Infrared light, too, is less affected by dust than visible light is. In the early 1980s the Infrared Astronomical Satellite (IRAS) surveyed the whole sky in far-infrared wavelengths (those closer to radio wavelengths). It tentatively identified infrared-bright galaxies, particularly spirals and starburst galaxies, in which stars are forming rapidly and plentifully. IRAS-selected galaxy candidates near the zone of avoidance are now being reexamined with images taken in the near-infrared wavelengths (those closer to visible light).

Another possible way to overcome the obscuration is to observe at very short wavelengths, such as x rays. Highly populated galaxy clusters emit copious x rays, which pass through the Milky Way almost unhindered. But an x ray investigation, which could draw on existing data from ROSAT and other satellites, has not been done yet.

In addition to direct observations, astronomers are exploring the zone of avoidance by indirect means. Signal-processing techniques, commonly applied by engineers to noisy and incomplete data, have been used successfully by researchers at the Hebrew University and one of us (Lahav) to predict the existence of clusters such as Puppis and Vela, as well as the continuity of the Supergalactic Plane across the zone. The galaxy velocities can also be used on both sides of the zone to predict the mass distribution in between. With this method the center of the Great Attractor was predicted to lie on a line connecting the constellations Centaurus and Pavo. These reconstruction methods, however, deduce only the largest-scale features across the zone; they miss individual galaxies and smaller clusters.

PREY OF THE MILKY WAY

A most surprising discovery came in 1994, when Rodrigo A. Ibata, then at the University of British Columbia, Gerard F. Gilmore of the University of Cambridge and Michael J. Irwin of the Royal Greenwich Observatory in Cambridge, England, who were studying stars in our Milky Way, accidentally found a galaxy right on our doorstep. Named the Sagittarius dwarf, it is now the closest known galaxy—just 80,000 light-years away from the solar system, less than half the distance of the next closest, the Large Magellanic Cloud. In fact, it is located well inside our galaxy, on the far side of the galactic center.

Because the Sagittarius dwarf lies directly behind the central bulge of the Milky Way, it cannot be seen in direct images. Its serendipitous detection was based on velocity measurements of stars: the researchers spotted a set of stars moving differently from those in our galaxy. By pinpointing the stars with this velocity, looking for others at the same distance and compensating for the light of known foreground stars, they mapped out the dwarf. It extends at least 20 degrees from end to end, making it the largest apparent structure in the sky after the Milky Way itself. Its angular size corresponds to a diameter of at least 28,000 light-years, about a fifth of the size of our galaxy, even though the dwarf is only a thousandth as massive.

Sagittarius is one of many surprises to have surfaced from the zone of avoidance. In August 1994 we and the rest of the Dwingeloo Obscured Galaxy Survey team examined our first 21-centimeter spectra. We selected a region where many filaments are lost in the zone and where the nearby galaxy group IC 342 resides. Quite soon we came across an intriguing radio spectrum in the direction of the constellation Cassiopeia.

George K. T. Hau of the University of Cambridge identified an extremely dim visible-light object that matched the location of this radio signal. Before long, deeper images were obtained at various telescopes, which fully revealed the shape of the galaxy: a bar with spiral arms protruding at its ends. If it were not lying behind the plane of the Milky Way, the galaxy—named Dwingeloo 1—would be one of the 10 brightest in the sky. Judging from its rate of rotation it has about one third the mass of the Milky Way, making it comparable to M33, the third heaviest galaxy of the Local Group after the Milky Way and Andromeda.

Although astronomers have yet to explore the entire zone of avoidance, they can now rule out other Andromeda-size galaxies in our backyard. The Milky Way and Andromeda are indeed the dominant galaxies of the Local Group. Disappointing though the lack of another major discovery may be, it removes the uncertainties in the kinematics of our immediate neighborhood.

CLUSTERS AND SUPERCLUSTERS

Studies in the zone of avoidance have also upset astronomers' ideas of the more distant universe. Using the 100-meter radio telescope near Effelsberg, Germany, astronomers discovered a new cluster 65 million light-years away in the constellation Puppis. Several other lines of evidence, including an analysis of galaxies discovered by IRAS, have converged on the same conclusion: the inclusion of Puppis brings the expected motion of the Local Group into better agreement with the observed cosmic background dipole.

Could these searches demystify the Great Attractor? Although the density of visible galaxies does increase in the attractor's presumed direction, the core of this amorphous mass has eluded researchers. A cluster was identified in roughly the right location by George O. Abell in the 1980s, at which time it was the only known cluster in the zone of avoidance.

The true richness and significance of this cluster has become clear in the recent searches. Kraan-Korteweg, with Patrick A. Woudt of the European Southern Observatories in Garching, Germany, has discovered another 600 galaxies in the cluster. With colleagues in France and South Africa, we obtained spectral observations at various telescopes in the Southern Hemisphere. The observed velocities of the galaxies suggest that the cluster is very massive indeed—on par with the well-known Coma cluster, an agglomeration 10,000 times as massive as our galaxy. At long last, astronomers have seen the center of the Great Attractor. Along with surrounding clusters, this discovery could fully explain the observed galaxy motions in the nearby universe.

For generations of astronomers, the zone of avoidance has been an obstacle in investigating fundamental issues such as the formation of the Milky Way, the origin of the Local Group motion, the connectivity of chains of galaxies and the true number of galaxies in the universe. The efforts to lift this thick screen during the 1990s have turned the former zone of avoidance into one of the most exciting regions in the extragalactic sky. The mysterious Great Attractor is now well mapped; the discovery of the Sagittarius dwarf has shown how the Milky Way formed; and the vast cosmic filaments challenge theories of dark matter and structure formation. Step by step, the missing pieces of the extragalactic sky are being filled in.

Colossal Galactic Explosions

Sylvain Veilleux, Gerald Cecil and Jonathan Bland-Hawthorn

Millions of galaxies shine in the night sky, most made visible by the combined light of their billions of stars. In a few, however, a pointlike region in the central core dwarfs the brightness of the rest of the galaxy. The details of such galactic dynamos are too small to be resolved even with the Hubble Space Telescope. Fortunately, debris from these colossal explosions—in the form of hot gas glowing at temperatures well in excess of a million degrees—sometimes appears outside the compact core, on scales that can be seen directly from Earth.

The patterns that this superheated material trace through the interstellar gas and dust surrounding the site of the explosion provide important clues to the nature and history of the powerful forces at work inside the galactic nucleus. Furthermore, because such cataclysms appear to have been taking place since early in the history of the universe, they have almost certainly affected the environment in which our own Milky Way Galaxy evolved.

Astronomers have proposed two distinctly different mechanisms for galactic dynamos. The first was the brainchild of Martin J. Rees of the University of Cambridge and Roger D. Blandford, now at the California Institute of Technology. During the early 1970s, the two sought to explain the prodigious luminosity—thousands of times that of the Milky Way—and the spectacular "radio jets" (highly focused streams of energetic material) that stretch over millions of light-years from the centers of some hyperactive young galaxies known as quasars. They suggested that an ultramassive black hole—not much larger than the Sun but with perhaps a million times its mass—could power a quasar.

A black hole itself produces essentially no light, but the disk of accreted matter spiraling in toward the hole heats up and radiates as its density increases. The inner, hotter part of the disk produces ultraviolet and x-ray photons over a broad range of energies, a small fraction of which are absorbed by the surrounding gas and reemitted as discrete spectral lines of ultraviolet and visible light.

As the disk heats up, gas in its vicinity reaches temperatures of millions of degrees and expands outward from the galactic nucleus at high speed. This flow, an enormous cousin to the solar wind that streams away from the Sun or other stars, can sweep up other interstellar gases and expel them from the nucleus. The resulting luminous shock waves can span thousands of light-years—comparable to the visible sizes of the galaxies themselves—and can be studied from space or ground-based observatories. Some of these galaxies also produce radio jets: thin streams of rapidly moving gas that emit radio waves as they traverse a magnetic field that may be anchored within the accretion disk.

Black holes are not the only engines that drive violent galactic events. Some galaxies apparently undergo short episodes of rapid star formation in their cores: so-called nuclear starbursts. The myriad new stars produce strong stellar winds and, as the stars age, a rash of supernovae. The fast-moving gas ejected from the supernovae strikes the background interstellar dust and gas and heats it to millions of degrees.

The pressure of this hot gas forms a cavity, like a steam bubble in boiling water. As the bubble expands, cooler gas and dust accumulate in a dense shell at the edge of the bubble, slowing its expansion. The transition from free flow inside the bubble to near stasis at its boundary gives rise to a zone of turbulence that is readily visible from Earth. If the energy injected into the cavity is large enough, the bubble bursts out of the galaxy's gas disk and spews the shell's fragments and hot gas into the galaxy halo or beyond, thousands of light-years away from their origins.

IDENTIFYING THE ENGINE

Both the starburst and the black-hole explanations appear plausible, but there are important differences between the two that may reveal which one is at work in a given galaxy. A black hole can convert as much as 10 percent of the infalling matter to energy. Starbursts, in contrast, rely on nuclear fusion, which can liberate only 0.1 percent of the reacting mass. As a result, they require at least 100 times as much matter, most of which accumulates as unburned fuel. Over the lifetime of a starburst-powered quasar, the total mass accumulated in the nucleus of the galaxy could reach 100 billion times the mass of the Sun, equivalent to the mass of all the stars in the Milky Way Galaxy.

The more mass near the nucleus, the more rapidly the orbiting stars must move. Recent ground-based near-infrared observations have revealed the presence of a dark compact object with a mass two million times that of the Sun at the center of our own Milky Way. And recent radio-telescope findings have revealed an accretion disk with an inner radius of half a light-year spinning rapidly around a mass 20 million times that of the Sun at the center of a nearby spiral galaxy called NGC 4258.

Starbursts and black holes also differ in the spectra of the most energetic photons they produce. Near a black hole, the combination of a strong magnetic field and a dense accretion disk creates a soup of very fast particles that collide with one another and with photons to generate x rays and gamma rays. A starburst, in contrast, produces most of its high-energy radiation from collisions between supernova ejecta and the surrounding galactic gas and dust. This impact heats gas to no more than about a billion degrees and so cannot produce any radiation more energetic than x rays. The large numbers of gamma rays detected recently from some quasars by the Compton Gamma Ray Observatory imply that black holes are at their centers.

A final difference between black holes and starbursts lies in the forces that focus the flow of outrushing gas. The magnetic-field lines attached to the accretion disk around a black hole direct outflowing matter along the rotation axis of the disk in a thin jet. The material expelled by a starburst bubble, in contrast, simply follows the path of least resistance in the surrounding environment. A powerful starburst in a spiral galaxy will spew gas perpendicular to the plane of the galaxy's disk of stars and gas, but the flow will be distributed inside an hourglass-shaped region with a wide opening. The narrow radio jets that extend millions of light-years from the core of some active galaxies clearly suggest the presence of black holes.

SPECTRAL SIGNATURES

Thanks to the Doppler shift, which changes the frequency and wavelength of light emitted by moving sources, this analysis also reveals how fast the gas is moving. Approaching gas emits light shifted toward the blue end of the spectrum, and receding gas emits light shifted toward the red end.

Until recently, astronomers unraveled gas behavior by means of two complementary methods: emission-line imaging and long-slit spectroscopy. The first produces images through a filter that selects light of a particular wavelength emitted by an element such as hydrogen. Such images often dramatically reveal the filamentary patterns of explosions, but they cannot tell observers anything about the speed or direction of the gases' motions, because the filter does not discriminate finely enough to measure redshifts or blueshifts. Long-slit spectrometers, which disperse light into its constituent colors, provide detailed information about gas motions but only over a tiny region.

For almost a decade, our group has used an instrument that combines the advantages of these two methods without the main drawbacks. The Hawaii Imaging Fabry-Perot Interferometer (HIFI) yields detailed spectral information over a large field of view.

The HIFI takes its pictures atop the 4,200-meter dormant volcano Mauna Kea, using the 2.2-meter telescope owned by the University of Hawaii and the 3.6-meter Canada-France-Hawaii instrument. The smooth airflow at the mountaintop produces sharp images. Charge-coupled devices, which are very stable and sensitive to faint light, collect the photons. In a single night, this powerful combination can generate records of up to a million spectra across the full extent of a galaxy.

NEARBY ACTIVE GALAXIES

We have used the HIFI to explore NGC 1068, an active spiral galaxy 46 million light-years away. As the nearest and brightest galaxy of this type visible from the Northern Hemisphere, it has been studied extensively. At radio wavelengths, NGC 1068 looks like a miniature quasar: two jets extend about 900 light-years from the core, with more diffuse emission from regions farther out. Most likely, emission from gaseous plasma moving at relativistic speeds creates the radio jets, and the "radio lobes" arise where the plasma encounters matter from the galactic disk. As might a supersonic aircraft, the leading edge of the northeast jet produces a V-shaped shock front.

The same regions also emit large amounts of visible and ultraviolet light. We have found, however, that only 10 percent of the light comes from the nucleus. Another 5 percent comes from galaxy-disk gas that has piled up on the expanding edge of the northeast radio lobe. All the rest comes from two fans of high-velocity gas moving outward from the center at speeds of up to 1,500 kilometers per second.

The effects of the activity within the nucleus reach out several thousand light-years, well beyond the radio lobes. The diffuse interstellar gas exhibits unusually high temperatures and a large fraction of the atoms have lost one or more electrons and become ionized. At the same time, phenomena in the disk appear to influence the nucleus. Infrared images reveal an elongated bar of stars that extends more than 3,000 light-years from the nucleus. The HIFI velocity measurements suggest that the bar distorts the circular orbit of the gas in the disk, funneling material toward the center of the galaxy. This inflow of material may in fact fuel the black hole.

Another tremendous explosion is occurring in the core of one of our nearest neighbor galaxies, M82, just a few million light-years away. In contrast to NGC 1068, this cataclysm appears to be an archetypal starburst-driven event. Images exposed through a filter that passes the red light of forming hydrogen atoms reveal a web of filaments spraying outward along the galactic poles. Our spectral grids of emission from filaments perpendicular to the galactic disk reveal two main masses of gas, one receding and the other approaching. The difference in velocity between the two increases as the gas moves outward from the core, reaching about 350 kilometers per second at a distance of 3,000 light-years. At a distance of 4,500 light-years from the core, the velocity separation diminishes.

The core of M82 is undergoing an intense burst of star formation, possibly triggered by a recent orbital encounter with its neighbors M81 and NGC 3077. Its infrared luminosity is 30 billion times the total luminosity of the Sun, and radio astronomers have identified the remnants of large numbers of supernovae. The upcoming launch of the Advanced X-ray Astrophysics Facility (AXAF), the third of NASA's four planned Great Observatories, should open up exciting new avenues of research in the study of this hot-wind component.

AMBIGUOUS ACTIVITY

Unfortunately, the identity of the principal source of energy in active galaxies is not always so obvious. Sometimes a starburst appears to coexist with a black-hole engine. Like M82, many of these galaxies are abnormally bright at infrared wavelengths and

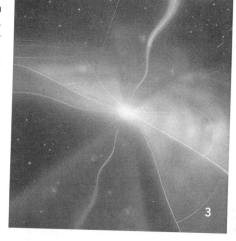

Colossal forces at work in the center of an active galaxy can make themselves felt half a million light years away or more, as jets of gas moving at relativistic speeds plow into the intergalactic medium and create enormous shock waves (1). Closer to the center of the galaxy(2, 3), dense equatorial disk of dust and molecular gas feeds matter to the active nucleus, while hot gas and radiation spill out along the poles. The high density of the infalling gas within a few dozen light years of the center causes a burst of star formation (4). Even closer to the center (5), the disk, glowing at ultraviolet and x-ray wavelengths, tapers inward to feed what astronomers believe is a black hole containing millions of stellar masses—but still so small as to be invisible on this scale.

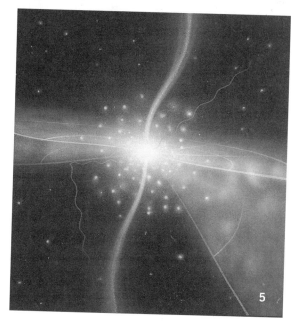

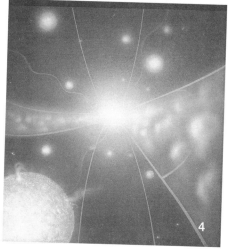

rich in molecular gas, the raw material of stars. Radio emission and visual spectra resembling those of a quasar, however, suggest that a black hole may also be present.

Such ambiguity plagues interpretations of the behavior of the nearby galaxy NGC 3079. This spiral galaxy appears almost edge-on from Earth—an excellent vantage point from which to study the gas expelled from the nucleus. Like galaxy M82, NGC 3079 is anomalously bright in the infrared, and it also contains a massive disk of molecular gas spanning 8,000 light-years around its core. At the same time, the core is unusually bright at radio wavelengths, and the linear shape of radio-emitting regions near the core suggests a collimated jet outflow.

Our spectral observations imply that the total energy of this violent outflow is probably 10 times that of the explosions in NGC 1068 or M82. The alignment of the bubble along the polar axis of the host galaxy implies that galactic dust and gas, rather than a central black hole, are collimating the outflow. Nevertheless, the evidence is fairly clear that NGC 3079 contains a massive black hole at its core.

Is the nuclear starburst solely responsible for such a gigantic explosion? We have tried to answer this question by analyzing the infrared radiation coming from the starburst area. Most of the radiation from young stars embedded in molecular clouds is absorbed and reemitted in the infrared, so the infrared luminosity of NGC 3079's nucleus may be a good indicator of the rate at which supernovae and stellar winds are injecting energy at the center of the galaxy. When we compare the predictions of the starburst model with our observations, we find that the stellar ejecta appears to have enough energy to inflate the bubble. Although the black hole presumed to exist in the core of NGC 3079 may contribute to the outflow, there is no need to invoke it as an energy source.

HOW ACTIVE GALAXIES FORM

Although astronomers now understand the basic principles of operation of the engines that drive active galaxies, many details remain unclear. There is a vigorous debate about the nature of the processes that ignite a starburst or form a central black hole. Researchers are also divided on which comes first, nuclear starburst or black hole.

The anomalous gas flows that we and others have observed are almost certainly only particularly prominent examples of widespread, but more subtle, processes that affect many more galaxies. Luminous infrared galaxies are common, and growing evidence is leading astronomers to believe that many of their cores are also the seats of explosions. These events may profoundly affect the formation of stars throughout the galactic neighborhood. Nuclear reactions in the torrent of supernovae unleashed by the starburst enrich this hot wind in heavy chemical elements. As a result, the wind will not only heat its surroundings but also alter the environment's chemical composition.

The full impact of this "cosmic bubble bath" over the history of the universe is difficult to assess accurately because we currently know very little of the state of more distant galaxies. Images of distant galaxies taken by the Hubble will help clarify some of these questions. Indeed, as the light that left those galaxies billions of years ago reaches our instruments, we may be watching an equivalent of our own galactic prehistory unfolding elsewhere in the universe.

Dark Matter in the Universe

Vera Rubin

Today, as we have done for centuries, we gaze into the night sky from our planetary platform and wonder where we are in this cavernous cosmos. Flecks of light provide some clues about great objects in space. And what we do discern about their motions and apparent shadows tells us that there is much more that we cannot yet see.

From every photon we collect from the universe's farthest reaches, we struggle to extract information. Astronomy is the study of light that reaches Earth from the heavens. Our task is not only to collect as much light as possible—from ground- and space-based telescopes—but also to use what we can see in the heavens to understand better what we cannot see and yet know must be there.

Based on 50 years of accumulated observations of the motions of galaxies and the expansion of the universe, most astronomers believe that as much as 90 percent of the stuff constituting the universe may be objects or particles that cannot be seen. First posited some 60 years ago by astronomer Fritz Zwicky, this so-called missing matter was believed to reside within clusters of galaxies. Nowadays we prefer to call the missing mass "dark matter," for it is the light, not the matter, that is missing.

Astronomers and physicists offer a variety of explanations for this dark matter. On the one hand, it could merely be ordinary material, such as ultrafaint stars, large or small black holes, cold gas, or dust scattered around the universe—all of which emit or reflect too little radiation for our instruments to detect. It could even be a category of dark objects called MACHOs (MAssive Compact Halo Objects) that lurk invisibly in the halos surrounding galaxies and galactic clusters. On the other hand, dark matter could consist of exotic, unfamiliar particles that we have not figured out how to observe. A third possibility is that our understanding of gravity needs a major revision—but most physicists do not consider that option seriously.

In some sense, our ignorance about dark matter's properties has become inextricably tangled up with other outstanding issues in cosmology—such as how much mass the universe contains, how galaxies formed and whether or not the universe will expand forever. So important is this dark matter to our understanding of the size, shape and ultimate fate of the universe that the search for it will very likely dominate astronomy for the next few decades.

OBSERVING THE INVISIBLE

Understanding something you cannot see is difficult—but not impossible. Not surprisingly, astronomers currently study dark matter by its effects on the bright matter that we do observe. For instance, when we watch a nearby star wobbling predictably, we infer from calculations that a "dark planet" orbits around it. Applying similar principles to spiral galaxies, we infer dark matter's presence because it accounts for the otherwise inexplicable motions of stars within those galaxies.

When we observe the orbits of stars and clouds of gas as they circle the centers of spiral galaxies, we find that they move too quickly. These unexpectedly high velocities signal the gravitational tug exerted by something more than that galaxy's visible matter. From detailed velocity measurements, we conclude that large amounts of invisible matter exert the gravitational force that is holding these stars and gas clouds in high-speed orbits. We deduce that dark matter is spread out around the galaxy, reaching beyond the visible galactic edge and bulging above and below the otherwise flattened, luminous galactic disk. As a rough approximation, try to envision a typical spiral galaxy, such as our Milky Way, as a relatively flat, glowing disk embedded in a spherical halo of invisible material—almost like an extremely diffuse cloud.

Looking at a single galaxy, astronomers see within the galaxy's radius (a distance of about 50,000 light-years) only about one tenth of the total gravitating mass needed to account for how fast individual stars are rotating around the galactic hub.

In trying to discover the amount and distribution of dark matter in a cluster of galaxies, x-ray astronomers have found that galaxies within clusters float immersed in highly diffuse clouds of 100-million-degree gas—gas that is rich in energy yet difficult to detect. Observers have learned to use the x ray–emitting gas's temperature and extent in much the same way that optical astronomers use the velocities of stars in a single galaxy. In both cases, the data provide clues to the nature and location of the unseen matter.

In a cluster of galaxies, the extent of the x ray–emitting region and temperature of the gas enable us to estimate the amount of gravitating mass within the cluster's radius, which measures almost 100 million light-years. In a typical case, when we add together the luminous matter and the x ray–emitting hot gas, we are able to sense roughly 20 to 30 percent of the cluster's total gravitating mass. The remainder, which is dark matter, remains undetected by present instruments.

Subtler ways to detect invisible matter have recently emerged. One clever method involves spotting rings or arcs around clusters of galaxies. These "Einstein rings" arise from an effect known as gravitational lensing, which occurs when gravity from a massive object bends light passing by. For instance, when a cluster of galaxies blocks our view of another galaxy behind it, the cluster's gravity warps the more distant galaxy's light, creating rings or arcs, depending on the geometry involved.

Using computer models, we can calculate the mass of the intervening cluster, estimating the amount of invisible matter that must be present to produce the observed geometric deflection. Such calculations confirm that clusters contain far more mass than the luminous matter suggests.

Even compact dark objects in our own galaxy can gravitationally lens light. When a foreground object eclipses a background star, the light from the background star is distorted into a tiny ring, whose brightness far exceeds the star's usual brightness. Consequently, we observe an increase, then a decrease, in the background star's brightness. Careful analysis of the light's variations can tease out the mass of the dark foreground lensing object.

WHERE IS DARK MATTER?

Given the strong evidence that spiral and elliptical galaxies lie embedded in large dark-matter halos, astronomers now wonder about the location, amount and distribution of the invisible material.

To answer those questions, researchers compare and contrast observations from specific nearby galaxies. For instance, we learn from the motions of the Magellanic Clouds, two satellite galaxies gloriously visible in the Southern Hemisphere, that they orbit within the Milky Way Galaxy's halo and that the halo continues beyond the clouds, spanning a distance of almost 300,000 light-years. In fact, motions of our galaxy's most distant satellite objects suggest that its halo may extend twice as far—to 600,000 light-years.

Because our nearest neighboring spiral galaxy, Andromeda, lies a mere two million light-years away, we now realize that our galaxy's halo may indeed span a significant fraction of the distance to Andromeda and its halo. We have also determined that clusters of galaxies lie embedded in even larger systems of dark matter. At the farthest distances for which we can deduce the masses of galaxies, dark matter appears to dwarf luminous matter by a factor of at least 10, possibly as much as 100.

Overall, we believe dark matter associates loosely with bright matter, because the two often appear together. Yet, admittedly, this conclusion may stem from biased observations, because bright matter typically enables us to find dark matter.

By meticulously studying the shapes and motions of galaxies over decades, astronomers have realized that individual galaxies are actively evolving, largely because of the mutual gravitational pull of galactic neighbors. Within galaxies, stars remain enormously far apart relative to their diameters, thus little affecting one another gravitationally. For example, the separation between the Sun and its nearest neighbor, Proxima Centauri, is so great that 30 million Suns could fit between the two. In contrast, galaxies lie close together, relative to their diameters—nearly all have neighbors within a few diameters. So galaxies do alter one another gravitationally, with dark matter's added gravity a major contributor to these interactions.

As we watch many galaxies—some growing, shrinking, transforming or colliding—we realize that these galactic motions would be inexplicable without taking dark matter into account. Right in our own galactic neighborhood, for instance, such interactions are under way. The Magellanic Clouds, our second nearest neighboring galaxies, pass through our galaxy's plane every billion years. Indeed, on every passage, they lose energy and spiral inward. In less than 10 billion years, they will fragment and merge into the Milky Way.

Recently [1994] astronomers identified a still nearer neighboring galaxy, the Sagittarius dwarf, which lies on the far side of the Milky Way, close to its outer edge. (Viewed from Earth, this dwarf galaxy appears in the constellation Sagittarius.) As it turns out, gravity from our galaxy is pulling apart this dwarf galaxy, which will cease to exist as a separate entity after several orbits. Our galaxy itself may be made up of dozens of such previous acquisitions.

In many ways, our galaxy, like all large galaxies, behaves as no gentle neighbor. It gobbles up nearby companions and grinds them into building blocks for its own growth. By studying the spinning, twisting and turning motions and structures of many galaxies as they hurtle through space, astronomers can figure out the gravitational forces required to sustain their motions—and the amount of invisible matter they must contain.

How much dark matter does the universe contain? The destiny of the universe hinges on one still unknown parameter: the total mass of the universe. If we live in a high-density, or "closed," universe, then mutual gravitational attraction will ultimately halt the universe's expansion, causing it to contract—culminating in a big crunch, followed perhaps by reexpansion. If, on the other hand, we live in a low-density, or "open," universe, then the universe will expand forever.

Observations thus far suggest that the universe—or, at least, the region we can observe—is open, forever expanding. When we add up all the luminous matter we can detect, plus all the dark matter that we infer from observations, the total still comes to only a fraction—perhaps 20 percent—of the density needed to stop the universe from expanding forever.

Another complicating factor to take into account is that totally dark systems may exist—that is, there may be agglomerations of dark matter into which luminous matter has never penetrated. At present, we simply do not know if such totally dark systems exist because we have no observational data either to confirm or to deny their presence.

WHAT IS DARK MATTER?

Whatever dark matter turns out to be, we know for certain that the universe contains large amounts of it. For every gram of glowing material we can detect, there may be tens of grams of dark matter out there. Currently the astronomical jury is still out as to exactly what constitutes dark matter. In fact, one could say we are still at an early stage of exploration. Many candidates exist to account for the invisible mass, some relatively ordinary, others rather exotic.

Nevertheless, there is a framework in which we must work. Nucleosynthesis, which seeks to explain the origin of elements after the big bang, sets a limit to the number of baryons—particles of ordinary, run-of-the-mill matter—that can exist in the universe. This limit arises out of the Standard Model of the early universe, which has one free parameter—the ratio of the number of baryons to the number of photons.

From the temperature of the cosmic microwave background—which has been measured—the number of photons is now known. Therefore, to determine the number of baryons, we must observe stars and galaxies to learn the cosmic abundance of light nuclei, the only elements formed immediately after the big bang.

Without exceeding the limits of nucleosynthesis, we can construct an acceptable model of a low-density, open universe. In that model, we take approximately equal amounts of baryons and exotic matter (nonbaryonic particles), but in quantities that add up to only 20 percent of the matter needed to close the universe. This model universe matches all our actual observations. On the other hand, a slightly different model of an open universe in which all matter is baryonic would also satisfy observations. Unfortunately, this alternative model contains too many baryons, violating the limits of nucleosynthesis. Thus, any acceptable low-density universe has mysterious properties: most of the universe's baryons would remain invisible, their nature unknown, and in most models much of the universe's matter is exotic.

EXOTIC PARTICLES

Theorists have posited a virtual smorgasbord of objects to account for dark matter, although many of them have fallen prey to observational constraints. As leading possible candidates for baryonic dark matter, there are black holes (large and small), brown dwarfs (stars too cold and faint to radiate), Sun-size MACHOs, cold gas, dark galaxies and dark clusters, to name only a few.

What we fail to see with our eyes, or detectors, we can occasionally see with our minds, aided by computer graphics. Computers now play a key role in the search for dark matter. Historically, astronomers have focused on observations; now the field has evolved into an experimental science. Today's astronomical experimenters sit neither at lab benches nor at telescopes but at computer terminals. They scrutinize cosmic simulations in which tens of thousands of points, representing stars, gas and dark matter, interact gravitationally over a galaxy's lifetime. A cosmologist can tweak a simulation by adjusting the parameters of dark matter and then watch what happens as virtual galaxies evolve in isolation or in a more realistic, crowded universe.

Computer models can thus predict galactic behavior. For instance, when two galaxies suffer a close encounter, violently merging or passing briefly in the night, they sometimes spin off long tidal tails. Yet from the models, we now know these tails appear only when the dark matter of each galaxy's halo is three to 10 times greater than its luminous matter. Heavier halos produce stubbier tails.

New tools, no less than new ways of thinking, give us insight into the structure of the heavens. Less than 400 years ago Galileo put a small lens at one end of a cardboard tube and a big brain at the other end. In so doing, he learned that the faint stripe across the sky, called the Milky Way, in fact comprised billions of single stars and stellar clusters. Suddenly, a human being understood what a galaxy is. Perhaps in the coming century, another—as yet unborn—big brain will put her eye to a clever new instrument and definitively answer, What is dark matter?

What's New in the Milky Way?

Michael V. Magee

When I was about 12 years old my father took me outside one summer night in August and told me we were going to look for meteors in the sky. It was a dark, moonless night far out in the countryside where we lived at that time in the mountains of eastern Tennessee. I had no idea what a meteor shower was and all I knew about the night sky was that there were stars scattered about in what I thought were random patterns. Occasionally, the Moon appeared in different phases, and there were sometimes "stars" that appeared brighter than the rest, and after a short period of weeks or months they would be gone. It made no sense to me at that time but if my father said that there would be "shooting stars" then I believed him. And so, we watched. And so, I began to wonder what the night sky was all about.

It wasn't until I moved to Tucson, Arizona, that I really began to take an interest in astronomy, and even then it was from an amateur's level of understanding. I took a couple of courses in beginning astronomy at the University of Arizona and joined the local amateur astronomy club, where I began to absorb as much about the subject as I could. I even went to work as a student employee at the local planetarium thinking, "What better way to keep on the cutting edge of what's new in my favorite subject?" I didn't realize just how lucky those decisions would make me until I attended a lecture, as part of an ongoing lecture series at the planetarium, by none other than Bart J. Bok. Many astronomers noted for their expertise in their fields lectured as part of this series, but it was Bok's lecture on the Milky Way that truly began to open my eyes to the over-whelming vastness and complexity of the universe. I was fortunate to attend at least one more lecture by Bok and even had the chance to get to know him as an acquaintance before his passing. In that time he inspired me to learn much more about the Milky Way, and I have continued to do so to this day. I am now Director of Planetarium Operations at the University of Arizona's Flandrau Science Center in Tucson. Flandrau is the public outreach arm of the university's College of Science department. Through exhibits, star shows, science demonstrations and outreach programs to local schools, Flandrau's mission is to bring the excitement and enjoyment of science to our audience. As I continue to learn more about the Milky Way, I also get the honor of passing that knowledge on to schoolchildren and the general public through planetarium programs. I'm humbled by the ever-increasing knowledge professional astronomers pass along as they make new discoveries daily about the known universe. I only hope that I can convey some of that awe and inspiration along to you in this discussion of the latest findings about our home, the Milky Way.

In Bart Bok's fifth and final edition of his book *The Milky Way* (1981), he states in the opening chapter, "In this book we invite you to join us on a brief tour along the road to the heaven of the Greeks." What a beautiful and intriguing way to welcome someone to a discussion about the Milky Way! I immediately imagined peering through a peephole into the forbidden realm of the gods, where mysteries both amazing and frightening would await. The rest of the book did not let me down, and to this day the latest discoveries about the Milky Way continue to fascinate me. We live on a blue planet that is alive with oceans and storm systems, mountains and rivers, and countless forms of animal and vegetable life. It is our home in a neighborhood we call the solar system, with an average yellow star, called the Sun, at its center. The solar system is but one of countless other systems making up a larger city we call the Milky Way. The analogy could continue on to the distant galaxies, but it is the "city" and its inner workings that are at the forefront.

The process of galactic evolution is important in understanding how our Milky Way galaxy formed, how old it is and what its fate will be ultimately. Current understanding of stellar evolution goes a long way in refining that process. The Milky Way has been observed for thousands of years but it has only been in the past century that astronomers came to model our home galaxy as a flattened disk with differentially rotating arms and a central bulge. During that same time we realized that our Sun is actually a part of the Milky Way. A lot has happened in the decades since those realizations and we now have a much clearer picture of our galaxy. Much of what Bok discovered in his research during the middle part of the 20th century still holds true, but much has also changed, or at least been updated. New observing techniques and instrumentation for ground-based astronomy, along with the new age of satellite observatories such as the Hubble Space Telescope, the Advanced X-ray Astrophysical Facility (named Chandra after the late Nobel laureate Subrahmanyan Chandrasekhar), Hipparcos, the Space Infrared Telescope Facility and others have contributed to a huge increase in our knowledge base. With it comes an everchanging understanding of the universe in which we live.

STAR FORMATION

Estimates place the formation of our galaxy at about 1 billion years after the Big Bang. As astronomers study how stars are born, live out their lives and die, they combine those stellar evolution models with observations of distant galaxies to make a model of how our Milky Way Galaxy was formed. Simply put, stars form out of large clouds of gas and dust that collapse under their own gravity until the pressures and temperatures reach a level where nuclear fusion can begin. The Milky Way probably also began as a huge cloud of mostly hydrogen gas that collapsed, either as a whole or as several gas cloud fragments to form a protogalaxy. As star formation began, the resulting collapse of small pockets of gas within the galaxy, combined with explosions of unstable stars, gradually started to shape the Milky Way into what we observe today.

As Bok studied the Milky Way more than 50 years ago he came across small dark regions (dark nebulae), roughly circular in shape, within brighter emission nebulae. They became known as Bok globules and are thought to be areas of dust and gas that are collapsing to form new stars. More than 250 have been catalogued in our galaxy to date. Researchers using the Hubble Space Telescope's (HST) and its Wide Field and Planetary Camera 2 (WFPC2) have "accidentally" found five new Bok globules, but this time they are located in the Large Magellanic Cloud, a companion galaxy to the Milky Way. Four of these new globules have masses less than our Sun, but the fifth has an estimated mass of about 80 Suns. If these globules are truly the beginnings of star formation then we now know it is not a process peculiar to just the Milky Way.

Looking toward the constellation Orion, astronomers using the 4-meter Mayall tele-scope at Kitt Peak National Observatory have produced high-resolution images of Herbig-Haro objects, thought to be the products of protostar formation. They appear as bright nebulae on the edges of darker nebulae (similar to Bok globules), and in this new image narrow fountains of ionized gas can be seen ejecting from their parent stars. These young solar-mass stars are evolving onto the main sequence, and jets from the stars produce shock waves within the surrounding material, probably the result of a series of eruptive mass-loss events. Studying these jets will hopefully provide insight into the process by which cold gas clouds gradually collapse into new stars.

Star formation is certainly a complex process and many parts of the process are not well understood. Astronomers have thought for a long time now that stars begin their lives as dark clouds of dust and gas that, by some unknown process, begin to coalesce into a protostellar clump. Finding details of this initial process had been almost impossible until the Hubble Space Telescope came along. New images from the WFPC2 are shed-ding important information on this process. Taking a look deep into the huge nebula surrounding NGC 3603, a star cluster some 20,000 light-years away in the constella-tion Carina, HST reveals not only many clues about star formation but clues about all of the stages of stellar evolution. Protostars within the nebula eject jets of material from their poles, which produce shock waves throughout the nearby medium. Proplyds, or protoplanetary disks, surrounding newly formed stars may not last long enough for planets to form. The ionizing radiation from the newer stars that survive the protostar stage strips other nearby protostars of the very material they need to continue their formation process, robbing them of their chance to live. Ionizing radiation from the main cluster of hot blue-white young stars that have already formed is too overpowering for most of the other protostellar systems to survive. Some of the first stars to form end up as blue supergiants, such as Sher 25, and in their enthusiasm to gather up as much mass as they could in the early stages of development, they end up signing their own death warrants. At 120 solar masses, Sher 25 is only about three million years old (very young for a star) and is already showing signs of dying due to its excess mass. Observations of its spectrum show large amounts of nitrogen in the gas and dust clouds that the star has already blown off. This means that the star is rapidly burning through its hydrogen and helium and moving on to produce much heavier elements, a sure sign of old age. As it reaches the iron-burning stage it will most likely go supernova, and the resulting expansion of material may contribute to the beginnings of new star formation elsewhere in the nebula.

Another active region of star formation can be found in the Trifid Nebula (M20) in Sagittarius. Similar to images taken by HST of the Eagle Nebula, the Trifid shows many of the same types of features. Evaporating Gaseous Globules (called EGGs) stretch out almost a light-year in length looking like pillars of dust and gas that have been ejected from some powerful hidden object. In fact, the hidden object is a protostar and its jets have created these pillars as they push their way through the surrounding medium. In some cases the parent protostar may be spinning or wobbling over a period of many years to produce a corkscrew appearance of the ejected material. Once again the radiation from newly formed stars within the nebula may help star formation along or may extinguish many of these protostars before they get a chance to evolve into a more stable form. These observed jets and pillars of material may in fact be the last breaths of some of the protostars before they die of starvation, their supply of gas and dust being cut off by the more better-established stars of the system. Our Sun and most stars like it were probably created in situations just like this. It is no small miracle that any of these stars survive, much less produce planets, much less produce planets like Earth. We can certainly take some pride in the fact that we (meaning our Sun and system of planets) beat the odds in the game of star formation.

MULTIPLE STAR SYSTEMS

As astronomers study the processes of stellar evolution, one fact has become very clear. The majority of stars observed exist in multiple star systems. And of those systems well over half are binary, two stars orbiting around each other. Binary star systems are popping up everywhere. Well, not popping up in the sense of being formed overnight, but as observational and theoretical techniques improve, astronomers are finding that multiple star systems are quite the norm in our galaxy. In some cases they are discovering binary protostar systems, lending evidence to the understanding of how binary stars form early on. As they observe younger and younger star populations through the help of instruments like the HST, researchers are gaining ground in understanding the early stages of star formation. One area under study is how multiple star systems form in the early stages of the collapse of a protostellar cloud. New computer modeling techniques combined with faster computer speeds have produced simulations of binary star system formation that are being confirmed by observational evidence. Using a process called speckle imaging, the relatively young stars of both the Hyades and Pleiades star clusters show populations that consist of about half as multiple stars, mostly binaries. Even younger stars in the Taurus and Ophiuchus star-forming regions show similar proportions. Using the Near Infrared Camera and Multi-Object Spectrometer (NICMOS) aboard HST, researchers have found an object in Taurus called TMR 1, confirmed as a binary protostar system. This system is only a few hundred thousand years old, which implies that binaries must form very early in the development of protostellar systems. Current computer simulations using a process called smoothed-particle-hydrodynamics can be applied to the initial conditions of the dense clouds on the verge of collapsing into a protostar. In the early stages magnetic fields control the collapse of the cloud, but at a certain point the collapse becomes much more rapid and then the rotational rate of the cloud takes over to determine the outcome. If the system rotates fast it will most likely turn into a binary; if the rotation rate is too slow, it will become a single protostar. The formation of planets in binary star systems is far less understood much less than that of the binaries themselves. About a third of the extrasolar planets discovered so far have been found around binary stars, and most of those planets are giants, much bigger than our planet Jupiter.

The basic life force of ordinary stars is nuclear fusion. Deep in their cores hydrogen is converted into helium in huge quantities every second. Energy is released by this fusion process in the form of x rays and gamma rays. It takes time for this high-energy radiation to reach the surface of the star, and by the time it gets there it has dispersed much of its energy into longer wavelengths of visible and infrared light. But recent evidence suggests that this is not the only way that fusion generates energy that we can observe. Binary star systems made up of white dwarf stars and ordinary companion stars emit large amounts of "soft" x-ray energy, radiation bordering on the ultraviolet part of the electromagnetic spectrum. As the white dwarf orbits its companion, it gradually aquires hydrogen gas from its partner and accumulates the gas on its surface until it reaches the critical mass needed for nuclear fusion to occur. Hence, x-ray emissions are created by nuclear fusion at the surface of the dwarf star rather than in the core like ordinary stars. This type of system now provides an answer for the cause of one of the types of supernovae that have been observed for many years. Type Ia supernovae provide one of the tools cosmologists use to measure the distance to the most distant galaxies, which in turn helps define the cosmic expansion rate and ultimately the age of the universe. In addition, supersoft x-ray stars provide an explanation for the cause of many of the variable stars, novae and supernovae that have been observed for many years but are not well understood. They are all unified as variations of the same white dwarf–companion star theme.

STRANGER STILL

Brown dwarfs are unusual objects in that they never quite made it as a star but are much more than just a gas-giant planet. They are kind of the missing link that astronomers were looking for to help interpret how stars and their planetary systems form and evolve. Since the process of formation for brown dwarfs is thought to be virtually the same as for stars, it lends even more support to the idea of binary system formation in the early stages of star formation. Most astronomers refer to them as failed stars, but they offer a great deal of insight into the processes by which planets like Jupiter and Saturn form and live out their lives. Up until about 1995, brown dwarfs were only theoretical objects, but now several dozen are known. One of the most prolific projects for discovering these objects is the 2 Micron All Sky Survey (2MASS). By 1999, the survey was only about half complete but had already discovered 39 new L-type objects, of which about a third are brown dwarfs. (The L spectral type follows M in the stellar spectral sequence and represents the first addition to the classification scheme in more than 50 years.) Young L-type brown dwarfs have been discovered around several nearby stars including one in the Pleiades star cluster. Called Roque 25, this brown dwarf is about 35 Jupiter masses, and as a member of the Pleiades is thought to be of the same age as the cluster, about 100 million years old. From this observation astronomers infer that brown dwarfs may form out of the collapse of a large cloud of dust and gas in the same way that stars do, rather than from the planetary disks surrounding stars that have already formed.

The Ring Nebula (M57) is one of the easiest to find and most notable of all the planetary nebulas. Anyone with a reasonably-sized amateur telescope can view this wonder and discern its ring shape. For many of us that have viewed this nebula over the years, we have usually been told that it is not really ring-shaped but is in fact more of a sphere. But in recent years astronomers have posed the notion that this nebula has probably more of a cylinder shape and that we are looking at it edge-on. New research using the HST has yielded imaging results that lend support to the cylindrical shape theory. In the edges of the nebula are dark dust globules that partially obscure light from behind, but toward the center of the nebula no such globules can be seen blocking the view of the inside. This gives support to the idea that the view toward the center is through a much less dense region than previously thought. The nebula is formed of the gaseous remains of a Sunlike star that ejected most of its atmosphere, and the remaining core of the star ionizes the gases with ultraviolet radiation causing the glow that we see. If in fact the ring is not spherical in shape, then it might have been formed from an earlier stellar system with a high angular momentum containing another star or even massive planets.

New imaging techniques in every field of astronomy are leading to clarification about many objects that have not always been well understood. In radio astronomy one of the big challenges is how to achieve the best resolution of an object over a given area. A small area can be imaged with high resolution or a much larger area with less resolution. Now, techniques combining data from observations made of both smaller and larger areas allow objects to be studied in more detail. An interesting case in point is the supernova remnant W50 and its starlike companion SS 433. Previous estimates of the distance to these objects often varied from about 7,000 to 17,000 light-years giving different values for the distance to each object. But analysis of the new composite images formed from the Very Large Array in Socorro, New Mexico, leads astronomers to a distance estimate of 9,800 light years for the system. In addition, studies of SS 433 show it to be a close binary system consisting of a normal star and a neutron star. Mass transfer between the two has resulted in a pair of jets that shoot outward in opposite directions tracing a corkscrew pattern that is visible in the two extensions of W50. These factors led astronomers to view the two objects as part of an integrated system rather than just a line-of-sight anomaly.

New imaging and observational techniques are also helping to add more pieces to the puzzle of stellar evolution. Many stars once thought to be fairly normal are turning out to be anything but. Most of us are familiar with Polaris, the North Star. As steady a beacon as Polaris seems to be, it is in fact a variable star classified as a Cepheid variable. This is not news, but the amplitude and period of Polaris's light variations has been changing over the years in ways atypical of Cepheids, and that is news. Studies of Polaris will modify our understanding of variable stars and stellar evolution overall. Then there are "blue stragglers," stars in globular clusters that are a little bluer than the cluster's hydrogen-burning main-sequence stars. When looking for pieces to the puzzle of a star's life cycle astronomers have long used globular star clusters as an example of the oldest stars in our galaxy. But blue stragglers seem to be a little too old. They should have moved off the main sequence to become red giants but they haven't. Astronomers think this is due to the infusion of new hydrogen fuel from collisions with other stars. Blue stragglers have been found outside the Milky Way in the globular cluster NGC 121 in the Small Magellanic Cloud (SMC). Using the WFPC2 on the HST, astronomers singled out 23 blue stragglers. At the same time they determined that while this cluster is the oldest of those found in the SMC, it is also some 2 billion years younger than most of the globular clusters of the Milky Way. Continuing studies will focus on how these stars fit into the stellar age scale and what they tell us about the evolution of stars.

We have already discussed the significance of multiple star systems. But in the case of Eta Carinae, a star in the southern constellation Carina, astronomers have yet to determine if it is just one star or more. Eta Carina is unusual in that it varies its x-ray output as well as exhibiting changes in its spectrum in a period of every 5.5 years. Research conducted with Hubble's visible-light and ultraviolet spectrograph showed that the star doubled in brightness between December 1997 and February 1999. Researchers say that this is the most rapid and largest brightening of Eta Carinae in the past 50 years and it is now brighter than at any time in the past 130 years. Data from HST's near infrared camera (dormant at the end of the 1990s) will be used in the future to construct a three-dimensional view of the structure of the nebula surrounding the star. Continued observations of this star will be very valuable in studies of stellar life cycles, and it is also one area where amateur observations will be quite useful. Eta Carinae is one of the most massive known stars and will probably explode as a supernova in the near astronomical future.

To catch a supernova in the past you either had to be lucky enough to see it visually as it happened (a rare event) or you could catch it on a photographic plate several days or weeks after it happened, in another galaxy. The last time anyone saw a supernova in the Milky Way was when Johannes Kepler saw one in 1604, and luckily it was very bright. Then there was SN 1987A discovered in the Large Magellanic Cloud, but it was not all that bright and ended up being somewhat unusual for a supernova. It was a blue supergiant of about 6 solar masses, whereas most of the likely supernova candidates are heavy stars like red giants of more than 8 solar masses. So, how can anyone reasonably predict when the next supernova is going to appear? The answer lies in how supernovae happen. When a giant star runs out of fuel its core collapses, becoming very dense and almost instantly releasing huge quantities of neutrinos. The neutrinos travel outward through space long before the core of the star begins to rebound pushing shock waves out toward the surface. The visual explosion we recognize as a supernova may occur as much as a day after the initial neutrino release. Particle physicists using neutrino observatories can now detect these neutrino bursts and warn observers, both professional and amateur alike. With as much as a day's advance notice observers can prepare to photograph and analyze the resulting supernova. Rates of supernova occurrence have been estimated without much precision over the years, and some observations of x-ray remnants and strong radio sources suggest that many supernovae may be optically unexciting. One very special case is in the constellation Cassiopeia. Cas A is an example of a supernova that was not very bright optically, but it is one of the strongest radio sources known. Why was it not bright? Astronomers say that dimming by interstellar gas and dust seems unlikely. The Chandra Advanced X-Ray Astrophysical Facility satellite has now taken its first light picture of this unusual object and the star that caused the supernova may have been imaged at last. The brightness question has not been answered, but future studies of this object may help provide an answer. Improved chances of observing supernovae gives hope to increasing our database of supernova sightings and with that increase will come a better understanding of the early stages of star death.

THE FUTURE MILKY WAY

To aid astronomers in their quest to understand the evolutionary processes within the Milky Way, it is of utmost importance to know precise distances to the stars. The Hipparcos astrometry mission of the European Space Agency has yielded a catalog of accurate distances to 118,218 stars using parallax motions produced by the Earth's annual motion around the Sun. From nearby stars such as Vega and Fomalhaut to the stars of the Hyades cluster our understanding of stellar distances has improved at least tenfold. Especially important is the distance to the Hyades cluster, because astronomers have long used this cluster as a yardstick for calibrating the cosmic distance scale to the edge of the known universe. Also important is a revision of the local stellar mass density, which attempts to calculate the mass in a representative sample of a "typical" region of the galaxy. In this case the typical example is the region around our Sun out to a few hundred light-years. This mass density estimate is important in determining the amount of dark matter in not only our galaxy but the rest of the universe as well. Current interpretations using data from Hipparcos show that observational estimates of the amount of gas and dust in interstellar space combined with the mass of stars in the same region indicate that there is no significant dark matter in the disk of the Milky Way. This leads to a new problem in explaining why the outer parts of the Milky Way rotate at such high speeds, because there is no longer enough mass to account for it.

There is one mystery, however, that may now have an answer. Astronomers have thought for some time that the Milky Way Galaxy is powered by a black hole engine at its center. But attempts to see any details of this region in Sagittarius have been stalled by an inability to resolve anything out of the wide range of radio waves being emitted from the area. Using the Very Long Baseline Array (VLBA) astronomers now think they have resolved the size and possibly the shape of Sagittarius A*. As you look at Sgr A* in shorter and shorter radio wavelengths, it seems to get smaller and smaller, suggesting that it has not been resolved. But at a wavelength of 7mm, the shortest observed so far, Sgr A* does not appear as small as expected. Current data suggest that at a distance of 27,000 light-years Sgr A* appears to have a physical diameter of 3.6 astronomical units, or slightly larger than the orbit of Mars. More studies are needed to confirm these results and new observations already made at a wavelength of 3mm may confirm or refute these results.

Finally, there is the question of where the Milky Way will go from here. Part of this answer may come from astronomers using the 0.6-meter Burrell Schmidt telescope at Kitt Peak in Arizona. They have found a plume of luminous material to the side of NGC 4874, one of the two supermassive elliptical galaxies at the core of the Coma cluster. The plume resembles computer models showing tidal disruptions of one galaxy by another. The mechanism for the formation of these stellar plumes is still not understood, but astronomers suggest that as galaxies move past one another the tidal forces that result may rip some of the stars away from their parent galaxy or possibly an entire galaxy may break apart distributing its stars along its orbital path. This galactic cannibalism has been observed repeatedly in many galaxies outside of our own Milky Way, and it has been known for many years that the Milky Way is interacting with the Large and Small Magellanic Clouds. The gravitational attraction between the Milky Way and the clouds was thought to keep them in orbit around our galaxy. But at some point in the distant future the Milky Way may actually cannibalize the Large Magellanic Cloud. A lesser-known interaction is of a recently discovered dwarf spheroidal galaxy in Sagittarius that is orbiting the Milky Way at roughly the distance from the center as our own solar system. But, this galaxy is on the opposite side of the center from us, which makes it difficult to see. It has been in orbit around the Milky Way for billions of years passing through more and less dense regions of our galaxy with every orbit. Researchers are prone to wonder whether or not this galaxy will eventually gather up many of the stars of our Milky Way for its own or will it gradually fall apart, yielding to the more powerful gravity of the Milky Way, blending its stars with our own.

The universe may be on the order of 10 to 15 billion years old or older. If our Milky Way was formed about 1 billion years after the big bang, then is it a young galaxy or an old one? Using the trip our Sun takes once around the Milky Way as a birthday measure, then one cosmic year for our galaxy equals 240 million of our standard years. That would make the Milky Way about 58 cosmic years old. I do not know if that answers the question but I prefer to think that the Milky Way still has a lot of life left in it. We have come a long way in uncovering its secrets, but it is a large and bustling city, and from our home here in the solar system we have so much more to learn before we can truly understand its nature.

Chapter Four:
The Milky Way

Introduction

As we continue our journey inward through space and forward through time, we now focus on the Milky Way. I first saw the broad circle of faint light that represents the more distant stars of our galaxy, the Milky Way, while away at camp as a child in the mid-1950s. At that time Bart and Priscilla Bok were just completing the third edition of their classic book *The Milky Way* (1957), and the fact that our galaxy was spiral in shape had just been discovered. "The authors could hardly have guessed in 1945," they wrote, "that in the third edition they would have to find room for chapters on radio astronomy and on the spiral structure of the galaxy."

It seems that while I was lazily gazing at the Milky Way from summer camp, giant optical and radio telescopes were defining the shape of our galactic home as a giant spiral galaxy with arms. One laces through Cygnus, Cepheus, Cassiopeia, Perseus, and all the way through Orion. A second arm also stretches along through the sky, featuring the famous double cluster in Perseus. More recently we have become aware of an arm through Sagittarius and the Carina arm as well.

This chapter explores the myriad structures that exist within our galaxy, from the stellar nurseries that surround the formation of stars, to the binary suns that often interact with each other, to the novae, supernovae, neutron stars and black holes that mark the end of a star's life cycle. With this chapter, we leave the large-scale structures of the universe and start our focus toward home. The first step in that process is to learn about our home galaxy, which Bart Bok loved to call the "Bigger and Better Milky Way."

How the Milky Way Formed

Sidney van den Bergh and James E. Hesser

Attempts to reconstruct how the Milky Way formed and began to evolve resemble an archaeological investigation of an ancient civilization buried below the bustling center of an ever-changing modern city. Like archaeologists, astronomers, too, look at small, disparate clues to determine how our galaxy and others like it were born about a billion years after the big bang and took on their current shapes. The clues consist of the ages of stars and stellar clusters, their distribution and their chemistry—all deduced by looking at such features as color and luminosity. The shapes and physical properties of other galaxies can also provide insight concerning the formation of our own.

The evidence suggests that our galaxy, the Milky Way, came into being as a consequence of the collapse of a vast gas cloud. Yet that cannot be the whole story. Recent observations have forced workers who support the hypothesis of a simple, rapid collapse to modify their idea in important ways. This new information has led other researchers to postulate that several gas cloud fragments merged to create the protogalactic Milky Way, which then collapsed. Other variations on these themes are vigorously maintained. Investigators of virtually all persuasions recognize that the births of stars and supernovae have helped shape the Milky Way. Indeed, the formation and explosion of stars are at this moment further altering the galaxy's structure and influencing its ultimate fate.

Much of the stellar archaeological information that astronomers rely on to decipher the evolution of our galaxy resides in two regions of the Milky Way: the halo and the disk. The halo is a slowly rotating, spherical region that surrounds all the other parts of the galaxy. The stars and star clusters in it are old. The rapidly rotating, equatorial region constitutes the disk, which consists of young stars and stars of intermediate age, as well as interstellar gas and dust. Embedded in the disk are the sweepingly curved arms that are characteristic of spiral galaxies such as the Milky Way. Among the middle-aged stars is our Sun, which is located about 25,000 light-years from the galactic center. (When you view the night sky, the galactic center lies in the direction of Sagittarius.) The Sun completes an orbit around the center in approximately 200 million years.

That the Sun is part of the Milky Way was discovered less than 80 years ago. At the time, Bertil Lindblad of Sweden and the late Jan H. Oort of the Netherlands hypothesized that the Milky Way system is a flattened, differentially rotating galaxy. A few years later John S. Plaskett and Joseph A. Pearce of Dominion Astrophysical Observatory accumulated three decades' worth of data on stellar motions that confirmed the Lindblad-Oort picture.

In addition to a disk and a halo, the Milky Way contains two other subsystems: a central bulge, which consists primarily of old stars, and, within the bulge, a nucleus. Little is known about the nucleus because the dense gas clouds in the central bulge obscure it. The nuclei of some spiral galaxies, including the Milky Way, may contain a large black hole. A black hole in the nucleus of our galaxy, however, would not be as massive as those that seem to act as the powerful cores of quasars.

All four components of the Milky Way appear to be embedded in a large, dark corona of invisible material. In most spiral galaxies the mass of this invisible corona exceeds by an order of magnitude that of all the galaxy's visible gas and stars. Investigators are intensely debating what the constituents of this dark matter might be.

The clues to how the Milky Way developed lie in its components. Perhaps the only widely accepted idea is that the central bulge formed first, through the collapse of a gas cloud. The central bulge, after all, contains mostly massive, old stars. But determining when and how the disk and halo formed is more problematic.

In 1958 Oort proposed a model according to which the population of stars forming in the halo flattened into a thick disk, which then evolved into a thin one. Meanwhile further condensation of stars from the hydrogen left over in the halo replenished that structure. Other astronomers prefer a picture in which these populations are discrete and do not fade into one another. In particular, V. G. Berman and A. A. Suchkov of the Rostov State University in Russia have indicated how the disk and halo could have developed as separate entities.

These workers suggest a hiatus between star formation in the halo and that in the disk. According to their model, a strong wind propelled by supernova explosions interrupted star formation in the disk for a few billion years. In doing so, the wind would have ejected a significant fraction of the mass of the protogalaxy into intergalactic space. Such a process seems to have prevailed in the Large Magellanic Cloud, one of the Milky Way's small satellite galaxies. There an almost 10-billion-year interlude appears to separate the initial burst of creation of conglomerations of old stars called globular clusters and the more recent epoch of star formation in the disk. Other findings lend additional weight to the notion of distinct galactic components. The nearby spiral M33 contains a halo but no nuclear bulge. This characteristic indicates that a halo is not just an extension of the interior feature, as many thought until recently.

In 1962 a model emerged that served as a paradigm for most investigators. According to its developers—Olin J. Eggen, now at the National Optical Astronomical

Observatories, Donald Lynden-Bell of the University of Cambridge and Allan R. Sandage of the Carnegie Institution—the Milky Way formed when a large, rotating gas cloud collapsed rapidly, in about a few hundred million years. As the cloud fell inward on itself, the protogalaxy began to rotate more quickly; the rotation created the spiral arms we see today. At first, the cloud consisted entirely of hydrogen and helium atoms, which were forged during the hot, dense initial stages of the big bang. Over time the protogalaxy started to form massive, short-lived stars. These stars modified the composition of galactic matter, so that the subsequent generations of stars, including our Sun, contain significant amounts of elements heavier than helium.

Although the model gained wide acceptance, observations made during the past three decades have uncovered a number of problems with it. In the first place, investigators found that many of the oldest stars and star clusters in the galactic halo move in retrograde orbits—that is, they revolve around the galactic center in a direction opposite to that of most other stars. Such orbits suggest that the protogalaxy was quite clumpy and turbulent or that it captured sizable gaseous fragments whose matter was moving in different directions. Second, more refined dynamic models show that the protogalaxy would not have collapsed as smoothly as predicted by the simple model; instead the densest parts would have fallen inward much faster than more rarefied regions.

Several investigators have attempted to develop scenarios consistent with the findings. In 1977 Alar Toomre of the Massachusetts Institute of Technology postulated that most galaxies form from the merger of several large pieces rather than from the collapse of a single gas cloud. Once merged in this way, according to Toomre, the gas cloud collapsed and evolved into the Milky Way now seen. Leonard Searle of Carnegie Institution and Robert J. Zinn of Yale University have suggested a somewhat different picture, in which many small bits and pieces coalesced. In the scenarios proposed by Toomre and by Searle and Zinn, the ancestral fragments may have evolved in chemically unique ways. If stars began to shine and supernovae started to explode in different fragments at different times, then each ancestral fragment would have its own chemical signature. Work by one of us (van den Bergh) indicates that such differences do indeed appear among the halo populations.

Discussion of the history of galactic evolution did not advance significantly beyond this point until the 1980s. At that time, workers became able to record more precisely than ever before extremely faint images. This ability is critically important because the physical theories of stellar energy production—and hence the lifetimes and ages of stars—are most secure for so-called main-sequence stars. Such stars burn hydrogen in their cores; in general, the more massive the star, the more quickly it completes its main-sequence life. Unfortunately, this fact means that within the halo the only remaining main-sequence stars are the extremely faint ones. The largest, most luminous ones, which have burned past their main-sequence stage, became invisible long ago. Clusters are generally used to determine age. They are crucial because their distances from the earth can be determined much more accurately than can those of individual stars.

The technology responsible for opening the study of extremely faint halo stars is the charge-coupled device (CCD). This highly sensitive detector produces images electronically by converting light intensity into current. CCDs are far superior in most respects to photographic emulsions, although extremely sophisticated software, such as that developed by Peter B. Stetson of Dominion Astrophysical Observatory, is required to take full advantage of them. So used, the charge-coupled device has yielded a tenfold increase in the precision of measurement of color and luminosity of the faint stars in globular clusters.

Among the most important results of the CCD work done so far are more precise age estimates. Relative age data based on these new techniques have revealed that clusters whose chemistries suggest they were the first to be created after the big bang have the same age to within 500 million years of one another. The ages of other clusters, however, exhibit a greater spread.

The ages measured have helped researchers determine how long it took for the galactic halo to form. For instance, Michael J. Bolte, now at Lick Observatory, carefully measured the colors and luminosities of individual stars in the globular clusters NGC 288 and NGC 362. Comparison between these data and stellar evolutionary calculations shows that NGC 288 is approximately 15 billion years old and that NGC 362 is only about 12 billion years in age. This difference is greater than the uncertainties in the measurements. The observed age range indicates that the collapse of the outer halo is likely to have taken an order of magnitude longer than the amount of time first envisaged in the simple, rapid collapse model of Eggen, Lynden-Bell and Sandage.

Of course, it is possible that more than one model for the formation of the galaxy is correct. The Eggen–Lynden-Bell–Sandage scenario may apply to the dense bulge and inner halo. The more rarefied outer parts of the galaxy may have developed by the merger of fragments, along the lines theorized by Toomre or by Searle and Zinn. If so, then the clusters in the inner halo would have formed before those in the more tenuous outer regions. The process would account for some of the age differences found for the globular clusters.

Knowing the age of the halo is, however, insufficient to ascertain a detailed formation scenario. Investigators need to know the age of the disk as well and then to compare that age with the halo's age. Whereas globular clusters are useful in determining the age of the halo, another type of celestial body—very faint white dwarf stars—can be used to determine the age of the disk. The absence of white dwarfs in the galactic disk near the Sun sets a lower limit on the disk's age. White dwarfs, which are no longer producing radiant energy, take a long time to cool, so their absence means that the population in the disk is fairly young—less than about 10 billion years. This value is significantly less than the ages of clusters in the halo and is thus consistent with the notion that the bulk of the galactic disk developed after the halo.

It is, however, not yet clear if there is a real gap between the time when formation of the galactic halo ended and when creation of the old thick disk began. To estimate the duration of such a transitional period between halo and disk, investigators have compared the ages of the oldest stars in the disk with those of the youngest ones in the halo. The oldest known star clusters in the galactic disk, NGC 188 and NGC 6791, have ages of nearly eight billion years, according to Pierre Demarque and David B. Guenther of Yale and Elizabeth M. Green of the University of Arizona. Stetson and his colleagues and Roberto Buonanno of the Astronomical Observatory in Rome and his coworkers examined globular clusters in the halo population. They found the youngest globulars—Palomar 12 and Ruprecht 106—to be about 11 billion years old. If the few billion years' difference between the disk objects and the young globulars is real, then young globulars may be the missing links between the disk and halo populations of the galaxy.

Unfortunately, the relative ages of only a few globular clusters have been precisely estimated. As long as this is the case, one can argue that the Milky Way could have tidally captured Palomar 12 and Ruprecht 106 from the Magellanic Clouds. This scenario, proposed by Douglas N. C. Lin of the University of California at Santa Cruz and Harvey B. Richer of the University of British Columbia, would obviate the need for a long collapse time. Furthermore, the apparent age gap between disk and halo might be illusory. Undetected systematic errors may lurk in the age-dating processes. Moreover, gravitational interactions with massive interstellar clouds may have disrupted the oldest disk clusters, leaving behind only younger ones.

Determining the relative ages of the halo and disk reveals much about the sequence of the formation of the galaxy. On the other hand, it leaves open the question of how old the entire galaxy actually is. The answer would provide some absolute framework by which the sequence of formation events can be discerned. Most astronomers who study star clusters favor an age of some 15 to 17 billion years for the oldest clusters (and hence the galaxy).

Confidence that those absolute age values are realistic comes from the measured abundance of radioactive isotopes in meteorites. The ratios of thorium 232 to uranium 235, of uranium 235 to uranium 238 or of uranium 238 to plutonium 244 act as chronometers. According to these isotopes, the galaxy is between 10 and 20 billion years old. Although ages determined by such isotope ratios are believed to be less accurate than those achieved by comparing stellar observations and models, the consistency of the numbers is encouraging.

Looking at the shapes of other galaxies alleviates to some extent the uncertainty of interpreting the galaxy's evolution. Specifically, the study of other galaxies presents a perspective that is unavailable to us as residents of the Milky Way—an external view. We can also compare information from other galaxies to see if the processes that created the Milky Way are unique.

The most immediate observation one can make is that galaxies come in several shapes. In 1925 Edwin P. Hubble found that luminous galaxies could be arranged in a linear sequence according to whether they are elliptical, spiral or irregular. From an evolutionary point of view, elliptical galaxies are the most advanced. They have used up all (or almost all) of their gas to generate stars, which probably range in age from 10 to 15 billion years. Unlike spiral galaxies, ellipticals lack disk structures. The main differences between spiral and irregular galaxies is that irregulars have neither spiral arms nor compact nuclei.

The morphological types of galaxies can be understood in terms of the speed with which gas was used to create stars.

Star formation in spirals began less rapidly than it did in ellipticals but continues to the present day. Spirals are further subdivided into categories Sa, Sb and Sc. The subdivisions refer to the relative size of the nuclear bulges and the degree to which the spiral arms coil. Objects of type Sa have the largest nuclear bulges and the most tightly coiled arms. Such spirals also contain some neutral hydrogen gas and a sprinkling of young blue stars. Sb spirals have relatively large populations of young blue stars in their spiral arms. The central bulge, containing old red stars, is less prominent than is the central bulge in spirals of type Sa. Finally, in Sc spirals the light comes mainly from the young blue stars in the spiral arms; the bulge population is inconspicuous or absent. The Milky Way is probably intermediate between types Sb and Sc.

Information from other spirals seems consistent with the data obtained for the Milky Way. Like those in our galaxy, the stars in the central bulges of other spirals arose early. The dense inner regions of gas must have collapsed first. As a result, most of the primordial gas initially present near the centers has turned into stars or has been ejected by supernova-driven winds.

There is an additional kind of evidence on which to build our understanding of how the Milky Way came into existence: the chemical composition of stars. This information helps to pinpoint the relative ages of stellar populations. According to stellar models, the chemistry of a star depends on when it formed. The chemical differences exist because first-generation stars began to "pollute" the protogalaxy with elements heavier than helium. Such so-called heavy elements, or "metals," as astronomers refer to them, were created in the interiors of stars or during supernova explosions. Examining the makeup of stars can provide stellar evolutionary histories that corroborate or challenge age estimates.

Despite the quantity of data, information about metal content has proved insufficient to settle the controversy concerning the time scale of disk and halo formation.

Such differences in interpretation often reflect nearly unavoidable effects arising from the way in which particular samples of stars are selected for study. For example, some stars exhibit chemical compositions similar to those of "genuine" halo stars, yet they have kinematics that would associate them with one of the subcomponents of the disk. As vital as it is, chemical information alone does not resolve ambiguities about the formation of the galactic halo and disk.

As well as telling us about the past history of our galaxy, the disk and halo also provide insight into the Milky Way's probable future evolution. One can easily calculate that almost all of the existing gas will be consumed in a few billion years. This estimate is based on the rate of star formation in the disks of other spirals and on the assumption that the birth of stars will continue at its present speed. Once the gas has been depleted, no more stars will form, and the disks of spirals will then fade. Eventually the galaxy will consist of nothing more than white dwarfs and black holes encapsulated by the hypothesized dark matter corona.

Several sources of evidence exist for such an evolutionary scenario. In 1978 Harvey R. Butcher of the Kapteyn Laboratory in the Netherlands and Augustus Oemler, Jr., of Yale found that dense clusters of galaxies located about six billion light-years away still contained numerous spiral galaxies. Such spirals are, however, rare or absent in nearby clusters of galaxies. This observation shows that the disks of most spirals in dense clusters must have faded to invisibility during the past six billion years. Even more direct evidence for the swift evolution of galaxies comes from the observation of so-called blue galaxies. These galaxies are rapidly generating large stars. Such blue galaxies seem to be less common now than they were only a few billion years ago. Of course, the life of spiral galaxies can be extended. Copious infall of hydrogen from intergalactic space might replenish the gas supply. Such infall can occur if a large gas cloud or another galaxy with a substantial gas reservoir is nearby. Indeed, the Magellanic Clouds will eventually plummet into the Milky Way, briefly rejuvenating our galaxy. Yet the Milky Way will not escape its ultimate fate. Like people and civilizations, stars and galaxies leave behind only artifacts in an evolving, ever dynamic universe.

Black Holes and the Information Paradox

Leonard Susskind

Somewhere in outer space, Professor Windbag's time capsule has been sabotaged by his arch rival, Professor Goulash. The capsule contains the only copy of a vital mathematical formula, to be used by future generations. But Goulash's diabolical scheme to plant a hydrogen bomb on board the capsule has succeeded. Bang! The formula is vaporized into a cloud of electrons, nucleons, photons and an occasional neutrino. Windbag is distraught. He has no record of the formula and cannot remember its derivation.

Later, in court, Windbag charges that Goulash has sinned irrevocably: "What that fool has done is irreversible. Why, the fiend has destroyed my formula and must pay. Off with his tenure!"

"Nonsense," says an unflustered Goulash. "Information can never be destroyed. It's just your laziness, Windbag. Although it's true that I've scrambled things a bit, all you have to do is go and find each particle in the debris and reverse its motion. The laws of nature are time symmetric, so on reversing everything, your stupid formula will be reassembled. That proves, beyond a shadow of a doubt, that I could never have destroyed your precious information." Goulash wins the case.

Windbag's revenge is equally diabolical. While Goulash is out of town, his computer is burglarized, along with all his files, including his culinary recipes. Just to make sure that Goulash will never again enjoy his famous matelote d'anguilles [eel stew] with truffles, Windbag launches the computer into outer space and straight into a nearby black hole.

At Windbag's trial, Goulash is beside himself. "You've gone too far this time, Windbag. There's no way to get my files out. They're inside the black hole, and if I go in to get them I'm doomed to be crushed. You've truly destroyed information, and you'll pay."

"Objection, Your Honor!" Windbag jumps up. "Everyone knows that black holes eventually evaporate. Wait long enough, and the black hole will radiate away all its mass and turn into outgoing photons and other particles. True, it may take 1070 years, but it's the principle that counts. It's really no different from the bomb. All Goulash has to do is reverse the paths of the debris, and his computer will come flying back out of the black hole."

"Not so!" cries Goulash. "This is different. My recipe was lost behind the black hole's boundary, its horizon. Once something crosses the horizon, it can never get back out without exceeding the speed of light. And Einstein taught us that nothing can ever do that. There is no way the evaporation products, which come from outside the horizon, can contain my lost recipes even in scrambled form. He's guilty, Your Honor."

Her Honor is confused. "We need some expert witnesses. Professor Hawking, what do you say?"

Stephen W. Hawking of the University of Cambridge comes to the stand. "Goulash is right. In most situations, information is scrambled and in a practical sense is lost. For example, if a new deck of cards is tossed in the air, the original order of the cards vanishes. But in principle, if we know the exact details of how the cards are thrown, the original order can be reconstructed. This is called microreversibility. But in my 1976 paper I showed that the principle of microreversibility, which has always held in classical and quantum physics, is violated by black holes. Because information cannot escape from behind the horizon, black holes are a fundamental new source of irreversibility in nature. Windbag really did destroy information."

Her Honor turns to Windbag: "What do you have to say to that?" Windbag calls on Professor Gerard 't Hooft of Utrecht University.

"Hawking is wrong," 't Hooft begins. "I believe black holes must not lead to violation of the usual laws of quantum mechanics. Otherwise the theory would be out of control. You cannot undermine microscopic reversibility without destroying energy conservation. If Hawking were right, the universe would heat up to a temperature of 10^{31} degrees in a tiny fraction of a second. Because this has not happened, there must be some way out of this problem." The problem is known as the information paradox.

When something falls into a black hole, one cannot expect it ever to come flying back out. The information coded in the properties of its constituent atoms is, according to Hawking, impossible to retrieve. The problem, 't Hooft points out, is that if the information is truly lost, quantum mechanics breaks down. Despite its famed indeterminacy, quantum mechanics controls the behavior of particles in a very specific way: it is reversible. When one particle interacts with another, it may be absorbed or reflected or may even break up into other particles. But one can always reconstruct the initial configurations of the particles from the final products.

If this rule is broken by black holes, energy may be created or destroyed, threatening one of the most essential underpinnings of physics. The conservation of energy is ensured by the mathematical structure of quantum mechanics, which also guarantees reversibility; losing one means losing the other. As Thomas Banks, Michael Peskin and I showed in 1980 at Stanford University, information loss in a black hole leads to enormous amounts of energy being generated. For such reasons, 't Hooft and I believe the information that falls into a black hole must somehow become available to the outside world.

Some physicists feel the question of what happens in a black hole is academic or even theological, like counting angels on pinheads. But it is not so at all: at stake are the future rules of physics. Processes inside a black hole are merely extreme examples of interactions between elementary particles. At the energies that particles can acquire in today's largest accelerators (about 10^{12} electron volts), the gravitational attraction between them is negligible. But if the particles have a "Planck energy" of about 10^{28} electron volts, so much energy—and therefore mass—becomes concentrated in a tiny volume that gravitational forces outweigh all others. The resulting collisions involve quantum mechanics and the general theory of relativity in equal measure.

The physics at Planck energies may be revealed by the known properties of matter. Elementary particles have a variety of attributes that lead physicists to suspect they are not so elementary after all: they must actually have a good deal of undiscovered internal machinery, which is determined by the physics at Planck energies. We will recognize the right confluence of general relativity and quantum physics—or quantum gravity—by its ability to explain the measurable properties of electrons, photons, quarks or neutrinos.

A black hole is born when so much mass or energy gathers in a small volume that gravitational forces overwhelm all others and everything collapses under its own weight. The material squeezes into an unimaginably small region called a singularity, the density inside of which is essentially infinite. But it is not the singularity itself that will interest us.

Surrounding the singularity is an imaginary surface called the horizon. For a black hole with the mass of a galaxy, the horizon is 10^{11} kilometers from the center—as far as the outermost reaches of the solar system are from the Sun. For a black hole of solar mass, the horizon is roughly a kilometer away; for a black hole with the mass of a small mountain, the horizon is 10^{13} centimeters away, roughly the size of a proton.

The horizon separates space into two regions that we can think of as the interior and exterior of the black hole. Suppose that Goulash, who is scouting for his computer near the black hole, shoots a particle away from the center. If he is not too close and the particle has a high velocity, then it may overcome the gravitational pull of the black hole and fly away. It will be most likely to escape if it is shot with the maximum velocity—that of light. If, however, Goulash is too close to the singularity, the gravitational force will be so great that even a light ray will be sucked in. The horizon is the place with the (virtual) warning sign: Point of No Return. No particle or signal of any kind can cross it from the inside to the outside.

AT THE HORIZON

An analogy inspired by William G. Unruh of the University of British Columbia, one of the pioneers in black hole quantum mechanics, helps to explain the relevance of the horizon. Imagine a river that gets swifter downstream. Among the fish that live in it, the fastest swimmers are the "lightfish." But at some point, the river flows at the fish's maximum speed; clearly, any lightfish that drifts past this point can never get back up.

What happens to Goulash, who in a careless moment gets too close to the black hole's horizon? Like the freely drifting fish, he senses nothing special: no great forces, no jerks or flashing lights. He checks his pulse with his wristwatch—normal. His breathing rate—normal. To him the horizon is just like any other place.

But Windbag, watching Goulash from a spaceship safely outside the horizon, sees Goulash acting in a bizarre way. Windbag has lowered to the horizon a cable equipped with a camcorder and other probes, to better keep an eye on Goulash. As Goulash falls toward the black hole, his speed increases until it approaches that of light. Einstein found that if two persons are moving fast relative to each other, each sees the other's clock slow down; in addition, a clock that is near a massive object will run slowly compared with one in empty space. Windbag sees a strangely lethargic Goulash. As he falls, the latter shakes his fist at Windbag. But he appears to be moving ever more slowly; at the horizon, Windbag sees Goulash's motions slow to a halt. Although Goulash falls through the horizon, Windbag never quite sees him get there.

In fact, not only does Goulash seem to slow down, but his body looks as if it is being squashed into a thin layer. Einstein also showed that if two persons move fast with respect to each other, each will see the other as being flattened in the direction of motion. More strangely, Windbag should also see all the material that ever fell into the black hole, including the original matter that made it up—and Goulash's computer—similarly flattened and frozen at the horizon. With respect to an outside observer, all of that matter suffers a relativistic time dilation. To Windbag, the black hole consists of an immense junkyard of flattened matter at its horizon. But Goulash sees nothing unusual until much later, when he reaches the singularity, there to be crushed by ferocious forces.

Black hole theorists have discovered over the years that from the outside, the properties of a black hole can be described in terms of a mathematical membrane above the horizon. This layer has many physical qualities, such as electrical conductivity and viscosity. Perhaps the most surprising of its properties was postulated in the early 1970s by Hawking, Unruh and Jacob D. Bekenstein of the Hebrew University in Israel. They found that as a consequence of quantum mechanics, a black hole—in particular, its horizon—behaves as though it contains heat. The horizon is a layer of hot material of some kind.

The temperature of the horizon depends on just where it is measured. Suppose one of the probes that Windbag has attached to his cable is a thermometer. Far from the horizon he finds that the temperature is inversely proportional to the black hole's mass. For a black hole of solar mass, this "Hawking temperature" is about 10^{-8} degree—far colder than intergalactic space. As Windbag's thermometer approaches the horizon, however, it registers higher temperatures. At a distance of a centimeter, it measures about a thousandth of a degree; at a nuclear diameter it records 10 billion degrees. The temperature ultimately becomes so high that no imaginable thermometer can measure it.

Hot objects also possess an intrinsic disorder called entropy, which is related to the amount of information a system can hold. Think of a crystal lattice with N sites; each site can house one atom or none at all. Thus, every site holds one "bit" of information, corresponding to whether an atom is there or not; the total lattice has N such bits and can contain N units of information. Because there are two choices for each site and N ways of combining these choices, the total system can be in any one of 2^N states (each of which corresponds to a different pattern of atoms). The entropy (or disorder) is defined as the logarithm of the number of possible states. It is roughly equal to N—the same number that quantifies the capacity of the system for holding information.

Bekenstein found that the entropy of a black hole is proportional to the area of its horizon. The precise formula, derived by Hawking, predicts an entropy of 3.2×10^{64} per square centimeter of horizon area. Whatever physical system carries the bits of information at the horizon must be extremely small and densely distributed: their linear dimensions have to be $1/10^{20}$ the size of a proton's. They must also be very special for Goulash to completely miss them as he passes through.

The discovery of entropy and other thermodynamic properties of black holes led Hawking to a very interesting conclusion. Like other hot bodies, a black hole must radiate energy and particles into the surrounding space. The radiation comes from the region of the horizon and does not violate the rule that nothing can escape from within. But it causes the black hole to lose energy and mass. In the course of time an isolated black hole radiates away all its mass and vanishes.

BLACK HOLE COMPLEMENTARITY

Is it possible that Goulash and Windbag are in a sense both correct? Can it be that Windbag's observations are indeed consistent with the hypothesis that Goulash and his computer are thermalized and radiated back into space before ever reaching the horizon, even though Goulash discovers nothing unusual until long after, when he encounters the singularity? The idea that these are not contradictory but complementary scenarios was first put forward as the principle of black hole complementarity by Lárus Thorlacius, John Uglum and me at Stanford. Very similar ideas are also found in 't Hooft's work. Black hole complementarity is a new principle of relativity. In the special theory of relativity we find that although different observers disagree about the lengths of time and space intervals, events take place at definite space-time locations. Black hole complementarity does away with even that.

How this principle actually comes into play is clearer when applied to the structure of subatomic particles. Suppose that Windbag, whose cable is also equipped with a powerful microscope, watches an atom fall toward the horizon. At first he sees the atom as a nucleus surrounded by a cloud of negative charge. The electrons in the cloud move so rapidly they form a blur. But as the atom gets closer to the black hole, its internal motions seem to slow down, and the electrons become visible. The protons and neutrons in the nucleus still move so fast that its structure is obscure. But a little later the electrons freeze, and the protons and neutrons start to show up. Later yet, the quarks making up these particles are revealed. (Goulash, who falls with the atom, sees no changes.)

Many physicists believe elementary particles are made of even smaller constituents. Although there is no definitive theory for this machinery, one candidate stands out as being the most promising—namely, string theory. In this theory, an elementary particle does not resemble a point; rather it is like a tiny rubber band that can vibrate in many modes. The fundamental mode has the lowest frequency; then there are higher harmonics, which can be superimposed on top of one another. There are an infinite number of such modes, each of which corresponds to a different elementary particle.

A string is a minute object, $1/10^{20}$ the size of a proton. But as it falls into a black hole, its vibrations slow down, and more of them become visible. Mathematical studies done at Stanford by Amanda Peet, Thorlacius, Arthur Mezhlumian and me have demonstrated the behavior of a string as its higher modes freeze out. The string spreads and grows, just as if it were being bombarded by particles and radiation in a very hot environment. In a relatively short time the string and all the information that it carries are smeared over the entire horizon.

This picture applies to all the material that ever fell into the black hole—because according to string theory, everything is ultimately made of strings. Each elementary string spreads and overlaps all the others until a dense tangle covers the horizon. Each minute segment of string, measuring 10^{-33} centimeters across, functions as a bit. Thus, strings provide a means for the black hole's surface to hold the immense amount of information that fell in during its birth and thereafter.

STRING THEORY

It seems, then, that the horizon is made of all the substance in the black hole, resolved into a giant tangle of strings. The information, as far as an outside observer is concerned, never actually fell into the black hole; it stopped at the horizon and was later radiated back out. String theory offers a concrete realization of black hole complementarity and therefore a way out of the information paradox. To outside observers—that is, us—information is never lost. Most important, it appears that the bits at the horizon are minute segments of string.

Just as quantum mechanics describes the radiation of an atom by showing how an electron jumps from a high-energy "excited" state to a low-energy "ground" state, quantum strings seem to account for the spectrum of radiation from an excited black hole.

Quantum mechanics, I believe, will in all likelihood turn out to be consistent with the theory of gravitation; these two great streams of physics are merging into a quantum theory of gravity based on string theory. The information paradox, which appears to be well on its way to being resolved, has played an extraordinary role in this ongoing revolution in physics. And although Goulash would never admit it, Windbag will probably turn out to be right: the recipe for matelote d'anguilles is not forever lost to the world.

Collapse and Formation of Stars

Alan P. Boss

What are the early stages in the formation of a star? What determines whether a cloud of star-forming matter will evolve into one, two or several stars? Because clouds of gas, dust and debris largely obscure all but the initial and final stages of the birth of a star, these questions have so far not been answered by direct observation. Theoretical modeling offers a way to circumvent this obstacle, although not an easy one. Each model requires the execution of more basic calculations than were performed by the entire human race before 1940.

Stars form when nebulas (interstellar clouds of gas and dust), or parts of nebulas, collapse. Although these clouds are too dense for optical telescopes to penetrate, the more diffuse clouds are transparent to millimeter-wavelength radiation. A telescope sensitive to millimeter radiation can therefore be used to observe nebulas in which stars are on the verge of forming. The clouds are also partially transparent to infrared light, and so observations in the infrared wavelengths can be made of newly formed stars within the parent nebulas. These observations yield the basic data with which any theory of stellar formation must reckon: the initial conditions under which stars form and the characteristics of the newly formed stars.

Unfortunately there is still a difference of a factor of almost 10^{20} between the density of a star-forming cloud and that of the young stars that can be observed with infrared radiation: it has been impossible to date to view the cloud as it collapses through this range of densities. Consequently stars cannot be observed as they form.

Since the late 1960s, astrophysicists have developed increasingly sophisticated computer models of the events that take place between the two observable stages of stellar formation. Such models are based on systems of equations that describe the behavior of nebular gas and dust under the influence of many different forces; the solution of these equations can require roughly one million million basic operations.

Among the most important advances has been the use of increasingly realistic descriptions of the parent clouds. The early models pictured a spherically symmetric cloud with no rotation; at the next stage of complexity it was assumed that the cloud rotates but remains symmetric about its axis of rotation; in the most recent models the initial cloud rotates and is completely asymmetric.

These models have shown that a collapsing cloud will generally pass through two phases of rapid contraction (called dynamic collapse); phases during which outlying matter accumulates around a stable core follow each dynamic-collapse phase. In either phase of dynamic collapse the cloud might fragment into two or more protostars; whether or not the cloud fragments depends on such variables as its size and rate of rotation. It might also collapse in a way that produces a single protostar.

In fact such single stars are rare: in spite of the appearance of the night sky to the unaided eye, most stars are actually binary. (A binary system consists of two stars orbiting about each other.) Our Sun, as a single star, is part of a minority population. The clouds that do not fragment are thus particularly interesting: they may represent models for the formation of our own solar system.

How a cloud fragments is one of the two fundamental characteristics of stellar evolution that a theory must be able to describe. Interstellar clouds can be as large as 100,000 times the mass of the Sun—quite massive compared with stars, which are seldom larger than about 10 times the solar mass. In addition most stars in the disk of our galaxy seem to form in clusters containing about 100 stars. These two observations suggest that interstellar clouds fragment into many protostars.

The second fundamental characteristic that must be described concerns angular momentum. In rough terms, the angular momentum of a spinning body is a measure of how much mass in the body is spinning, how fast that mass is spinning and how large the body is. According to observational evidence, interstellar clouds have up to 10^5 times as much angular momentum per unit of mass as their progeny stars. Any theory of star formation must therefore describe how a cloud disposes of a considerable amount of angular momentum before it collapses to form a star or several stars.

One of the first sophisticated computer models of star formation was produced in 1968 by Richard B. Larson of Yale University. He developed a detailed model for the contraction of a spherically symmetric, nonrotating cloud. An important product of his work was a picture of the so-called dynamic-collapse phase of star formation. The dynamic-collapse phase is a period of rapid contraction that can be explained by the interplay of two major forces: gravity, which tends to contract the cloud, and thermal pressure, which is the tendency of hot gas within the cloud to expand. Larson showed that the dynamic-collapse phase is due in part to the way the relation between these two forces changes because of the flow of radiation within the cloud.

The outer shell of a very diffuse dust cloud is transparent to ultraviolet radiation from neighboring stars, and hence it tends to be heated substantially by such radiation. After gravity has compressed the cloud to the density of a dark cloud, the cloud becomes opaque to ultraviolet light, eliminating this source of heating. It is still transparent to

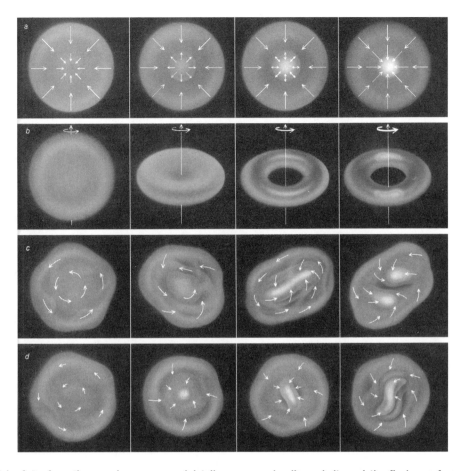

Models of star formation grow in accuracy and detail as astrophysicists employ increasingly realistic pictures of a star's parent dust cloud. The earliest model (A) pictures the cloud as a perfect sphere that does not rotate. The first panel shows the dynamic-collapse phase, in which gas and dust fall rapidly toward the cloud's center. When the center becomes so dense that it is opaque to infrared radiation (second panel), the compressional energy produced by the collapse can no longer radiate from the cloud. Instead it adds to the thermal energy, increasing gas pressure and stopping the dynamic collapse; a first core is formed. Eventually (third panel) the first core becomes hot enough so that diatomic hydrogren molecules break up into single atoms. As it breaks up, the hydrogen absorbs heat and so the core's temperature falls. Pressure inside the core drops rapidly until it is no longer able to withstand the force of gravity, and a second dynamic collapse occurs. Once all the hydrogen is dissociated (fourth panel) the second collapse halts and the final core forms. In a more sophisticated model (B) a cloud that is symmetric about an axis rotates. Matter along the axis collapses faster than matter away from the axis, and so the cloud assumes a lozenge shape (second panel), which eventually turns into a ring (third panel). If the rotating cloud is even slightly irregular, the ring may break up into two or more fragments (fourth panel). In C, a fully asymmetric, rotating cloud flattens out (second panel). The cloud becomes increasingly distorted about its rotation axis, so that it forms a bar shape (third panel). The bar becomes denser and more elongated as the cloud collapses, and then it breaks up into a binary protostellar system (fourth panel). If an asymmetric cloud rotates very slowly (D), it may collapse to form a single protostar. The collapse (first, second and third panels) is similar to that shown in C, but the cloud in D does not become as elongated and hence will not fragment; a single protostar forms.

infrared radiation, however, and so dust grains in its interior are able to radiate heat energy out of the cloud in the infrared portion of the spectrum.

Thus as the density of the cloud increases, its temperature drops, down to a minimum of about 10 degrees Kelvin (degrees Celsius above absolute zero). The cloud then enters an "isothermal phase" during which the temperature remains at 10 degrees as the cloud collapses through a wide range of densities, from approximately 10^5 to 10^{11} atoms per cubic centimeter. As the cloud grows smaller and denser, the gravitational force becomes stronger, eventually overwhelming the thermal pressure. The result is a dynamic collapse, in which the gas and dust fall into the center at rapidly increasing velocities. As the gas and dust fall in, the density of the cloud's center increases.

When the central regions of the cloud become dense enough to be opaque to infrared radiation, the dynamic-collapse phase ends. The collapse of the cloud has generated a great deal of heat because of the compressional work gravitational forces perform on the gas. During the isothermal phase this heat was radiated out of the cloud as infrared light; once the radiation can no longer easily escape from the cloud, the temperature and pressure begin to rise. When the center of the cloud reaches a temperature of about 100 degrees K and a density of about 10^{14} atoms per cubic centimeter, thermal pressure becomes great enough to overcome the gravitational force and to stop the cloud's dynamic collapse. The region in which the dynamic collapse halts has a radius of about five astronomical units (one astronomical unit, roughly 93 million miles, is the mean distance from Earth to the Sun). This region is called the first core; matter in the outer regions, still transparent to infrared radiation, continues to fall inward, accumulating at the core.

The first core is in a quasi-equilibrium state: the matter deepest within the core flows alternately inward and outward, producing a periodic increase and decrease in density.

As matter from the envelope continues to build up at the core, the center of the core becomes progressively denser and hotter. Eventually the central temperature and density become high enough so that the diatomic hydrogen molecules dissociate into single atoms of hydrogen. At this stage the temperature of the cloud is about 2,000 degrees K and its density is roughly 10^{16} atoms per cubic centimeter.

Because hydrogen absorbs energy as it dissociates, the temperature of the first core drops and there is less thermal pressure to support the mass of the cloud. Consequently the first core enters a second phase of dynamic collapse. The innermost regions fall in rapidly until the core reaches densities of approximately 10^{24} atoms per cubic centimeter (roughly the density of water) and temperatures of about 100,000 degrees K. At this point thermal pressure again becomes sufficient to counteract the force of gravity that has been pulling matter inward. A second core, smaller than the first, is therefore able to form. Initially this core contains only a small fraction of the total cloud and is a few times the size of the Sun. The remainder of the cloud, however, continues to fall inward and to enter the second core. As this falling matter accumulates, the second core replaces the first core, which disappears.

After the second core is formed and the remainder of the cloud collapses around it, the protostar enters the main sequence of stellar evolution. The entire dynamic collapse has occupied roughly 100,000 years.

Rotation and inhomogeneity within the cloud, neither of which is included in Larson's model, have important effects on the rate and type of collapse. It is worth noting that the nonrotational, spherically symmetric model cannot explain either fragmentation or the question of excess angular momentum.

The next step toward theoretical accuracy was taken in 1972 by Larson and in 1976 by David C. Black of the National Aeronautics and Space Administration's Ames Research Center and Peter H. Bodenheimer of the Lick Observatory of the University of California at Santa Cruz. These investigators studied the collapse of a rotating cloud; to keep the model relatively simple, they assumed that the cloud was symmetric with respect to its axis. They found that a rapidly rotating dense cloud may collapse, in several stages, to form a ring. Under certain conditions that ring may fragment into a system of several protostars.

In the first stage, matter along the axis of rotation collapses toward the center in the same way as the matter of a nonrotating cloud collapses. Matter distant from the axis collapses more slowly, because much of the gravitational force that would ordinarily pull it toward the center is needed simply to hold it in orbit about the axis of rotation. In other words, because the cloud is spinning, matter in it feels an apparent "centrifugal force": the matter would normally tend to fly off along a straight trajectory; gravity overcomes that tendency and bends the trajectory into a circle. The faster the material is moving and the smaller the orbit is, the more gravitational force is needed to maintain that orbit and the less gravitational force is available to pull the matter inward.

Since matter along the axis collapses faster than matter farther from the axis, the once spherical cloud flattens out, forming a lozenge-shaped cloud, which grows progressively flatter and more disklike as it collapses. Eventually, for reasons that were first described by Joel E. Tohline (now at Louisiana State University at Baton Rouge) and me, the disk forms a ring.

Tohline and I showed that the ring develops because of an interplay between the forces of gravity and the law of the conservation of angular momentum. The angular momentum of a spinning body depends in part on the distance between the rotating matter within that body and the axis of rotation. Since the angular momentum of an isolated spinning body must remain constant, matter that falls in toward the center must orbit faster as it falls. This means that orbiting matter cannot fall all the way to the center: as it accelerates, more of the gravitational force is necessary to keep it from flying off. Eventually the falling matter will reach "centrifugal equilibrium," where the force of gravity is precisely sufficient to maintain the matter's orbit and hold it at a constant radius.

During the collapse of a rotating cloud some of the matter falling toward the center reaches and passes the radius of centrifugal equilibrium. Since the force of gravity is not strong enough to hold this matter in a small orbit, the matter stops collapsing inward and begins to flow away from the center (under the influence of an apparent "centrifugal force").

Meanwhile other material that is farther from the axis is still falling inward. In the resulting collision between the mass falling inward and that flowing outward a significant amount of mass accumulates away from the axial center of the cloud. If the accumulation is large enough, the gravitational force it exerts will attract the rest of the falling matter as well as the matter from the central regions. The result is a growing ring of gas and dust around an empty central region.

Thomas L. Cook of the Los Alamos National Laboratory and Michael L. Norman, now at the Max Planck Institute for Physics and Astrophysics in Munich, have shown that such a ring might eventually fragment: if the ring is not perfectly symmetric about its axis, accumulations will form along the circumference, which eventually break the ring into a system of many protostars. Their model shows that the rest of the cloud's angular momentum goes into the fragments' orbital motion about one another.

The next theoretical advance took place in 1979, when it became possible to model completely asymmetric, rapidly rotating clouds. Bodenheimer, Tohline, Black and I found that some collapsing clouds will fragment without forming a ring; instead irregularities in the cloud can grow large enough to fragment it. The process would take roughly the same amount of time that an axially symmetric cloud requires to form a ring. We also found that those clouds that tend to fragment without forming rings usually evolve into binary systems rather than systems consisting of three or more members; apparently the first two accretions that form will pull the rest of the gas and dust toward themselves.

The initial fragments that form from rapidly rotating clouds typically have a mass of approximately one-tenth of the initial cloud mass and, as in the case of ring fragments, their spin angular momentum is much less per unit of mass than that of the original cloud. Furthermore, each of the fragments is likely to undergo a second dynamic collapse. As each one collapses, it breaks up into yet another set of fragments. These subfragments are themselves likely to collapse and fragment.

As useful as they are, computer models of asymmetric clouds have until recent years suffered one major flaw. Unlike Larson's model of the perfectly symmetric cloud, these models did not take thermodynamic factors into account. That is, they have not modeled the heating and cooling of various portions of the cloud as it is controlled by the flow of electromagnetic radiation. The flow of radiation is in turn dependent on the opacity and density of gas and dust particles within the cloud, factors that change as the cloud collapses. Because of this flaw, computer models were able to consider only the isothermal phase (that part of the first dynamic-collapse phase when the temperature of the cloud remains constant), in which the effects of radiation can be neglected.

My most recent work remedies this shortcoming: it consists of a detailed analysis of the thermodynamics of asymmetric clouds. More advanced methods make it possible to follow collapsing clouds through the isothermal phase and into the next one. In this phase the opacity rises, the first core is formed and the first stage of dynamic collapse and fragmentation ends in the central region of the cloud.

These developing models of general star formation can be applied to a specific case— the formation of our own Sun and the solar system. There are three primary models of its preliminary stages. The first of these models, which suggests that the Sun was originally part of a multiple stellar system, is the least likely. The second and third models, which suggest respectively that the Sun evolved from a decayed binary system and that it formed out of a slowly rotating single protostar, coincide—that is, they predict essentially the same sequence of development once the protostar is formed.

At the same time that the Sun is forming at the center of the nebula, the dust in the outer regions will form a flattened layer and begin the process of accumulation into a planetary system. This surrounding gas and dust may be essential for forming a single star, because it provides the protostar with a way to disperse some of the angular momentum that would otherwise impede its collapse. The formation of a planetary system may thus be a natural consequence of the formation of a single star.

As astrophysicists have made increasingly realistic assumptions about the dust clouds, a clearer image of the process of stellar formation has developed. The next stage in my own research will be an attempt to extend the thermodynamic models of asymmetric clouds. Until now the model has been applied only to the first quasi-equilibrium phase, the forming of the first core; next I shall examine the second dynamic-collapse phase. I believe that no further fragmentation occurs after the first core has formed, and that the protostar contracts to stellar densities, but the definite answer can come only after rigorous modeling, which may take several more years.

Accretion Disks in Interacting Binary Stars

John K. Cannizzo and Ronald H. Kaitchuck

isks are one of the most common structures found in the heavens. In most cases, disks surround a massive central object such as a star or a black hole. Matter in a disk usually migrates inward and eventually accretes onto the central object; these objects therefore are called accretion disks. Accretion disks are thought to be involved in diverse phenomena ranging from the formation of stars and planets to the powering of quasars.

The best-studied disks reside in interacting binary star systems. We have focused our attention on a particularly intriguing, unstable class of interacting binaries, known as cataclysmic variables, that can brighten by a factor of 100 in just a few hours. These systems, fascinating in their own right, serve as a laboratory for understanding the physics of accretion disks.

One might wonder how highly organized disk shapes would develop so frequently throughout the universe. It turns out that the basic laws of physics favor such formations. Consider, for example, an irregularly shaped cloud consisting of particles moving in random orbits, the entire ensemble possessing net angular momentum. (If the cloud is gaseous, the "particle" can be thought of as a small parcel of gas.)

Each particle in the cloud responds to the combined gravitational tugs of all the other parts of the cloud. Particles passing very close to one another experience gravitational and pressure forces that deflect their paths from one of pure orbital motion. Such interactions among the particles dissipate the energy of random motions, whereas conservation of angular momentum preserves the cloud's rotational energy. The particles will eventually settle into the lowest-energy configuration: circular orbits all lying in a single plane.

Theoreticians have developed a number of models aimed at understanding the behavior of such accretion disks. The simplest models assume that the mass flow into the disk exactly balances the accretion onto the central object. Most of these so-called static accretion disk models are based on a pioneering 1973 paper by the Russian scientists Nikolai I. Shakura of the Sternberg Astronomical Institute in Moscow and Rashid A. Sunyaev, then at the Institute of Applied Mathematics, also in Moscow.

For simplicity, Shakura and Sunyaev pictured the disk as a thin, flat, gaseous object whose gravitational field is negligible compared with that of the central object. The gas therefore obeys Kepler's laws of motion, which means that the orbital velocity of each parcel of gas is inversely proportional to the square root of the distance from the central object.

Inner parts of the disk rotate faster than (and therefore slide past) outer parts. Friction between material at adjacent radii—a phenomenon known as viscosity—heats the gas and transports orbital angular momentum outward. The heated gas emits electromagnetic radiation, which escapes from the system. In the process, gravitational potential energy is converted into radiant energy. As a result, gas slowly drifts inward, at about one ten-thousandth of the orbital velocity.

The intensity of the coupling between adjacent radii in the disk, which determines the rate of inflow, is controlled by the strength of the viscosity of the gas. Unfortunately, there is no good theory to predict this viscosity, so Shakura and Sunyaev assumed the viscosity is proportional to the pressure in the disk. In the equations describing the physical state of the disk, the strength of the viscosity appears as an adjustable parameter called alpha. All the unknown physics associated with the viscosity is hidden in alpha. Setting alpha as a constant makes it possible to construct models that predict the physical conditions in the disk. Fortunately, the spectrum of radiation emitted by a disk in a steady state is independent of alpha.

The nature of the central object determines the depth of the gravitational potential well, which in turn determines the energy flux from the disk. When the central object is an ordinary star, the disk radiates primarily in the visible and in the infrared. Matter orbiting a collapsed star, such as a white dwarf or a neutron star, falls much farther before hitting the stellar surface. Disks around these objects release more total energy, and the emission peaks at ultraviolet or x-ray wavelengths.

Because of the great distances involved, it is impossible to observe directly the accretion disks of even the nearest binary star systems. Astronomers have been forced to infer the structure of stellar systems by analyzing the radiation that they produce. Light emitted by the part of a disk moving toward the viewer shifts toward the blue end of the spectrum, whereas light from the parts moving away shift in the red direction, an effect called Doppler shifting. The swirling gas in an accretion disk therefore produces a distinctive spectral signature.

In this way, Arthur Wyse of Lick Observatory detected the first known accretion disk in a double star system in 1934. He was studying the spectrum of RW Tauri, a binary consisting of a hot, main-sequence star and a large, cool red companion. The orbital plane of the two stars is viewed nearly edge-on, so that once during each orbit the cool star obscures its companion. Wyse obtained a spectrum of the red star during the 90-minute interval when the companion was completely blocked from view. He observed

Doppler-shifted emission lines, indicating the presence of hot, rapidly moving gas.

In the early 1940s Alfred H. Joy of Mount Wilson Observatory obtained additional spectra of RW Tauri. He found that at the beginning of an eclipse, the emission lines were Doppler shifted to the red; at the end of an eclipse, the lines were shifted to the blue by an equal amount. Joy inferred from his observations that the hot star is surrounded by a rapidly rotating ring of gas. As an eclipse begins, the side of the ring rotating away from the observer remains visible beyond the limb of the cool star, producing redshifted emission lines. As an eclipse draws to a close, the side rotating toward the observer is visible, yielding blueshifted lines.

Over the following years, astronomers came to suspect that Joy's gaseous ring was actually a disk whose inner edge lay at the surface of the central star. In the 1940s Otto Struve of McDonald Observatory and a few others suggested that matter flows from the companion star into the disk and then accretes onto the central star, a concept now accepted by most astronomers.

The flow of matter in binary systems is sometimes a direct result of stellar evolution. When a star reaches the end of its stable life, its core becomes depleted in hydrogen fuel and so begins to contract; as the core shrinks, it grows hotter and releases more energy; causing the outer parts of the star to swell. If a star in a close binary system expands beyond a certain volume, called the Roche lobe, matter moving outward from the star falls under the gravitational influence of its companion. The size and shape of the Roche lobe are largely determined by three forces felt by a particle at rest with respect to the orbiting stars: the gravitational attraction of each star and the centrifugal force produced by the orbital motion of each particle. These competing forces give the Roche lobe's surface a teardrop shape.

If the star expands beyond this surface, the gas follows the least energetic path and falls toward the companion star. The gas emerges through the point of the teardrop, called the inner Lagrangian (LI) point, where the companion's gravitational influence is greatest. Gas leaving the LI point forms a narrow stream that move toward the companion star. Because the gas still carries the momentum of the orbital motion of the star it just left, it does not fall directly onto the companion. Instead the stream follows a curved path toward the trailing side of the mass-gaining star.

The events that follow depend on the size of the accreting star in comparison to the orbital separation between the stars. If the accreting star is small, the stream of gas curves around it and forms a ring. The ring quickly spreads into a flat disk as a result of viscosity, which causes some gas to lose angular momentum and spiral inward while a smaller volume of gas gains the lost momentum and spirals outward.

If the accreting star is comparatively large, the stream of gas collides with the body of the mass-gaining star. Oddly enough, something resembling an accretion disk still manages to form. But the resulting disk is turbulent and unstable; it disappears quickly if the mass transfer temporarily stops, much the way the spray from a garden hose falls to the ground and disappears when the faucet is turned off. As it turns out, RW Tauri is a transient-disk binary system.

Accretion disks can act in more complicated and violent ways, as seen in a class of binary stars called cataclysmic variables. These stellar systems contain a dense, hot companion that accretes matter from its cooler neighbor. Here the mass transfer is driven by a loss of orbital angular momentum, which slowly draws the stars closer and causes the Roche lobe to shrink around the red star. The disks in cataclysmic variables seem far from stable. Because the flow of matter through these disks may not be at all steady, understanding cataclysmic variables poses a difficult challenge. But it is a challenge worth accepting: the erratic behavior of these objects holds important clues to the general nature of accretion disks.

As their names imply, some cataclysmic variables undergo severe outbursts, during which their brightness jumps as much as 100-fold in a matter of days or even hours. One subset of cataclysmic variables, called dwarf novae, flare up in a quasi-periodic fashion. Dwarf nova eruptions recur on intervals of weeks to years; each episode typically lasts from a few days to a few weeks. Dwarf novae are distinct from ordinary novae, which are thought to be powered by the nuclear fusion of hydrogen accreted onto the surface of a white dwarf star. Dwarf nova outbursts seem to draw on gravitational energy only, which explains why they are 1,000 times less energetic than ordinary novae.

Groundbreaking work by Robert P. Kraft of Mount Wilson Observatory in the 1960s revealed that cataclysmic variables are binary stars whose components orbit very close to each other. A typical cataclysmic binary has an orbital period of four hours, and a few have periods of less than 90 minutes. Such rapid orbits mean that the separation between the two stars, and hence the stars themselves, must be extremely small. In fact, the typical cataclysmic system could fit inside our Sun.

Studies of light spectra and radiation fluxes at a broad range of wavelengths indicate that there are three major components to a cataclysmic binary: a red dwarf star, a white dwarf star and an accretion disk surrounding the white dwarf. The red dwarf is a cool, faint, low-mass star that is losing matter through its L1 point. The white dwarf is much hotter, more luminous and more massive. White dwarfs are the remnant cores of evolved stars that have depleted their hydrogen fuel. Lacking an internal energy source, the core grows fantastically dense—about 10 million times as dense as water. White dwarfs are roughly as massive as the Sun but are only about the size of Earth.

Because a white dwarf is so small and massive, it has a very deep gravitational potential well. Matter falling onto it from the accretion disk releases a large amount of gravitational potential energy via viscous heating in the disk. So intense is the heating that the disk becomes brighter than the stars.

Observational astronomers have deduced that enhanced mass flow through the accretion disk powers the outbursts of dwarf novae. This was demonstrated by Brian Warner of the University of Cape Town, who conducted photometric observations of the dwarf nova Z Chamæleontis. As seen from the Earth, the red star in this binary eclipses the white dwarf and its surrounding accretion disk once every orbit. During an outburst, the light loss at mideclipse is much greater than during times of quiescence, indicating that the source of the outburst cannot be the red dwarf star. Furthermore, the duration of successive eclipses lengthens as the outburst progresses, implying that the eruption is spreading across the face of the disk.

The optical spectrum of a dwarf nova changes drastically during the course of an outburst. The quiescent spectrum shows a blue continuum on which are superimposed lines of radiation emitted by hydrogen and by singly ionized helium (helium that is missing one electron). The continuum originates in the dense inner regions of the disk, whereas the emission lines are formed in the more rarefied outer regions. In systems where the orbital plane is seen nearly edge-on, the emission-line profiles appear as double-peaked curves. The peaks appear because each profile consists of two Doppler-shifted components: one from the side of the disk rotating toward the observer and one from the side rotating away.

During an outburst, the continuum radiation brightens sharply; at the same time, the emission lines can become difficult or impossible to see. Broad hydrogen absorption lines often appear, the result of radiation being absorbed by cooler gas surrounding the bright part of the disk. As the outburst declines in brightness, the continuum fades, and the emission lines once again become prominent.

Dwarf nova outbursts seem to result from a sudden increase in the flow of mass through the accretion disk onto the white dwarf star. The speedier inflow could result from a surge from the mass-losing red star or from a change in the accretion disk itself. The first possibility, proposed some 20 years ago by Geoffrey T. Bath, who was then at the University of Oxford, requires that the red star periodically overflow its Roche lobe, dumping excess material into the accretion disk. The extra matter causes the rate of accretion to increase, which in turn makes the disk shine more brilliantly.

A few years later Jozef Smak of the Copernicus Astronomical Center in Poland and Yoji Osaki of the University of Tokyo independently presented a competing idea. The Smak-Osaki hypothesis holds that mass flows from the secondary star at a constant rate. Some mechanism in the accretion disk itself stores up matter and then dumps it onto the white dwarf whenever the disk's mass exceeds a critical level.

Astrophysicists largely favor the disk instability model that has grown out of the work of Smak and Osaki. Most researchers now think the instability is triggered when a critical local surface density is attained somewhere in the disk. Once the density in a local region reaches the critical value, viscosity increases tremendously, and the stored-up matter accretes onto the white dwarf star, producing an outburst. When the surface density drops below a certain level, disk viscosity plummets, and the accumulation process starts anew.

For years, theorists have labored to create mathematical models that could describe accretion disk instabilities. These models predict the local surface density and temperature when values are specified for the accretion rate and for alpha at a certain radius. An increase in the accretion rate leads to higher surface density, which in turn causes increased viscous heating. Disk temperature is therefore expected to vary in proportion to surface density; a graph of local temperature versus surface density at a particular radius in an accretion disk should produce an upward-sloping curve.

The outburst models of Osaki and Smak required that at a given location in the disk there exist two possible stable values for temperature for the same value of surface density. The outburst corresponds to the high-temperature solution, the quiescent state to the low-temperature solution. In this case, the curve in the temperature-surface density plot must be S-shaped. Osaki and Smak implicitly assumed that such an S curve relation was physically plausible, but they did not propose a specific mechanism that would produce it.

In 1979 Reiun Hoshi of Rikkyo University in Tokyo demonstrated the existence of such a mechanism. Hoshi worked with a model that considered the vertical structure of the accretion disk (the physical conditions that prevail in a thin perpendicular slice through the disk). Although his model for the vertical structure was crude, he found that an S curve relation naturally emerged when his model accounted for the temperatures at which hydrogen becomes partially ionized, that is, when the disk becomes so hot that some of the hydrogen atoms lose their surrounding electrons.

The disk instability model makes a number of predictions that can be tested by observations of dwarf novae. Smak delineated many of these predictions in a landmark article that appeared in 1984. Similar studies appeared by the other researchers mentioned above at about the same time. Using computer models to simulate the time-dependent behavior of an accretion disk, Smak was able to produce artificial light curves that closely resemble the observed ones.

Several aspects of dwarf nova eruptions seem quite consistent with the disk instability model. Light curves produced by disk instabilities can show either a slow or a rapid increase in brightness. Computer models show that instabilities that begin near the outer edge of the disk produce the fast increases. In this case, the radius of the disk should increase during an outburst, a phenomenon that has been observed in several dwarf novae.

If the disk instability model for dwarf nova eruptions proves correct, it should help answer many of the questions about the physics of accretion disks in general. The greatest mystery in accretion disk research is the nature of the viscosity. Clearly, a preliminary step in understanding the physical mechanism that produces viscosity involves determining empirically the magnitude of the effect and how it depends on surface density and on radius of the disk. Researchers using time-dependent computer models have tried adjusting the relation between temperature and surface density in order to make the model outbursts resemble those observed. In this way, it is possible to check theories of viscous heating to make sure they are consistent with the behavior of dwarf novae.

Two main mechanisms have been proposed to account for viscous heating of the disk. One focuses on turbulence associated with convective motions that accompany the vertical transport of energy out of the disk; the other invokes energy released by the breaking and reconnecting of magnetic field lines in the disk.

Steven Balbus and John Hawley of the Virginia Institute for Theoretical Astronomy have recently uncovered a mechanism by which motions in the ionized (and hence magnetic) gas in the disk would amplify the weak magnetic fields that should always be present. The next task for theorists is to incorporate the results from such studies into a global theory that follows the temporal evolution of the entire disk to see if the models can reproduce in detail the recorded behavior of dwarf nova eruptions.

The disk instability model may explain not only cataclysmic variables but also distant, energetic quasars. Astrophysicists have long speculated that these objects harbor accretion disks that channel matter onto central black holes having masses millions of times that of the Sun.

A better understanding of accretion disks will be essential to determining the true nature of quasars and other violent phenomena that occur in the nuclei of galaxies, including the mysterious object at the center of the Milky Way. The information should also lead to new insights into the disks that seem to surround infant stars, the likely birthplace of planets such as the Earth.

Chapter Five:
The Genesis of the Solar System

Introduction

From the universe, past its galaxies, to the Milky Way, we have at last arrived at our own solar system. Compared to these greater structures, it seems almost trivial. But the story of the genesis of the planetary system, to the many worlds we see today, is a fabulous one, and one with which we are familiar. While we have examined the rest of the universe with telescopes and equations, we have actually visited most of the worlds of our solar system and have studied them in situ.

From these visits, and from centuries of detailed observations, we have come up with a fair idea of how the solar system began and evolved. In 1994, in fact, Comet Shoemaker-Levy 9 gave us a direct insight of what the early solar system was like. The comet's 21 fragments collided with Jupiter in yet another replay of an old story: comets and planetesimals accrete to build planets; comets continue to collide with these worlds. On Earth, these collisions had the incalculable effect of launching and changing the pageant of life.

The Origins of the Asteroids

Richard P. Binzel, M. Antonietta Barucci and Marcello Fulchignoni

Once referred to as uninteresting "vermin of the skies," asteroids store a rich variety of information that is proving to be of scientific and pragmatic interest.

Astronomers used to believe that these small worlds were the debris of a shattered planet. Actually, asteroids are remnants of a planet that failed to form. As such, they offer significant evidence of the nature of the poorly understood process that creates planets. They also provide clues to conditions in the early solar system.

Moreover, a number of asteroids reside in the inner solar system, within the orbit of Mars, and many of their orbits intersect that of the Earth. Evidence of past catastrophic impacts pervades the geologic record, indicating that collisions can strike with a force exceeding that of a nuclear warhead.

Unlike planets and comets, which have been known since antiquity, asteroids are a relatively recent discovery. They were found as the consequence of an inquiry into the spacing of the planets, a phenomenon that mystified astronomers of the 17th and 18th centuries. In 1766 the German astronomer Johann D. Titius calculated that a planet should exist between Mars and Jupiter, about 2.8 astronomical units (AUs) from the Sun (one AU is defined as the Earth's average orbital distance, about 150 million kilometers). Johann E. Bode of the Berlin Observatory later popularized the calculations, which became known as Bode's Law.

The search for the "missing planet" began in earnest at the turn of the 19th century, when the Hungarian baron Franz X. von Zach organized a group of astronomers who called themselves the celestial police. On January 1, 1801, the search apparently ended: Giuseppe Piazzi of Palermo discovered an object and named it Ceres, after the tutelary goddess of Sicily.

Although the discovery delighted astronomers, Ceres posed a puzzle. Observed through a telescope, it showed no disk, implying that it was substantially smaller than anticipated. Astronomers had expected a more massive object. The first clue in the solution to the "missing mass" problem came in 1802, when Heinrich W. M. Olbers found a second small planet, later named Pallas. By 1807 observers discovered the third and fourth minor planets, Juno and Vesta. Olbers suggested that these bodies were fragments of a larger planet that ruptured. Sir William Herschel, who discovered Uranus, proposed that these minor planets be called asteroids—Greek for "starlike," a term justified by their telescopic appearance.

Observations over the past two centuries have identified a total of about 50,000 asteroids. Astronomers have determined the precise orbits of about 12,000 of them. Each of these asteroids has a permanent catalogue identification consisting of a number that denotes its order of entry, which is usually followed by a name proposed by the observer. For instance, "3 Juno" formally identifies the third asteroid discovered. Only preliminary data characterize the orbits of the other 38,000.

Most asteroids orbit within a confined area between Mars and Jupiter—called the main asteroid belt—at an inclination of about 10 degrees with respect to the plane of the solar system. Their average distances from the Sun (referred to as the semimajor axes because the orbits trace ellipses) range between 2.1 and 3.3 AUs. Ceres, Pallas and Vesta hold about one half of the total mass of the asteroid belt, equal to about 0.0005 the mass of the Earth. Their diameters are 933, 523 and 501 kilometers, respectively. About 1,000 asteroids are larger than 30 kilometers across, and of these, more than 200 asteroids are larger than 100 kilometers. Not all the asteroids smaller than 30 kilometers have been discovered, but calculations indicate there are about one million whose diameters measure one kilometer or more.

Such a vast number conjures images from popular films that show spacecraft weaving through fields of crashing boulders. The volume of space in the main belt is so large, however, that asteroids usually remain several million kilometers apart. Collisions are infrequent; significant events occur only over geologic time scales.

Gravitational effects from Jupiter dominate the structure of the asteroid belt. Such effects first became evident in 1867, when the American astronomer Daniel Kirkwood found breaks in the uniformity of the asteroid belt. Called Kirkwood gaps, they occur in regions where the orbital period of a body would be some exact integer ratio of Jupiter's orbital period. For instance, an object 2.5 AUs from the Sun is said to be at the 3:1 "resonance"; that is, the body would complete exactly three revolutions for every one that Jupiter completes.

The resonances prevented the formation of a planet between Mars and Jupiter. The inner planets formed when swarms of planetesimals a few kilometers in size collided at velocities low enough to permit the bodies to grow larger by accretion. Numerous resonances from the rapidly growing and massive planet Jupiter probably permeated the region between two and four AUs. These resonances may have pumped up the orbital eccentricities of the planetesimals there, accelerating the objects to velocities so high that successful accretion on a planetary scale was impossible. Today the asteroids remain in an environment dominated by collisions, encountering one another at about five kilometers per second.

Increasing eccentricities may have also led to the collisional pulverization of some asteroids. In addition, collisions with planets or interactions with their gravitational fields (especially that of Jupiter) can effectively remove or scatter wayward bodies from the solar system. These events may have cleared the asteroid zone of most of its original mass, leaving behind the remnants observed today.

Some clusters of main-belt asteroids seem to share specific orbital distances, eccentricities and inclinations. In 1918 the Japanese astronomer Kiyotsugu Hirayama called these clusters families. He identified and named a number of them, including Themis, Eos, Koronis and Flora.

Hirayama proposed that families result from the disruption of large parent bodies. If so, the fragments would provide a view of the interiors of the precursor asteroids. The size distribution of family members provides information on the outcomes of collisions. As a result, researchers can use these families to deduce information about how the asteroid belt evolved through collisions since its formation.

Not all asteroids are found in the main belt. An especially interesting group, called the Trojans, resides at the 1:1 resonance. These asteroids have the same heliocentric distance and orbital period as Jupiter does. Trojans orbit in two stable regions near equilibrium positions, known as Lagrange points, that lie in the orbital plane about 60 degrees east and west of Jupiter.

Three populations of asteroids reside within the inner solar system. One class, the Atens, has orbits that keep them consistently close to the Sun. The semimajor axes of the Aten orbits are less than one AU. A second group, known as Apollo asteroids, has orbital semimajor axes beyond the Earth. Some Atens and Apollos orbit so eccentrically that they cross the Earth's orbit. A third class is called the Amors. These asteroids travel around the Sun between the orbits of Mars and the Earth and often cross the path of Mars. Collectively, Atens, Apollos and Amors constitute the near-Earth asteroids.

Astronomers think that some source must resupply the near-Earth population visible today. Asteroids in the inner solar system live only about 10 million to 100 million years—far shorter than the 4.5-billion-year age of the solar system. Within this lifetime, they disappear, either because of collisions with the inner planets or near misses, which result in their gravitational ejection from the solar system.

Workers suspect that the main belt may be supplying near-Earth asteroids, perhaps accounting for as much as 80 percent of them. A description of the delivery mechanism has until recently proved elusive. The solution hinges on findings by Jack L. Wisdom of the Massachusetts Institute of Technology. Some orbits, such as those near the 3:1 resonance, can undergo "chaotic motion." Such motion may lead an asteroid to cross the path of Mars. Studies by George W. Wetherill of the Carnegie Institution have shown that gravitational interactions with Mars further aid the delivery of asteroids to the vicinity of Earth.

Other small bodies in the solar system may be asteroids. Phobos and Deimos, the two satellites of Mars; the eight outer moons of Jupiter; and the Saturnian satellite Phoebe appear to be captured asteroids.

Although cataloguing bodies and defining orbits occupied the first 150 years of asteroid work, systematic studies of their physical properties began only in the 1950s. Because asteroids display no visible disk, observers must infer physical characteristics from the intensities and spectral properties of the reflected sunlight. Gerard P. Kuiper of the University of Chicago and his students, most notably Tom Gehrels of the University of Arizona, pioneered such studies.

The first physical characteristic noted by astronomers was that most asteroids do not maintain a constant brightness. Short-term variations result from the rotation of an asteroid's irregular shape about the spin axis. The changing cross-sectional area reflects different amounts of sunlight to the Earth. Each rotation typically brings two broad and two narrow ends into view, yielding two maxima and two minima in the observed light curve.

The maximum amplitude of the variation gives an indication of the shape. For instance, an elongated body shows a greater variation than does a roughly spherical one. Asteroids typically rotate once every four to 20 hours, and the brightness varies by about 20 percent.

From the accumulated data on the rotation rates, astronomers can infer the kinds of collisions an asteroid has experienced. Each noncatastrophic collision adds rotational angular momentum in a random way. Small asteroids tend to spin quickly. Larger asteroids, which have higher moments of inertia and experience fewer collisions with significantly massive projectiles, spin more slowly.

Curiously, the trend reverses for asteroids larger than about 125 kilometers: spin rates tend to increase with increasing size. Many rapidly rotating and elongated large asteroids are believed to be gravitationally bound "rubble piles"—that is, bodies that are thoroughly shattered in their interiors.

Although light curves reveal important characteristics of asteroids, such as shape, they do not provide any absolute measure of size. Instead the most widely applied means for estimating size comes from measurements of the thermal radiation asteroids emit.

Typically surface temperatures are about 200 kelvins, but that value and the corresponding thermal flux depend on an asteroid's albedo, diameter and distance from the Sun. For instance, a low-albedo surface reflects very little solar light: most of the Sun's energy is absorbed and reemitted at thermal wavelengths. So for two equidistant asteroids having the same apparent visual brightness, the one showing the greater thermal emission would have the lower albedo and larger diameter.

Because astronomers can quantitatively calculate the incident solar flux on an asteroid, they can use visual and thermal infrared measurements to derive albedos and diameters. The most detailed survey came from the Infrared Astronomical Satellite (IRAS). Launched in 1983, the satellite provided measurements of nearly 2,000 catalogued and many more uncatalogued asteroids.

The IRAS results suggest that the largest asteroids show an albedo distribution different from those of the smallest ones. The fact that many smaller asteroids may be fragments originating in the interiors of larger parent bodies could explain the difference. The IRAS discovery of dust bands within the asteroid belt also confirms that collisions occur with relative frequency.

Another method to measure an asteroid's size requires a fortuitous juxtaposition but yields the most accurate results. An asteroid's apparent motion sometimes causes it to pass directly in front of a distant star, occulting it; the shadow of the asteroid then sweeps across the Earth's surface. The width of the shadow path measures the length of the asteroid along one dimension. The duration of the star's disappearance multiplied by the calculated shadow velocity yields the other dimension.

One of the most powerful and promising new techniques applied to asteroid studies is radar. Unlike most other astronomical instruments, which passively measure the energy from celestial sources, radar can be used to perform controlled experiments on the target. Observers simply tailor the polarization as well as the time and frequency modulation of the outgoing signal. The return signals contain information on the asteroid's distance, size, shape, spin rate, orientation and surface properties.

Although many techniques can examine the physical dimensions of asteroids, only spectroscopic measurements can reveal chemical compositions. Because elements and compounds absorb distinct regions of the electromagnetic spectrum, researchers can use spectroscopic data to deduce the chemistry of a body. Compositional information for a large sample of asteroids enables astronomers to determine patterns of formation and thermal evolution as a whole.

As a result of such measurements, researchers recognize many classes of asteroids based on composition. The two broadest groups, first determined in the 1970s, are the C and S classes. The C class consists of low-albedo objects that generally exhibit neutral or flat spectra and have a strong absorption in the ultraviolet. The S class consists of moderate-albedo objects that show a broad absorption band in the blue as well as in the ultraviolet.

Astronomers soon recognized several other classes. A cluster analysis by David J. Tholen of the University of Hawaii, together with surveys, led to a major expansion of the taxonomic system in 1984. The currently recognized classes are designated by the letters S, C, M, D, F, P, G, E, B, T, A, V, Q and R, in order of their observed relative abundances.

Although it seems to be a random alphabet soup, the taxonomy actually reveals a distinct structure in the compositional distribution of asteroids. The distribution pattern suggests some primordial process produced a steep temperature gradient that altered the composition of the asteroids.

The most likely explanation is heating initiated by a strong solar wind during an early phase in the Sun's formation. Some researchers, however, have suggested that the radioactive decay of aluminum 26 served as the heating mechanism. This element could have been injected into the condensing solar system by a nearby supernova. Impacts, too, could have provided some heat to mix and melt the compounds detected on asteroid surfaces.

Mineralogical studies of asteroid spectra do not necessarily have to be remote. Meteorites probably represent samples of material from the asteroid belt. The pieces of the puzzle, however, do not yet fit together neatly. The observed samples of the asteroids and meteorites appear to represent partially mismatched sets. The greatest discrepancy occurs with the most common type of meteorite, the ordinary chondrites.

Theorists have proposed, that the parent bodies may be the S class asteroids. Such asteroids have the same qualitative mineralogy. Still, their quantitative proportions are difficult to determine. Some researchers think the S types are more metal rich and therefore more closely resemble the stony-iron meteorites—objects that have undergone substantial heating.

One possible answer to the puzzle may be "space weathering." Solar radiation or micrometeoroid impacts could have changed the upper few millimeters of an ordinary chondrite asteroid's surface enough to alter its spectral characteristics. Another solution may come from the recent observations of the near-Earth asteroid 1862 Apollo. This asteroid, about two kilometers wide, appears to be spectrally consistent with ordinary chondrites. Cratering impacts may have hurled pieces of 1862 Apollo into Earth-crossing orbits. Still, researchers think ordinary chondrites derive from several parent bodies. Perhaps extended surveys will reveal more of this asteroid class (Q).

Discrepancies in the models of collisional evolution remain as well. For example, Vesta's basaltic crust probably resulted from lava flows produced during an acute heating episode in the early solar system. It is difficult to understand how the intense bombardment that completely disrupted many asteroids of similar size could have allowed Vesta to preserve its crust intact.

Partial solutions to such questions may be forthcoming. Preliminary plans are under way for dedicated missions to near-Earth asteroids, many of which are more accessible than the Moon.

Still, spacecraft will study relatively few asteroids through the first part of the 21st century. Instead astronomers will continue to rely heavily on observations from the ground and from Earth's orbit. The current decade promises to bring a handsome return of information, to celebrate January 1, 2001—the bicentennial discovery date of 1 Ceres.

The Kuiper Belt

Jane X. Luu and David C. Jewitt

After the discovery of Pluto in 1930, many astronomers became intrigued by the possibility of finding a tenth planet circling the Sun. Cloaked by the vast distances of interplanetary space, the mysterious "Planet X" might have remained hidden from even the best telescopic sight, or so these scientists reasoned. Yet decades passed without detection, and most researchers began to accept that the solar system was restricted to the familiar set of nine planets.

But many scientists began seriously rethinking their notions of the solar system in 1992, when we identified a small celestial body—just a few hundred kilometers across—sited farther from the Sun than any of the known planets. A host of similar objects is likely to be traveling with it, making up the so-called Kuiper belt, a region named for Dutch-American astronomer Gerard P. Kuiper, who, in 1951, championed the idea that the solar system contains this distant family.

What led Kuiper, nearly half a century ago, to believe the disk of the solar system was populated with numerous small bodies orbiting at great distances from the Sun? His conviction grew from a fundamental knowledge of the behavior of certain comets—masses of ice and rock that plunge from the outer reaches of the solar system inward toward the Sun.

Astronomers have long realized that comets must be relatively new members of the inner solar system. Comets were created during the formation of the solar system 4.5 billion years ago and should have completely lost their volatile constituents by now, leaving behind either inactive, rocky nuclei or diffuse streams of dust. Why then are so many comets still around—including the recent cosmic spectacle, Comet Hyakutake—to dazzle onlookers with their displays?

GUIDING LIGHTS

The comets that are currently active formed in the earliest days of the solar system, but they have since been stored in an inactive state—most of them preserved within a celestial deep freeze called the Oort cloud. The Dutch astronomer Jan H. Oort proposed the existence of this sphere of cometary material in 1950. He believed that this cloud had a diameter of about 100,000 astronomical units (or AU, a distance defined as the average separation between Earth and the Sun—about 150 million kilometers, or 93 million miles) and that it contained several hundred billion individual comets. In Oort's conception, the random gravitational jostling of stars passing nearby knocks some of the outer comets in the cloud from their stable orbits and gradually deflects their paths to dip toward the Sun.

For most of the past half a century, Oort's hypothesis neatly explained the size and orientation of the trajectories that the so-called long-period comets (those that take more than 200 years to circle the Sun) follow. Oort's hypothesis could not explain short-period comets that normally occupy smaller orbits tilted only slightly from the orbital plane of Earth—a plane that astronomers call the ecliptic.

Most astronomers believed that the short-period comets originally traveled in immense, randomly oriented orbits (as the long-period comets do today) but that they were diverted by the gravity of the planets—primarily Jupiter—into their current orbital configuration. Yet not all scientists subscribed to this idea. As early as 1949, Kenneth Essex Edgeworth, an Irish gentleman-scientist (who was not affiliated with any research institution) wrote a scholarly article suggesting that there could be a flat ring of comets in the outer solar system. In his 1951 paper, Kuiper also discussed such a belt of comets, but he did not refer to Edgeworth's previous work.

Kuiper and others envisioned a belt beyond Neptune and Pluto consisting of residual material left over from the formation of the planets. The density of matter in this outer region would be so low that large planets could not have accreted there, but smaller objects, perhaps of asteroidal dimensions, might exist. It seemed likely that these distant objects would be composed of water ice and various frozen gases—making them quite similar (if not identical) to the nuclei of comets.

Kuiper's hypothesis languished until the 1970s, when Paul C. Joss of the Massachusetts Institute of Technology (MIT) began to question whether Jupiter's gravity could in fact efficiently transform long-period comets into short-period ones. He noted that the probability of gravitational capture was so small that the large number of short-period comets that now exists simply did not make sense.

Eventually other astronomers started to question the accepted view. In 1988 Martin J. Duncan of the University of Toronto, Thomas Quinn and Scott D. Tremaine (both at the Canadian Institute for Theoretical Astrophysics) used computer simulations to investigate how the giant gaseous planets could capture comets. They found that the process worked rather poorly, raising doubts about the veracity of this well-established concept for the origin of short-period comets. Indeed, they noted that the few comets that could be drawn from the Oort cloud by the gravitational tug of the major planets should be traveling in a spherical swarm, whereas the orbits of the short-period comets tend to lie in planes close to the ecliptic.

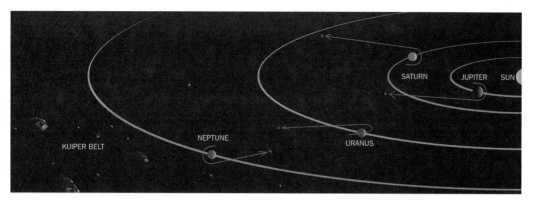

Gravity of the planets acted during the early stages of the solar system to sweep away small bodies within the orbit of Neptune. Some of these objects plummeted toward the Sun; others sped outward toward the distant Oort cloud (not shown).

Duncan, Quinn and Tremaine reasoned that short-period comets must have been captured from original orbits that were canted only slightly from the ecliptic, perhaps from a flattened belt of comets in the outer solar system.

WHY NOT JUST LOOK?

Even before Duncan, Quinn and Tremaine published their work, we wondered whether the outer solar system was truly empty or instead full of small, unseen bodies. In 1987 we began a telescopic survey intended to address exactly that question. Our plan was to look for any objects that might be present in the outer solar system using the meager amount of sunlight that would be reflected back from such great distances.

We conducted the bulk of our survey using the University of Hawaii's 2.2-meter telescope on Mauna Kea. Our strategy was to use a CCD (charge-coupled device) array with this instrument to take four sequential, 15-minute exposures of a particular segment of the sky. We then enlisted a computer to display the images in the sequence in quick succession—a process astronomers call "blinking." An object that shifts slightly in the image against the background of stars (which appear fixed) will reveal itself as a member of the solar system.

For five years, we continued the search with only negative results. On August 30, 1992, we were taking the third of a four-exposure sequence while blinking the first two images on a computer. We noticed that the position of one faint "star" appeared to move slightly between the successive frames. We both fell silent. The motion was quite subtle, but it seemed definite. When we compared the first two images with the third, we realized that we had indeed found something out of the ordinary. Its slow motion across the sky indicated that the newly discovered object could be traveling beyond even the outer reaches of Pluto's distant orbit. Still, we were suspicious that the mysterious object might be a near-Earth asteroid moving in parallel with Earth (which might also cause a slow apparent motion). But further measurements ruled out that possibility.

We observed the curious body again on the next two nights and obtained accurate measurements of its position, brightness and color. We then communicated these data to Brian G. Marsden, director of the International Astronomical Union's Central Bureau of Astronomical Telegrams at the Smithsonian Astrophysical Observatory in Cambridge, Mass. His calculations indicated that the object we had discovered was indeed orbiting the Sun at a vast distance (40 AU)—only slightly less remote than we had first supposed. He assigned the newly discovered body a formal, if somewhat drab, name based on the date of discovery: he christened it "1992 QB1."

Our observations showed that QB1 reflects light that is quite rich in red hues compared with the sunlight that illuminates it. This odd coloring matched only one other object in the solar system—a peculiar asteroid or comet called 5145 Pholus. Planetary astronomers attribute the red color of 5145 Pholus to the presence of dark, carbon-rich material on its surface. Perhaps the object we had just located was coated by some kind of red material abundant in organic compounds. How big was this ruddy new world? From our first series of measurements, we estimated that QB1 was between 200 and 250 kilometers across—about 15 times the size of the nucleus of Halley's Comet.

Some astronomers initially doubted whether our discovery of QB1 truly signified the existence of a population of objects in the outer solar system, as Kuiper and others had hypothesized. But such questioning began to fade when we found a second body in March 1993. This object is as far from the Sun as QB1 but is located on the opposite side of the solar system. During the past three years, several other research groups have joined the effort, and a steady stream of discoveries—including some confirming observations from the Hubble Space Telescope—has ensued. The current [1999] count of trans-Neptunian, Kuiper belt objects is 179.

The known members of the Kuiper belt share a number of characteristics. They are, for example, all located beyond the orbit of Neptune, suggesting that the inner edge of the belt may be defined by this planet. All these newly found celestial bodies travel in orbits that are only slightly tilted from the ecliptic—an observation consistent with the existence of a flat belt of comets. Each of the Kuiper belt objects is millions of times fainter than can be seen with the naked eye. The 32 objects range in diameter from 100 to 400 kilometers, making them considerably smaller than both Pluto (which is about 2,300 kilometers wide) and its satellite, Charon, (which measures about 1,100 kilometers across).

The current sampling is still quite modest, but the number of new solar system bodies

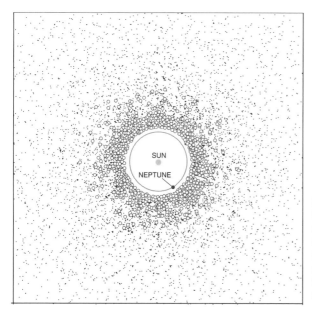

Countless objects in the Kuiper belt may orbit far from the Sun, but not all of those bodies can be seen from Earth. Objects (circles) that could reasonably be detected with the telescope on Mauna Kea in Hawaii typically lie near the inner border of the belt, as seen in this computer simulation of the distribution of distant matter.

found so far is sufficient to establish beyond doubt the existence of the Kuiper belt. It is also clear that the belt's total population must be substantial. The Kuiper belt probably has a total mass that is hundreds of times larger than the well-known asteroid belt between the orbits of Mars and Jupiter.

COLD STORAGE FOR COMETS

The Kuiper belt may be rich in material, but can it in fact serve as the supply source for the rapidly consumed short-period comets? Matthew J. Holman and Jack L. Wisdom, both then at M.I.T., addressed this problem using computer simulations. They showed that within a span of 100,000 years the gravitational influence of the giant gaseous planets (Jupiter, Saturn, Uranus and Neptune) ejects comets orbiting in their vicinity, sending them out to the farthest reaches of the solar system. But a substantial percentage of trans-Neptunian comets can escape this fate and remain in the belt even after 4.5 billion years. Hence, Kuiper belt objects located more than 40 AU from the Sun are likely to have held in stable orbits since the formation of the solar system.

Astronomers also believe there has been sufficient mass in the Kuiper belt to supply all the short-period comets that have ever been formed. So the Kuiper belt seems to be a good candidate for a cometary storehouse. And the mechanics of the transfer out of storage is now well understood. Computer simulations have shown that Neptune's gravity slowly erodes the inner edge of the Kuiper belt (the region within 40 AU of the Sun), launching objects from that zone into the inner solar system. Ultimately, many of these small bodies slowly burn up as comets.

If the Kuiper belt is the source of short-period comets, another obvious question emerges: Are any comets now on their way from the Kuiper belt into the inner solar system? The answer may lie in the Centaurs, a group of objects that includes the extremely red 5145 Pholus. Centaurs travel in huge planet-crossing orbits that are fundamentally unstable.

With orbital lifetimes that are far shorter than the age of the solar system, the Centaurs could not have formed where they currently are found. Yet the nature of their orbits makes it practically impossible to deduce their place of origin with certainty. Nevertheless, the nearest (and most likely) reservoir is the Kuiper belt. The Centaurs may thus be "transition comets." The strongest evidence supporting this hypothesis comes from one particular Centaur—2060 Chiron. Although its discoverers first thought it was just an unusual asteroid, 2060 Chiron is now firmly established as an active comet with a weak but persistent coma.

As astronomers continue to study the Kuiper belt, some have started to wonder whether this reservoir might have yielded more than just comets. Pluto, its satellite, Charon, and the Neptunian satellite Triton share similarities in their own basic properties but differ drastically from their neighbors. The Hubble Space Telescope has recently produced the first direct images of Pluto and Charon.

A PECULIAR TRIO

The densities of both Pluto and Triton, for instance, are much higher than any of the giant gaseous planets of the outer solar system. The orbital motions of these bodies are also quite strange. Triton revolves around Neptune in the "retrograde" direction—opposite to the orbital direction of all planets and most satellites. Pluto's orbit slants highly from the ecliptic, and it is so far from circular that it actually crosses the orbit of Neptune. Pluto is, however, protected from possible collision with the larger planet by a special orbital relationship known as a 3:2 mean-motion resonance. Simply put, for every three orbits of Neptune around the Sun, Pluto completes two.

These three bodies may have been swept up by Neptune, which captured Triton and locked Pluto—perhaps with Charon in tow—into its present orbital resonance.

Interestingly, orbital resonances appear to influence the position of many Kuiper belt objects as well. Up to one half of the newly discovered bodies have the same 3:2 mean-motion resonance as Pluto and, like that planet, may orbit serenely for billions of years. (The resonance prevents Neptune from approaching too closely and disturbing the orbit of the smaller body.) We have dubbed such Kuiper belt objects Plutinos—"little Plutos."

The recent discoveries of objects in the Kuiper belt provide a new perspective on the outer solar system. Pluto now appears special only because it is larger than any other member of the Kuiper belt. The many intriguing observations we have made of Kuiper belt objects remind us that our solar system contains countless surprises.

Migrating Planets

Renu Malhotra

In the familiar visual renditions of the solar system, each planet moves around the Sun in its own well-defined orbit, maintaining a respectful distance from its neighbors. The planets have maintained this celestial merry-go-round since astronomers began recording their motions, and mathematical models show that this very stable orbital configuration has existed for almost the entire 4.5-billion-year history of the solar system. It is tempting, then, to assume that the planets were "born" in the orbits that we now observe.

Certainly it is the simplest hypothesis. Modern-day astronomers have generally presumed that the observed distances of the planets from the Sun indicate their birthplaces in the solar nebula, the primordial disk of dust and gas that gave rise to the solar system. The orbital radii of the planets have been used to infer the mass distribution within the solar nebula. With this basic information, theorists have derived constraints on the nature and timescales of planetary formation. Consequently, much of our understanding of the early history of the solar system is based on the assumption that the planets formed in their current orbits.

It is widely accepted, however, that many of the smaller bodies in the solar system— asteroids, comets and the planets' moons—have altered their orbits over the past 4.5 billion years, some more dramatically than others. The demise of Comet Shoemaker-Levy 9 when it collided with Jupiter in 1994 was striking evidence of the dynamic nature of some objects in the solar system. Still smaller objects—micron- and millimeter-size interplanetary particles shaken loose from comets and asteroids—undergo a more gradual orbital evolution, gently spiraling in toward the Sun and raining down on the planets in their path.

Furthermore, the orbits of many planetary satellites have changed significantly since their formation. For example, Earth's moon is believed to have formed within 30,000 kilometers (18,600 miles) of Earth—but it now orbits at a distance of 384,000 kilometers. The Moon has receded by nearly 100,000 kilometers in just the past billion years because of tidal forces (small gravitational torques) exerted by our planet. Also, many satellites of the outer planets orbit in lockstep with one another: for instance, the orbital period of Ganymede, Jupiter's largest moon, is twice that of Europa, which in turn has a period twice that of Io. This precise synchronization is believed to be the result of a gradual evolution of the satellites' orbits by means of tidal forces exerted by the planet they are circling.

Until recently, little provoked the idea that the orbital configuration of the planets has altered significantly since their formation. But some remarkable developments during the past five years indicate that the planets may indeed have migrated from their original orbits. The discovery of the Kuiper belt has shown that our solar system does not end at Pluto. Approximately 100,000 icy "minor planets" (ranging between 100 and 1,000 kilometers in diameter) and an even greater number of smaller bodies occupy a region extending from Neptune's orbit—about 4.5 billion kilometers from the Sun—to at least twice that distance. The distribution of these objects exhibits prominent nonrandom features that cannot be readily explained by the current model of the solar system. Theoretical models for the origin of these peculiarities suggest the intriguing possibility that the Kuiper belt bears traces of the orbital history of the gas-giant planets—specifically, evidence of a slow spreading of these planets' orbits subsequent to their formation.

What is more, the recent discovery of several Jupiter-size companions orbiting nearby Sunlike stars in peculiarly small orbits has also focused attention on planetary migration. It is difficult to understand the formation of these putative planets at such small distances from their parent stars. Hypotheses for their origin have proposed that they accreted at more comfortable distances from their parent stars—similar to the distance between Jupiter and the Sun—and then migrated to their present positions.

PLUTO: OUTCAST OR SMOKING GUN?

Until just a few years ago, the only planetary objects known beyond Neptune were Pluto and its satellite, Charon. Pluto has long been a misfit in the prevailing theories of the solar system's origin: it is thousands of times less massive than the four gas-giant outer planets, and its orbit is very different from the well-separated, nearly circular and co-planar orbits of the eight other major planets. Pluto's is eccentric: during one complete revolution, the planet's distance from the Sun varies from 29.7 to 49.5 astronomical units (one astronomical unit, or AU, is the distance between the Earth and the Sun, about 150 million kilometers). Pluto also travels 8 AU above and 13 AU below the mean plane of the other planets' orbits. For approximately two decades in its orbital period of 248 years, Pluto is closer to the Sun than Neptune is.

In the decades since Pluto's discovery in 1930, the planet's enigma has deepened. Astronomers have found that most Neptune-crossing orbits are unstable—a body in such an orbit will either collide with Neptune or be ejected from the outer solar system in a relatively short time, typically less than 1 percent of the age of the solar system. But the particular Neptune-crossing orbit in which Pluto travels is protected from close

Migration of the outer planets is shown in an illustration of the solar system at the time when the planets formed (top) and in the present (bottom). The orbit of Jupiter is believed to have shrunk slightly, while the orbits of Saturn, Uranus and Neptune expanded. (The inner planetary region was not significantly affected by this process.) According to this theory, Pluto was originally in a circular orbit. As Neptune migrated outward, it swept Pluto into a 3:2 resonant orbit. Neptune's gravity forced Pluto's orbit to become more eccentric and inclined to the plane of the other planets' orbits.

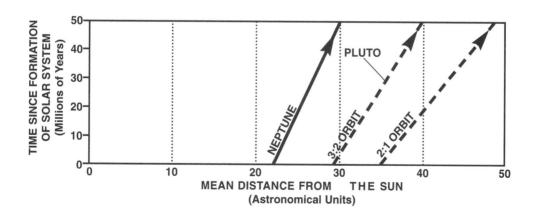

approaches to the gas giant by a phenomenon called resonance libration. Pluto makes two revolutions around the Sun during the time that Neptune makes three; Pluto's orbit is therefore said to be in 3:2 resonance with Neptune's. The relative motions of the two planets ensure that when Pluto crosses Neptune's orbit, it is far away from the larger planet. In fact, the distance between Pluto and Neptune never drops below 17 AU.

In addition, Pluto's perihelion—its closest approach to the Sun—always occurs high above the plane of Neptune's orbit, thus maintaining Pluto's long-term orbital stability. Computer simulations of the orbital motions of the outer planets, including the effects of their mutual perturbations, indicate that the relationship between the orbits of Pluto and Neptune is billions of years old and will persist for billions of years into the future. Pluto is engaged in an elegant cosmic dance with Neptune, dodging collisions with the gas giant over the entire age of the solar system.

How did Pluto come to have such a peculiar orbit? In the past, this question has stimulated several speculative and ad hoc explanations, typically involving planetary encounters. Recently, however, significant advances have been made in understanding the complex dynamics of orbital resonances and in identifying their Jekyll-and-Hyde role in producing both chaos and exceptional stability in the solar system. Drawing on this body of knowledge, I proposed in 1993 that Pluto was born somewhat beyond Neptune and initially traveled in a nearly circular, low-inclination orbit similar to those of the other planets but that it was transported to its current orbit by resonant gravitational interactions with Neptune. A key feature of this theory is that it abandons the assumption that the gas-giant planets formed at their present distances from the Sun. Instead it proposes an epoch of planetary orbital migration early in the history of the solar system, with Pluto's unusual orbit as evidence of that migration.

The story begins at a stage when the process of planetary formation was almost but not quite complete. The gas giants—Jupiter, Saturn, Uranus and Neptune—had nearly finished coalescing from the solar nebula, but a residual population of small planetesimals—rocky and icy bodies, most no larger than a few tens of kilometers in diameter—remained in their midst. The relatively slower subsequent evolution of the solar system consisted of the scattering or accretion of the planetesimals by the major planets. Because the planetary scattering ejected most of the planetesimal debris to distant or unbound orbits—essentially throwing the bodies out of the solar system—there was a net loss of orbital energy and angular momentum from the giant planets' orbits. But because of their different masses and distances from the Sun, this loss was not evenly shared by the four giant planets.

In particular, consider the orbital evolution of the outermost giant planet, Neptune, as it scattered the swarm of planetesimals in its vicinity. At first, the mean specific orbital energy of the planetesimals (the orbital energy per unit of mass) was equal to that of

Neptune itself, so Neptune did not gain or lose energy from its gravitational interactions with the bodies. At later times, however, the planetesimal swarm near Neptune was depleted of the lower-energy objects, which had moved into the gravitational reach of the other giant planets. Most of these planetesimals were eventually ejected from the solar system by Jupiter, the heavyweight of the planets.

Thus, as time went on, the specific orbital energy of the planetesimals that Neptune encountered grew larger than that of Neptune itself. During subsequent scatterings, Neptune gained orbital energy and migrated outward. Saturn and Uranus also gained orbital energy and spiraled outward. In contrast, Jupiter lost orbital energy; its loss balanced the gains of the other planets and planetesimals, hence conserving the total energy of the system. But because Jupiter is so massive and had so much orbital energy and angular momentum to begin with, its orbit decayed only slightly.

The possibility of such subtle adjustments of the giant planets' orbits was first described in a little-noticed paper published in 1984 by Julio A. Fernandez and Wing-Huen Ip, a Uruguayan and Taiwanese astronomer duo working at the Max Planck Institute in Germany. Their work remained a curiosity and escaped any comment among planet formation theorists, possibly because no supporting observations or theoretical consequences had been identified.

In 1993 I theorized that as Neptune's orbit slowly expanded, the orbits that would be resonant with Neptune's also expanded. In fact, these resonant orbits would have swept by Pluto, assuming that the planet was originally in a nearly circular, low-inclination orbit beyond Neptune. I calculated that any such objects would have had a high probability of being "captured" and pushed outward along the resonant orbits as Neptune migrated. As these bodies moved outward, their orbital eccentricities and inclinations would have been driven to larger values by the resonant gravitational torque from Neptune. (This effect is analogous to the pumping-up of the amplitude of a playground swing by means of small periodic pushes at the swing's natural frequency.) The final maximum eccentricity would therefore provide a direct measure of the magnitude of Neptune's migration. According to this theory, Pluto's orbital eccentricity of 0.25 suggests that Neptune has migrated outward by at least 5 AU. Later, with the help of computer simulations, I revised this to 8 AU and also estimated that the timescale of migration had to be a few tens of millions of years to account for the inclination of Pluto's orbit.

Of course, if Pluto were the only object beyond Neptune, this explanation of its orbit, though compelling in many of its details, would have remained unverifiable. The theory makes specific predictions, however, about the orbital distribution of bodies in the Kuiper belt, which is the remnant of the primordial disk of planetesimals beyond Neptune (see Chapter 5, "The Kuiper Belt"). Provided that the largest bodies in the primordial Kuiper belt were sufficiently small that their perturbations on the other objects in the belt would be negligible, the dynamical mechanism of resonance sweeping would work not only on Pluto but on all the trans-Neptunian objects, perturbing them from their original orbits. As a result, prominent concentrations of objects in eccentric orbits would be found at Neptune's two strongest resonances, the 3:2 and the 2:1. Such orbits are ellipses with semimajor axes of 39.5 AU and 47.8 AU, respectively. (The length of the semimajor axis is equal to the object's average distance from the Sun.)

More modest concentrations of trans-Neptunian bodies would be found at other resonances, such as the 5:3. The population of objects closer to Neptune than the 3:2 resonant orbit would be severely depleted because of the thorough resonance sweeping of that region and because perturbations caused by Neptune would destabilize the orbits of any bodies that remained. On the other hand, planetesimals that accreted beyond 50 AU from the Sun would be expected to be largely unperturbed and still orbiting in their primordial distribution.

Fortunately, recent observations of Kuiper belt objects, or KBOs, have provided a means of testing this theory. More than 174 KBOs had been discovered as of mid-1999. Most have orbital periods in excess of 250 years and thus have been tracked for less than 1 percent of their orbits. Nevertheless, reasonably reliable orbital parameters have been determined for about 45 of the known KBOs. Their orbital distribution is not a pattern of uniform, nearly circular, low-inclination orbits, as would be expected for a pristine, unperturbed planetesimal population. Instead one finds strong evidence of gaps and concentrations in the distribution. A large fraction of these KBOs travel in eccentric 3:2 resonant orbits similar to Pluto's, and KBOs in orbits interior to the 3:2 orbit are nearly absent—which is consistent with the predictions of the resonance sweeping theory.

Still, one outstanding question remains: Are there KBOs in the 2:1 resonance comparable in number to those found in the 3:2, as the planet migration theory would suggest? And what is the orbital distribution at even greater distances from the Sun? At present, the census of the Kuiper belt is too incomplete to answer this question fully. But on Christmas Eve 1998 the Minor Planet Center in Cambridge, Massachusetts, announced the identification of the first KBO orbiting in 2:1 resonance with Neptune. Two days later the center revealed that another KBO was traveling in a 2:1 resonant orbit. Both these objects have large orbital eccentricities, and they may turn out to be members of a substantial population of KBOs in similar orbits. They had previously been identified as orbiting in the 3:2 and 5:3 resonances, respectively, but new observations made last year strongly indicated that the original identifications were incorrect. This episode underscored the need for continued tracking of known KBOs in order to map their orbital distribution correctly. We must also acknowledge the dangers of overinterpreting a still small data set of KBO orbits.

In short, although other explanations cannot be ruled out yet, the orbital distribution of KBOs provides increasingly strong evidence for planetary migration. The data suggest that Neptune was born about 3.3 billion kilometers from the Sun and then moved about 1.2 billion kilometers outward—a journey of almost 30 percent of its present orbital radius. For Uranus, Saturn and Jupiter, the magnitude of migration was smaller, perhaps 15, 10 and 2 percent, respectively; the estimates are less certain for these planets because, unlike Neptune, they could not leave a direct imprint on the Kuiper belt population.

Most of this migration took place over a period shorter than 100 million years. That is long compared with the timescale for the formation of the planets—which most likely took less than 10 million years—but short compared with the 4.5-billion-year age of the solar system. In other words, the planetary migration occurred in the early history of the solar system but during the later stages of planet formation. The total mass of the scattered planetesimals was about three times Neptune's mass. The question arises whether even more drastic orbital changes might occur in planetary systems at earlier times, when the primordial disk of dust and gas contains more matter and perhaps many protoplanets in nearby orbits competing in the accretion process.

OTHER PLANETARY SYSTEMS?

In the early 1980s theoretical studies by Peter Goldreich and Scott Tremaine, both then at the California Institute of Technology, and others concluded that the gravitational forces between a protoplanet and the surrounding disk of gas, as well as the energy losses caused by viscous forces in a gaseous medium, could lead to very large exchanges of energy and angular momentum between the protoplanet and the disk. If the torques exerted on the protoplanet by the disk matter just inside the planet's orbit and by the matter just beyond it were slightly unbalanced, rapid and drastic changes in the planet's orbit could happen. But again, this theoretical possibility received little attention from other astronomers at the time. Having only our solar system as an example, planet formation theorists continued to assume that the planets were born in their currently observed orbits.

In the past five years, however, the search for extrasolar planets has yielded possible signs of planetary migration. By measuring the telltale wobbles of nearby stars—within 50 light-years of our solar system—astronomers have found evidence of more than a dozen Jupiter-mass companions in surprisingly small orbits around main-sequence stars. The first putative planet was detected orbiting the star 51 Pegasi in 1995 by two Swiss astronomers, Michel Mayor and Didier Queloz of the Geneva Observatory, who were actually surveying for binary stars. Their observations were quickly confirmed by Geoffrey W. Marcy and R. Paul Butler, two American astronomers working at Lick Observatory near San Jose, California. As of June 1999, 20 extrasolar planetary candidates have been identified, most by Marcy and Butler, in search programs that have surveyed almost 500 nearby Sunlike stars over the past 10 years (see Chapter 7: "Giant Planets Orbiting Faraway Stars"). The technique used in these searches—measuring the Doppler shifts in the stars' spectral lines to determine periodic variations in stellar velocities—yields only a lower limit on the masses of the stars' companions. Most of the candidate planets have minimum masses of about one Jupiter-mass and orbital radii shorter than 0.5 AU.

What is the relationship between these objects and the planets in our solar system? According to the prevailing model of planet formation, the giant planets in our solar system coalesced in a two-step process. In the first step, solid planetesimals clumped together to form a protoplanetary core. Then this core gravitationally attracted a massive gaseous envelope from the surrounding nebula. This process must have been completed within about 10 million years of the formation of the solar nebula itself, as inferred from astronomical observations of the lifetime of protoplanetary disks around young Sunlike stars.

At distances of less than 0.5 AU from a star, there is insufficient mass in the primordial disk for solid protoplanetary cores to condense. Furthermore, it is questionable whether a protoplanet in a close orbit could attract enough ambient gas to provide the massive envelope of a Jupiterlike planet. One reason is simple geometry: an object in a tight orbit travels through a smaller volume of space than one in a large orbit does. Also, the gas disk is hotter close to the star and hence less likely to condense onto a protoplanetary core. These considerations have argued against the formation of giant planets in very short-period orbits.

Instead several theorists have suggested that the putative extrasolar giant planets may have formed at distances of several AU from the star and subsequently migrated inward. Three mechanisms for planetary orbital migration are under discussion. Two involve disk-protoplanet interactions that allow planets to move long distances from their birthplaces as long as a massive disk remains.

With the disk-protoplanet interactions theorized by Goldreich and Tremaine, the planet would be virtually locked to the inward flow of gas accreting onto the protostar and might either plunge into the star or decouple from the gas when it drew close to the star. The second mechanism is interaction with a planetesimal disk rather than a gas disk: a giant planet embedded in a very massive planetesimal disk would exchange energy and angular momentum with the disk through gravitational scattering and resonant interactions, and its orbit would shrink all the way to the disk's inner edge, just a few stellar radii from the star.

The third mechanism is the scattering of large planets that either formed in or moved into orbits too close to one another for long-term stability. In this process, the outcomes would be quite unpredictable but generally would yield very eccentric orbits for both planets. In some fortuitous cases, one of the scattered planets would move to an eccentric orbit that would come so near the star at its closest approach that tidal friction would eventually circularize its orbit; the other planet, meanwhile, would be scattered to a distant eccentric orbit. All the mechanisms accommodate a broad range of final orbital radii and orbital eccentricities for the surviving planets.

These ideas are more than a simple tweak of the standard model of planet formation. They challenge the widely held expectation that protoplanetary disks around Sunlike stars commonly evolve into regular planetary systems like our own. It is possible that most planets are born in unstable configurations and that subsequent planet migration can lead to quite different results in each system, depending sensitively on initial disk properties. An elucidation of the relation between the newly discovered extrasolar companions and the planets in our solar system awaits further theoretical and observational developments. Nevertheless, one thing is certain: the idea that planets can change their orbits dramatically is here to stay.

Comet Shoemaker-Levy 9 Meets Jupiter

David H. Levy, Eugene Shoemaker and Carolyn S. Shoemaker

We were working underneath the dome of the small Schmidt telescope at Palomar Observatory in California, in a cramped room cluttered with papers, books and a laptop computer. It was May 22, 1993. Carolyn sat hunched over her stereomicroscope, an instrument she has used to examine photographs for asteroids and comets for more than a decade, since she joined her husband, Gene, in his survey of these small wanderers in the sky.

Gene has spent a significant part of his career examining such objects. More recently Gene, along with Carolyn, has been engaged in a systematic search for asteroids capable of striking the Earth.

Peering at his computer that day, David checked his e-mail to see whether any newly detected comets or asteroids needed to be added to the observing schedule. David's e-mail conveyed astonishing news from the International Astronomical Union's Central Bureau for Astronomical Telegrams—a kind of wire service for astronomers. A comet we had discovered two months earlier would strike Jupiter in July 1994. After a professional lifetime examining impact craters and the bodies that make them, Gene might actually get to see a collision.

THE IMPACT OF IMPACTS

As anyone who has gazed at the Moon through even the smallest telescope knows, the lunar surface is studded with impact craters. The Moon itself probably formed from the remains of a collision. During our planet's youth, a body the size of Mars may have struck the Earth, melting it and sending into orbit a stream of debris that eventually congealed to form the Moon. Tectonically static and without air or water, the moon can retain its crater-scarred face indefinitely. Erosion and the deposition of sediments constantly smooth the Earth's surface, which consequently shows few craters, although our world has been hit far more often than the Moon. For example, comets showered the Earth during its formative period between 3.9 and 4.6 billion years ago, bringing carbon, hydrogen, nitrogen and oxygen—critical elements that allowed life to evolve.

Such collisions have also taken life away. Sixty-five million years ago an object somewhat larger than Halley's Comet slammed into what is now the coast of Mexico's Yucatán peninsula. The impact gouged a crater 170 kilometers (105 miles) across and launched debris worldwide. As the multitude of tiny ballistic missiles fell back toward the Earth, meteors filled the sky, and the atmosphere became red-hot. Fires erupted over the Earth's surface, but the global inferno was soon followed by persistent darkness, as dust lifted into the atmosphere blocked the Sun's rays. Months of planet-wide cooling then gave way to centuries of greenhouse warming from the carbon dioxide released during the impact from the target rocks. Many species became extinct.

That ancient catastrophe demonstrates that projectiles from space can indeed affect this planet significantly. Our research program at Palomar was one of several designed to assess the rate at which interplanetary intruders of this kind collide with the planets and satellites. We did not, however, expect to witness such a colossal impact in the near future.

A LUCKY DISCOVERY

Our discovery of the comet began quietly enough. A typical comet has a nucleus several kilometers across composed of ices, rocky material and organic compounds. When it nears the Sun, the ices turn directly from solid to gas and release dust to form a light-scattering halo called a coma. The pressure of solar radiation then blows this material into an elongated tail. But instead of a single coma and tail, our new comet had a bar-shaped agglomeration of comae, with a composite tail stretching to the north. The strangest observation was that on either end of the bar was a pencil-thin line of light.

Our weird discovery needed confirmation with a better telescope. We contacted our colleague James V. Scotti of the University of Arizona, who was observing that night from the Spacewatch telescope atop Kitt Peak in Arizona. Jim agreed to take high-resolution television images of the comet. He was stunned. "There are at least five discrete comet nuclei side by side," Scotti explained to us over the telephone as he described the view, "but comet material exists between them. I suspect that there are more nuclei that I'll see when the sky clears."

We immediately reported this bizarre comet to Brian G. Marsden, director of the Central Bureau for Astronomical Telegrams at the Harvard-Smithsonian Center for Astrophysics, and Scotti followed with his observations. The next day Marsden's office announced the discovery. The description of the object was so unusual that astronomers around the world began to examine it at once. Jane Luu of Stanford University and David Jewitt of the University of Hawaii obtained a magnificent image using the 88-inch reflector at Jewitt's institution. They later resolved 21 separate nuclei strung out, they wrote, "like pearls on a string."

CLOSE ENCOUNTERS

By the middle of April 1993, Marsden, Syuichi Nakano in Japan and Donald K. Yeomans of the Jet Propulsion Laboratory in Pasadena, California, had determined that the comet we had uncovered was actually in an orbit about Jupiter. They also ascertained that the comet had passed very close to the planet about eight months before we located it. Such proximity would explain why there were multiple fragments.

On July 7, 1992, S-L 9 [Comet Shoemaker-Levy 9] had approached within about 20,000 kilometers (12,400 miles) of Jupiter's cloud tops. As it made a hairpin turn around the giant planet, it came apart because the pieces nearest Jupiter were deflected more sharply than those farther away. The difference in orbital paths resulted from the decrease in the strength of Jupiter's gravitational attraction between the near and far sides of the comet. The stress on S-L 9 was extremely weak, but it nonetheless broke up the comet easily. This behavior suggests that the original body was merely a pile of fragments held together by their weak gravitational attraction for one another.

Although astronomers had earlier established that comets have orbited Jupiter for brief periods in the past, S-L 9 is the first comet that anyone has seen in orbit about a planet. Jupiter indeed had not one but 21 tiny new moons. Yet these recently acquired satellites were not to last long. After further calculations, Marsden announced that the fractured comet would crash down on Jupiter in July of 1994.

The comets were predicted to strike well over on Jupiter's nightside, where they would be hidden from the Earth's view by the body of the planet. Jupiter would have to rotate eastward before any remains could be visible from the Earth. Nature was to put on the biggest impact extravaganza in history, and it seemed that our seat was to be behind a post.

A GLOBAL OBSERVING SESSION

As "impact week" approached in the summer of 1994, it became clear that this event was so extraordinary that it deserved observation time on as many telescopes as possible. Just as extraordinary was the good fortune that events had allowed astronomers a full 14 months to coordinate their programs. Heading the list of powerful telescopes to be aimed at Jupiter was the Hubble Space Telescope, whose newly corrected optics had already captured the comet nuclei with amazing clarity. For a team led by Harold A. Weaver of the Space Telescope Science Institute in Baltimore, Hubble's wide-field planetary camera would monitor the comet nuclei as they moved closer to Jupiter. A group led by Heidi B. Hammel of the Massachusetts Institute of Technology used the telescope to take detailed images of the entire planet on the day before the first collision, to compare with later views to come during the week. The telescope would also collect spectrographic signatures of elements and gases released during the explosions.

That is, of course, if something could still be seen when the impact sites rotated into the Earth's view.

But even if the nightside strikes were to be invisible from the Earth, there was another means to examine them. On its way toward a rendezvous with Jupiter, the Galileo space probe was in a position that would give its cameras and other instruments a direct view of the impact sites. Controllers at the Jet Propulsion Laboratory instructed the spacecraft to collect and return data on several of the impacts.

Many of the world's great telescopes were destined to play a vital role in recording the strikes and the related phenomena. The collisions would occur over a period lasting almost six days; thus, telescopes spread over the globe were needed. Palomar's venerable five-meter telescope, other large telescopes in Spain, Chile, Hawaii and Australia, and a host of smaller telescopes participated. The NASA Kuiper Airborne Observatory, flying out of Melbourne, Australia, captured key spectroscopic measurements. In addition, teams of radio astronomers monitored Jupiter for the effects of the impacts on the Jovian magnetosphere.

Using the Keck Observatory's giant 10-meter telescope atop Mauna Kea in Hawaii, Imke de Pater of the University of California at Berkeley and her colleagues planned to record infrared images in the wavelengths of light absorbed by cold methane gas. Because Jupiter's methane-rich atmosphere absorbs these wavelengths, filters that pass light only in the "methane band" would darken the face of the planet and highlight anything happening very high in or above the planet's atmosphere.

JULY 16, 1994: SHOW TIME

After 14 months of waiting, the first word was electrifying: Calar Alto Observatory in Spain had recorded the infrared signature of a collapsing plume from the first impact (Nucleus A). The detection was confirmed by the European Southern Observatory in Chile. Not only was the impact detectable, it was spectacular. As we were soon to find out, the plume shot some 3,000 kilometers above the clouds of Jupiter.

A BATTERED PLANET

From the outset the performance of the comets was intriguing. As Jupiter rotated, a large spot left by Nucleus A came into view. It was made up of three distinct parts: a central streak, an expanding ring and a peculiar crescent-shaped outer cloud. In the visible part of the spectrum, the markings looked extraordinarily dark, but in the infrared light of a methane absorption band, the spot appeared bright against the dark planet. The entire spot was as large as the Earth. Several hours later Nucleus B struck Jupiter with quite different effects. Even though B had been brighter than A, the plume that rose from its impact was so much smaller that only the largest telescope in the world,

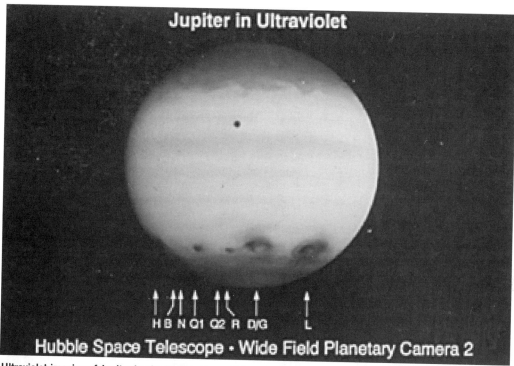

Ultraviolet imaging of Jupiter by the Hubble Space Telescope shows clearly the aftermath of comet-fragment hits. Those dark, smoky blemishes visible in the upper atmosphere of the southern part of the planet are each larger than the Earth itself. Dark spot visible in northern latitude is the shadow of one of Jupiter's moons.

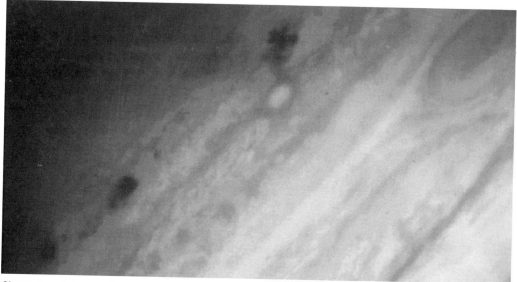

Closer view of damage done to Jupiter by fragments of Comet Shoemaker-Levy 9, taken by the Hubble Space Telescope.

the 10-meter Keck, recorded it easily. Nucleus B may have consisted of a swarm of small house-size subnuclei that split off from Nucleus C sometime after the initial breakup. An observer on Jupiter would have seen a fabulous storm of meteors, but little was detected from the Earth.

Nuclei C and E crashed with much the same effects as A. Two days later there was great anticipation as Nucleus G—which had a bright coma and presumably large mass—made its final descent. Hubble had a clear view of Jupiter, but that night all the big telescopes at Mauna Kea Observatories were closed because of fog and drizzle. Yet miraculously, only a minute before the impact, the clouds above Mauna Kea parted. The observatory domes raced open, and the telescopes captured images of the strike before more fog and rain forced them to close again only 10 minutes later. They were lucky to get a view: Nucleus G hit with such tremendous energy that the collapsing plume was much brighter than the entire planet in the infrared methane band. Nucleus G left the same imprint as the earlier major impacts of A, C and E, but the scar was much bigger. The great flash of energy was well recorded in Australia and at the South Pole.

At this point, Hubble had detected expanding rings from impacts A, E and G in the clear regions between the inner dark core clouds and the outer dark crescents. It was found that these were expanding outward at about 450 meters per second. Interpretation of these features fell to Andrew P. Ingersoll of Caltech. Soon after impact week, Ingersoll realized the rings were not moving out fast enough to be sound waves— they were not the "boom from the plume," as he had originally thought. But the speed of the waves was the same for all impacts. Ultimately, Ingersoll and Hiroo Kanamori, also at Caltech, found that an "internal gravity" wave had been produced, somewhat like the waves formed by a stone thrown into a pond.

As impact week continued, Nucleus L left the largest spot yet, once again complete with a central core and outer, crescent-shaped cloud. By this time amateur astronomers around the world had found that these dark features on Jupiter were so large and dense that they could be seen by using small telescopes. The nuclei of H, K and L were all preceded by a long train of particles whose entry into the atmosphere produced a rising infrared glow before the arrival of the main part of the nucleus. The Galileo spacecraft took an engaging series of "snapshots" of the brilliant meteor and incandescent rising plume from the impact of W, the final nucleus, as it tore into Jupiter. The Hubble image sequence of the same fall ended with a view of the plume collapsing directly on top of the spot made earlier by Nucleus K.

Comet S-L 9 probably began its wanderings in the outer solar system beyond the orbit of Neptune. A series of close encounters with Jupiter gradually altered its orbital period from one revolution about the Sun every several thousand years to about once a decade. The latest orbital calculations, by Paul W. Chodas of the Jet Propulsion

Laboratory, indicate that probably about 1929 (the year an unrelated crash hit the Earth's stock market) the comet made a slow approach to Jupiter that allowed the planet to capture the comet as a moon. The resulting two-year-long orbit about the planet was, however, unstable. Some revolutions followed narrow ellipses; others were roughly circular. In 1992, when the orbit was highly elliptical, the comet passed so close to Jupiter that it was broken apart.

The initial disintegration dispersed the cometary material into a long swarm of debris. Erik I. Asphaug of the NASA Ames Research Center and Willy Benz of the University of Arizona have shown that the loose string of rubble could have then coalesced into a set of distinct nuclei under the mutual gravitational attraction of the fragments. We suspect that large coherent pieces of fractured comet were present in some nuclei but not in others.

After the main disruption event, additional nuclei split from some of the earlier formed nuclei. Just how this later fracturing occurred is not understood. Possibly internal gas pressure ruptured large chunks, or perhaps the force of collisions between fragments traveling in the swarm knocked them apart. The largest individual nuclei in the entire train probably were no more than a kilometer or two across. These nuclei did not complete even one more orbit before striking Jupiter's flank. When they hit, the energy from each of the largest impacts probably equaled hundreds of thousands of large hydrogen bombs exploding simultaneously.

The great dark scars left on Jupiter gradually spread, merged and slowly faded in the months after the impacts. Such dark clouds have never been seen on Jupiter before, and one wonders just how rare such a dramatic event must be.

The frequency of impacts depends on the scale of the body involved, and we are still uncertain about the size of this comet before it broke apart. But by making some reasonable assumptions, we can estimate that the crash of a string of nuclei such as S-L 9 probably occurs less than once every few thousand years. Thus, we feel fortunate to be living at this moment, to have found the comet on its way toward Jupiter and to have witnessed its demise in a blaze of glory.

POSTSCRIPT

David H. Levy

More than two years after the impacts, Jupiter was still showing considerable changes. At a symposium on the Shoemaker-Levy 9 impacts held at Meudon, France, A. Marten reported continued presence of both hydrogen cyanide (HCN) and carbon sulfide (CS). In May of 1995, spectra recorded at the 30-meter IRAM telescope in Granada, Spain, and at the James C. Maxwell Telescope in Hawaii showed that the HCN and CS were still present.

Even more interesting, Both HCN and CS possibly have increased with time, spreading to the planet's northern hemisphere. Many new molecular species were detected either for the first time or in greater quantity after the impacts, including S_2, CS_2, OCS, NH_3, HCN, H_2O, and CO. Some, like S_2, OCS, H_2S, and NH_3 apparently did not survive more than a few months. NH_3 was still visible eight months after the impacts, and although CS_2 and CO weakened, they were still detected in the Jovian stratosphere as late as May 1996. Compared to these, the CS and HCN remained prominent.

What was the origin of these chemicals? Since the high temperatures of the fireballs would have destroyed most of the organic materials, Peter Wilson and Carl Sagan suggested that they were formed by the process of quench synthesis as cometary material that was vaporized and dissociated in the hot fireball recombined into the observed substances as the temperature plunged in the rising plume.

Besides these chemicals, an upper atmosphere haze was still visible as late as June 1995. By that time it had spread from latitude -70 degrees as far north as -20 degrees. During the two years following the impacts, the debris spread as a result of meridional and vertical shears of Jupiter's zonal winds.

Radio telescopes were as productive in observing the impacts and their aftermath as were optical instruments. Observations with telescopes like New Mexico's Very Large Array showed an east-west asymmetry in radio intensity that began just after the first impact. Although the main radiation peaks returned to normal after the last impact, emissions at high latitudes persisted for weeks.

Analyses of the impacts' results support the model by Eric Asphaug and Willy Benz that suggested a relatively small progenitor comet. Mordecai-Mark MacLow used the luminosity of the fireballs, chemistry of the impact spots and the lack of any seismic waves to conclude that the fragments did not exceed 750m in diameter, and that the parent comet had a diameter of 1.5 km and a density of 0.6 g cm^3. Although several groups attempted to detect seismic effects from the impacts, there was no confirmed evidence of any.

This lack of seismic effect also suggests a small impactor size. Other researchers agree on this lower size figure, including Kevin Zahnle, whose studies of the plumes showed that the terminal explosions took place at the 1 to 4 bar level of Jupiter's atmosphere.

However, Zdenek Sekanina and his colleagues concluded that the observed position angle of the fragment train after the comet's discovery in March 1993 cannot be explained by the comet's splitting apart precisely at perijove 8.5 months earlier. They suggest that a body about 10 kilometers in diameter and with 10^{17} g of mass at an assumed density of 0.2 g/cm^3 broke apart due to tidal interactions with Jupiter, and that this breakup was assisted somewhat by the comet's own rapid rotation. In their 1998 model Sekanina and colleagues maintained that the progenitor was large and that it did not begin to disrupt until after its July 1992 perijove. Jovian tidal forces caused the comet to suffer extensive cracks and then begin to split along the planes of the tidally induced cracks due to the comet's rapid rotation. Secondary fragmentation took place later, and the velocities imparted to these fragments are almost all along a great circle, indicating that these later disruptions were induced by rapid rotation.

"We still have a huge amount of unreduced data," noted Glenn Orton at the Meudon meeting in 1996. By 1998 most of these data were still unreduced; it is possible that some of the answers may lie in all these data.

As Richard West noted in his closing remarks at the Meudon conference, the only thing that everyone agrees on is that a series of objects struck Jupiter in July 1994. Although more observations were made of this astronomical event than of any other in the history of astronomy, much of the data still remains to be reduced and analyzed. Hopefully more answers will be forthcoming as the data set continues to take shape, and modelers continue to work with it.

On Comets

James V. Scotti

We humans are on the verge of making our first detailed *in situ* studies of cometary nuclei by sending robotic spacecraft to several comets during the first decade of the 21st century. Where do we stand in our knowledge of comets at this unique threshold of cometary exploration? And how have comets affected the development of human civilization and of life on Earth?

Modern comet research can be traced back to two landmark papers published half a century ago in the same year of 1950. In the first, Jan H. Oort examined the original orbits of a small sample of long-period comets and concluded that they originated in a spherical, isotropic reservoir between about 50,000 and 150,000 astronomical units (AU) from the Sun, now called the Oort Cloud. In the second, the "father" of modern comet research, Fred L. Whipple, theorized that comet nuclei had a structure he likened to a "dirty snowball." Our study of the structure of cometary nuclei continues to support this hypothesis today.

The last decade of the 20th century saw a number of significant cometary events and discoveries. Perhaps the most important was the discovery of Comet Shoemaker-Levy 9 in March 1993 and its subsequent impact on Jupiter in July 1994. Another important discovery was that in 1992 of the first of the so-called Edgeworth-Kuiper belt (EKB) objects, 1992 QB1, as well as an explosion in the discovery rate of such objects by the end of the decade. These objects are thought to be the possible source of the short-period comets. The decade also saw the discovery of several bright comets, including Comet Hale-Bopp, discovered almost two years before it reached perihelion in early 1997.

The differences between comets and asteroids have become fewer in many ways with time. We find most asteroids orbiting the Sun between the orbits of Mars and Jupiter. But we also find asteroids on planet-crossing orbits. Their orbits are "chaotic" in that they suffer frequent orbit changes during close approaches to the planets, and because of those close approaches, their orbits cannot be predicted over very long periods of time. Other asteroids have been found orbiting among the outer planets, the so-called Centaur asteroids, the first discovered by C. T. Kowal in 1977 and called 2060 Chiron. Comets are nearly always found on planet-crossing orbits, but visually they differ from asteroids that appear stellar in their appearance with comae and tails when they approach close to the Sun.

There are two classes of comets: the long-period comets, with orbital periods greater than 200 years and orbits that are isotropically distributed on the sky, and the short-period comets, with orbital periods less than 200 years. The short-period comets can be separated into two different classes as well. Those with orbital periods less than about 20 years and low inclinations are called the Jupiter family of short-period comets while those with longer periods are called Halley-type comets and are thought to be derived from the isotropically distributed long-period comets. We now think that the Jupiter-family comets are derived from a low inclination source, the aforementioned Edgeworth-Kuiper belt.

THE ORIGIN OF COMETS

Our understanding of the origin of the comets has advanced greatly since Oort first proposed their origin in a cloud at the fringes of our solar system. According to Oort's original work, he found an abundance of comets with aphelia between 50,000 and 150,000 AU. He also estimated the amount of velocity change that passing stars would impart on members of this cloud and found that the mechanism would easily supply long-period comets to the inner solar system. Based on his assumptions, he was able to estimate the number of comets required to be in this cloud as 1.9×10^{11}. Oort suggested that this cloud of comets might be minor planets that escaped from an early stage of the planetary system and brought into large and stable orbits through the perturbations of stars and of Jupiter.

With a larger number of long-period comet orbits available today, the Oort cloud's structure is still largely unknown, but the idea of the Oort cloud is still sound. Several outstanding problems are still being debated. One is the problem of cometary fading. A simple Monte Carlo simulation of the arrival of new Oort cloud comets results in an excess of comets making their first entry into the solar system as compared to the observed distribution. The heliocentric distribution of the comets is still largely unknown, with the vast majority of comets only being seen very near perihelion, and no long-period comets have been seen with perihelia beyond about 7 AU. This is due to large selection affects in the discovery of comets. They need to be active, and they are too faint to be seen while far from the Sun and are not active.

Debates have also ensued over the actual distribution of the directions of the aphelia of comets. Is the clumpiness that is seen due to discovery selection affects, or the result of some other affect, such as the recent passage of a star through the Oort cloud or simply due to their small numbers? Numerical experiments since Oort's work have confirmed the basic idea of the ejection of comets into the Oort cloud during solar system formation. Jupiter is far too efficient at ejecting bodies from its vicinity, and it tends to eject the would-be comets completely from the solar system. Instead, it is the action of the outer planets, principally Uranus and Neptune, that are most likely the largest contributors to the Oort cloud population. Objects ejected into large enough orbits will have their orbits strongly perturbed by external forces, such as the galactic tide and nearby stars, so that they are preserved for longer times outside the solar system, safe for billions of years from planetary encounters. Once a long-period comet reenters the solar system, it is quickly doomed to impact or, more likely, ejection or physical destruction on timescales much shorter than the age of the solar system.

Comets at the outer fringes of the Oort cloud orbit at low speeds around the Sun, taking millions of years to orbit the Sun at speeds of less than 0.1 kilometers per second. Many of the external forces acting on comets at this distance can impart a comparable change in velocity Δv, to its orbital speed, which can either eject a comet or send it inward. Objects whose velocities are almost completely negated fall in nearly straight toward the Sun to within the inner solar system so that we might see them from our vantage point on Earth. Objects orbiting closer to the Sun move with significantly higher velocity so that the same Δv hardly changes the comet's orbit at all.

Long-period comets entering the solar system are perturbed by the planets. Nearly half leave the solar system on slightly hyperbolic orbits, removing them completely from the solar system. Most of the rest have their orbit changed so that it is more tightly bound to the Sun. This so-called diffusion process continues with the comets randomly walking in orbital energy, some moving outward, others becoming still more tightly bound to the Sun. By this mechanism along with the stochastic interaction with the planets, especially Jupiter, it was thought that the supply of short-period comets could be generated from the long-period comets.

Numerical experiments by Edgar (1972, 1977) suggested that this capture mechanism was plausible. More recent work, however, has called this mechanism into doubt. The Jupiter-family comets, with orbital periods less than 20 years, are on low-inclination orbits. Early work by Joss (1973) threw doubt on the ability of Jupiter to transform the isotropic distribution of the long-period comets onto low inclination orbits in the plane of the solar system. More recently, Martin Duncan, Thomas Quinn and Scott Tremaine (1987, 1988) confirmed this problem by direct integration of test particles in a solar system composed of planets with enhanced masses to speed up the process. They suggested that the Jupiter family of short-period comets had to come from a low-inclination source.

Kenneth Edgeworth (1943, 1949) and Gerard Kuiper (1951) both suggested that there was no reason to believe that accretion of small bodies did not continue beyond Neptune and that perhaps material might still exist in orbits outside that of Neptune. Julio Fernandez suggested in 1980 that a comet belt might exist between 35 and 50 AU from the Sun and could provide a source of short-period comets. Confirmation of the existence of this belt of trans-Neptunian objects (TNOs), or the Edgeworth-Kuiper belt, began with their discovery of 1992 QB1 by Dave Jewitt and Jane Luu, followed shortly by the discovery of 1993 FW.

Objects in the Edgeworth-Kuiper belt are found in regions of resonance with Neptune, similar to the orbit of Pluto, which is protected from close approaches to Neptune despite having a perihelion distance inside the orbit of Neptune by a mean-motion resonance called the 2:3 resonance. Pluto orbits the Sun twice in the same time it takes Neptune to orbit the Sun three times. Other objects orbit far enough outside of Neptune that they remain relatively unperturbed. Long-term orbit integrations of objects in these types of orbits have shown that they are chaotic on long timescales. While their dynamical lifetimes are comparable to or longer than the age of the solar system, some of them leak out, become Neptune-crossing. Eventually they encounter Neptune and are perturbed out of their stable orbits.

In 1977 Charles T. Kowal discovered an object that orbits entirely beyond the orbit of Jupiter. It was numbered and named 2060 Chiron and is the prototypical Centaur asteroid. It crosses the orbits of both Saturn and Uranus and has a low-inclination orbit so that it is dynamically unstable. Integrations of objects on Chironlike orbits by Gerhard Hahn and Mark E. Bailey in 1990 found that Chiron is likely to be ejected from the solar system in less than one million years. It also has a substantial probability of evolving a short-period comet orbit or even an Earth-crossing orbit. Chiron was found to have cometary activity with a coma and later a tail as it approached perihelion. Fifteen Centaur asteroids have been found as of 1999 and appear to be an intermediate population between the short-period comets and their source.

THE STRUCTURE AND PHYSICAL EVOLUTION OF COMETS

Comets are transient bodies. Not only are they dynamically short lived, but they also clearly show signs of decay and physical destruction. Even the appearance of a bright comet displays signs of its limited life span. The appearance of comets usually includes structures such as comae, tails, and often, multiple nuclei.

Comets are fragile bodies. We have seen comets split into multiple pieces quite frequently. Most comet-splitting events occur when a small piece of the comet calves off the nucleus of the comet and rapidly dissipates. More rarely, the comet splits into two pieces that can survive for long times.

The relative motion of split comets was studied by Zdenek Sekanina (1982). They were found to separate because of their differential nongravitational activity. There appears to be little correlation with a comet's location in its orbit and the time of splitting. Some comets appear to split far from the Sun, while others split closer to the Sun. Estimates have been made for the physical lifetime of comets based on their various decay mechanisms, splitting and outgassing, as well as collisions with debris and even planets during their orbits around the Sun. Comets may also become dormant or extinct when their surfaces are crusted over with layers of debris that lack the volatile substances that drive their visible activity.

Some comets have split because of tidal disturbances of planets, especially Jupiter. One pair of comets, P/Van Biesbroeck and P/Neujmin 3, which are presently on significantly different orbits, can be traced back to the vicinity of Jupiter in 1887, when they were both in the same area. Integrating their orbits back a little further, we find them in similar orbits and can conclude that these two separate comets were once part of one parent comet.

The most famous tidally disrupted comet is Shoemaker-Levy 9, discovered on March 24, 1993, as an unusual appearing "squashed comet" by Carolyn Shoemaker on films obtained by herself, her husband Eugene Shoemaker, and David Levy. When viewed under better conditions with larger telescopes, it was found to be a string of as many as 24 individual nuclei, all in a straight line. Each comet was surrounded by a coma and had tails. Long dust wings extended off the ends of the train of nuclei. H. J. Melosh and I (1993) found it to be consistent with the tidal disruption of a weak progenitor with a diameter only 2 kilometers as it passed its previous perijove in July of 1992. A more detailed model later suggested that it might have had a structure like a loose rubble pile, which not only explained the number of nuclei that were visible but also some of the debris complex around the comet. Several of the nuclei appeared to break up and some dissipated after discovery. The danger of comet impacts on planets was clearly demonstrated between July 16 and July 22 of 1994 as the bits of Shoemaker-Levy 9 crashed into Jupiter, leaving large impact scars that slowly dissipated. The impact scars were Earth-sized at first and very obvious despite estimates of the sizes of the individual impactors being smaller than 1.0 kilometers in diameter.

Other comets have dissipated completely during perihelion passage, some never to be seen again, and others seen as a large diffuse cloud that slowly disappeared. Comet Biela was a short-period comet seen in the 19th century during several orbits about the Sun. It was seen to be split into two pieces during its return to perihelion in 1846 and both pieces were seen to return in 1852.

Neither was seen again. Their 1859 perihelion passage was poorly placed in the sky and the comets were missed. The Earth experienced a large meteor shower at the time the comets would have been nearby during its return in 1866, but the comets themselves remained unseen despite their relatively favorable viewing conditions.

Other short-period comets have failed to be seen at subsequent returns. Many are simply poorly observed and lost, but some have been well observed and have apparently vanished. The most recent case is that of periodic Comet Kohoutek, discovered in

1973 and seen during returns in 1980 and 1987 but missed at its next return despite several searches.

One well-known example of split comets is the Kreutz family of sungrazing comets. These comets graze the surface of the Sun, within about half a million kilometers of its surface, some of them even impacting the surface of the Sun. Heinrich Kreutz, for whom they are named, first studied them in 1888. The sungrazing comets are often seen to break up into multiple pieces as they pass perihelion—pulled apart by the gravitational and thermal stresses of their passage so close to the Sun.

Brian G. Marsden has carried out the most extensive study of these comets, finding in 1967 that the appearance of Comet Ikeya-Seki, a bright comet seen even in daylight when at perihelion and that of the great comet of 1882. These two comets were found to have orbital periods that suggested they might have split from a sungrazing comet seen in 1106. More recently, in 1989, Marsden found that a relatively large progenitor probably split into two significant pieces that have followed similar but differing dynamical evolution, each return generating fragments that also evolve on independent orbits.

THE DANGER OF COMETS

Humans have looked on comets throughout history as omens and heavenly signs. Given their frequent but random appearances, often at times of monumental historical events, whether it be a war, the death of a king, or a time of plague, it is easy to see how our ancestors might come to fear bright comets. But bad things happen quite frequently and so do bright comets. Was Hale-Bopp, which reached perihelion in May of 1997, a portent of doom for Princess Diana, killed in a terrible automobile accident just three months later? In olden days, these events may well have been associated thusly. But our modern understanding of comets has removed the primordial fear of comets for most of the general public. By the end of the 20th century, however, we have come to realize that perhaps comets are worthy of at least some of this fear. We have found that the threat of asteroid and comet impacts on Earth and the corresponding terrible consequences is a real one. Nonetheless, there is some evidence that disasters in the past and in relatively recent history may well be related to the breakup of Earth-crossing comets.

The threat of impact of a large comet is extremely small, although the consequences of such an impact would be devastating. The impact of a comet or asteroid whose diameter was between 10 and 20 kilometers occurred in a shallow sea that is now under the Yucatán peninsula on the margin of the Gulf of Mexico about 65 million years ago. The side affects of that impact were so terrible that about two-thirds of all species that inhabited the Earth at the time became extinct. Most of the individuals on the planet at that moment died as a result of the impact, either from the direct affects of the blast, the fallback of impact ejecta into the atmosphere or in the long nuclear winter that followed and enveloped the planet. Luckily, such large, extinction-level events occur very rarely, perhaps every 100 million years.

It is now thought that the impacts of objects larger than about 1 kilometer in diameter are large enough to cause global consequences such as a short term climate change that could kill most inhabitants of the planet but would probably not cause mass extinctions. These objects hit Earth every few hundred thousand years and maybe 20 percent of these impacts, once or twice every million years, are due to comets. Unlike near-Earth asteroids and short-period comets that make repeated orbits in the inner solar system and can be discovered many years ahead of an impact, the long-period comets enter the inner solar system from the depths of space. They are likely not to be seen until weeks or at most a couple years before they might hit the Earth. But at rates of about once every million years, the chance we will be hit anytime soon is small.

Much smaller objects, however, could cause the short-term disruption of life on Earth. Objects that might only cause regional or local damage impact more frequently—for example, objects a few meters in diameter hit the Earth about every month. Dust-sized bits hit Earth almost continuously, as evidenced by the sporadic meteors we can see in the nighttime sky. Asteroidal objects that are around 30–50 meters in diameter that could damage an area the size of a city are likely to hit the Earth about every century. Objects a few hundred meters in diameter, which are large enough to cause large tsunami that might devastate a coastline along an ocean, the object impacts hit Earth perhaps every 10,000 years.

On the peaceful morning of June 30, 1908, Russian Siberia was disrupted when a fireball as bright as the early morning Sun streaked northward, frightening many eyewitnesses. Moments later, over the Tunguska region, a huge explosion knocked down eyewitnesses there and flattened about 2000 square kilometers of forest. Later analysis would suggest that the explosion had the equivalent of 10–20 megatons of TNT and was caused by the explosion of a 50-meter stony asteroid at an altitude of about 8–10 kilometers. If the object had arrived just a few hours earlier, it might have exploded over St. Petersberg, destroying that city in an instant of devastation only demonstrated decades later after the invention of nuclear weapons.

The timing of the Tunguska event coincides with the Earth's passage through the Taurid meteor complex that has been associated with Comet Encke and a number of near-Earth asteroids. Researchers such as Victor Clube and Bill Napier have led a group of "cosmic catastrophists" who believe that this debris complex is the result of a larger comet that had disintegrated while on an Earth-crossing orbit. The significance of such a mechanism is profound. Instead of having just one chance of impact, if a large object were to break up into a hundred smaller objects, the chance of impact is 100 times greater. Though the damage due to each impactor is much less, it may still be significant. If the Taurid complex progenitor fragmented into many smaller objects, the impact rate of objects capable of causing local or regional devastation might rise from the background rate of once per 10,000 years to once per 1,000 years, for example.

Dendrochronologist Mike Baillie at Queens University in Belfast, Northern Ireland, has found evidence of multiple short-term catastrophic climate changes over the past 6,000 years. Dendrochronology uses the study of tree rings to determine chronology and the climate of a region. Baillie has found episodes of low growth rates at sites around the world in tree rings that coincide with upheaval in human affairs. The most recent corresponds to the Dark Ages in Europe around A.D. 530 to 540. Similar global climate changes have been found around 2345 B.C., 1628 B.C., 1159 B.C., 207 B.C. and 44 B.C. Baillie suggests that the impact of small objects exploding in the atmosphere or in the oceans would inject enough dust into the atmosphere to reduce sunlight—similar to the dark impact scars caused by Comet Shoemaker-Levy 9 when it hit Jupiter in 1994. This would cause years without summers, crop failures, famine and other side effects. These small objects, he believes, are part of the Taurid complex and have impacted during periods when the core of the debris complex intersects the orbit of the Earth. Many of the legends and mythology of mankind may well be linked to the appearance of bright comets in the sky overhead. Is the primordial fear of comets founded in actual prehistoric events?

FUTURE EXPLORATION

Several missions to short-period comets have been planned. Our first spacecraft explorations of comets occurred in the 1980s, when we sent spacecraft on flybys of comets Giacobini-Zinner and Halley. These spacecraft flew past the comets at very high speeds, taking measurements like a tourist would take photographs while driving past a museum. Our next explorations of comets will be much more detailed—we will crack the museum door open and perhaps even look inside. Although budget cuts have canceled several missions, others are moving ahead. For example, the Stardust spacecraft, launched on February 7, 1999, will encounter Comet Wild 2 in 2004, actually sampling the comet's coma for return to Earth. Another mission is planned to send a spacecraft to Comet Tempel 1. Called Deep Impact, it will crash a 500-kilogram copper ball into the comet so that material from within the comet will be exploded out for study by instruments on the spacecraft and from the ground. Launch of Deep Impact is planned for January 2004, with arrival at the comet in July 2005. The Contour spacecraft is scheduled for launch in July 2002. Its mission is to fly past Comet Encke at a distance of just 100 kilometers, followed by encounters with Comet Schwassmann-Wachmann 3 in 2006 and Comet d'Arrest in 2008.

The new century and millenium promises to advance our knowledge of comets well beyond our present level. We should have a better understanding of the composition and structure of comets and will be able to decipher their clues to the origin of our solar system.

Solar System Origins

Mark Washburn

If there were only one kind of rock in the world, geologists would have a tough time deciphering the Earth's history. If there were only one species of plant in the world, biologists would be hard-pressed to explain where it came from. Until recently, scientists probing the origin of the planets were stuck with only one kind of solar system to study. Working from this necessarily limited perspective, they nevertheless managed to develop a very tidy and satisfying theory of how the solar system came to be.

The Sun, the planets, and associated debris such as comets and asteroids all formed from the gas and dust residing in a flat, spinning disk known as the solar nebula. The protosun, at the center of the nebula, pulled in so much material that eventually the pressure and temperature in its interior rose to the point where thermonuclear reactions could begin. Meanwhile, chunks of rock and ice called planetesimals duplicated the process on a smaller scale, sweeping up enough material to form planetary cores.

In the hot inner regions of the nebula, ice was unavailable for planet-building, limiting the size of the four rocky inner planets. Beyond a theoretical "frost line," ices (including water and simple compounds such as methane and ammonia) were added to the mix, allowing the assembly of planetary cores 10 to 20 times more massive than the Earth. These cores, in turn, gravitationally attracted the nebula's abundant hydrogen and helium gases, building up the huge gas giant worlds of the outer solar system. Then the Sun turned on and blew away the leftover nebula material, leaving the solar system as we see it today.

The theory was especially satisfying because it explained the apparent order in our solar system. Only rocky planets could form in a nebula's inner zone, and only the outer zone could be home to Jovian-style worlds. Theory neatly matched observation—the ideal state of scientific equilibrium.

That pleasant state of affairs was upset when, beginning in the 1990s, new observational techniques and better technology began turning up evidence of planetary bodies orbiting other stars. These discoveries offered a broad confirmation of the solar nebula theory—it happened here, and we could see it happening elsewhere—but the surprising details of these neighboring solar systems sent theorists scurrying back to the drawing board.

Extrasolar planets are almost impossible to observe directly, since they tend to be lost in the glare of their star's glow. But every planet exerts a small gravitational tug on the star it orbits. For example, Jupiter and the Sun both revolve around their common center of mass, which lies about 50,000 kilometers above the Sun's surface. To a meticulous observer 33 light-years away, the Sun would appear to sway back and forth around a circle 0.001 arcseconds in diameter. The period of this wobble is 11.8 years, equal to the time it takes Jupiter to complete one orbit around the Sun. This motion would be detectable by careful observation of the Doppler shifting of the Sun's light, toward the blue end of the spectrum as it approaches and toward the red as it recedes. The observer would thus be able to infer the presence of Jupiter without ever seeing it. A simple application of the laws of gravitation and planetary motion would then lead to an accurate description of Jupiter's size, mass and orbit.

Now consider the case of the otherwise unremarkable star HD 114762, 90 light-years from Earth. A wobble in the star's motion implies the presence of another body, 11 times as massive as Jupiter, orbiting HD 114762 at a distance closer to the star than Mercury is to the Sun. The mystery body is so massive that it may, in fact, be a brown dwarf—a substellar body that never got quite big enough to become a full-fledged star.

Then there is 70 Virginis, 59 light-years from the Sun. There, a body 6.8 times as massive as Jupiter (too small to be a brown dwarf) orbits the star at a distance about half that of the Earth to the Sun. Or take Tau Bootis; it has an orbital companion 3.7 times the mass of Jupiter that completes a circuit around the star in just three days and eight hours, with an orbital radius of less than 10 million kilometers!

The presence of such huge bodies so close to their stars challenged prevailing theory. How could gas giants form so close to their suns? Could they have formed elsewhere? If so, how did they get to their present locations? Are their orbits unstable? And what does all of this say about our own solar system? Is it a freak whose apparent orderliness is illusory? The discovery of more than a dozen extrasolar planets (the number is steadily rising) has forced a serious rethinking of many details of the solar nebula theory and, in particular, the subtleties of orbital dynamics. One cornerstone of the standard theory has been that the planets first formed at or near their present locations relative to the Sun. But the news from afar, combined with more sophisticated computer modeling techniques, has suggested a more complicated and—literally—chaotic scenario for planetary formation.

A number of recent theories have proposed that young planets are acted upon by a variety of different forces. Friction and drag produced by the material in the nebula can slow a protoplanet's motion and alter its orbit. Spiral density waves—which control the structure of spiral galaxies and Saturn's rings—are produced by the gravitational interaction of the protoplanets and the gas and dust of the nebula. The waves, in turn, affect the protoplanets, driving them out of circular orbits. And the protoplanets also gravitationally interact with one another, turning the nebulae into cosmic pinball machines, ejecting planets from the system entirely or hurling them into the fires of their sun. The highly eccentric orbits of at least three extrasolar planets suggest that such instabilities are common.

There is strong evidence that our own seemingly orderly solar system has undergone episodes of chaos and planetary ping-pong. Renu Malhotra, of the Lunar and Planetary Institute, has proposed that Pluto—that notorious maverick world—initially formed in a more or less circular orbit in the plane of the ecliptic. However, in the later stages of planetary formation, the gas giant worlds tended to "clean up" the nebula through gravitational interactions with the remaining planetesimals. The net effect of these interactions was that the planetesimals were flung out of the solar system, while the orbits of Saturn, Uranus and Neptune expanded slightly. In compensation, the orbit of Jupiter would have become slightly smaller. As Neptune's orbit expanded, it may have reached a gravitational resonance with Pluto, giving the smaller world a small but decisive kick on every second or third orbit. Over tens of millions of years, Neptune would have booted Pluto entirely out of its circular orbit and into the tilted, elliptical orbit we observe today.

The action in a typical solar nebula may be even more dramatic than that. If too many "superplanets" (Jovian size or larger) form within a system, they may actually knock each other into their sun. Perhaps some of the extrasolar superplanets observed are in the midst of such a process; or they may simply be the survivors of a violent winnowing in their youth. Indeed, we may owe our own existence to Jupiter, whose massive influence has, over billions of years, served to stabilize the orbits of our planetary neighbors, making the solar system safe for life as we know it. If Jupiter's orbit was as eccentric as that of some extrasolar planets, Earth might never have had a chance to evolve a stable climate.

As for terrestrial planets elsewhere, the chances now seem good that they exist in profusion throughout the cosmos. Earth-sized planets are virtually undetectable with current techniques and technology, but a proposed new generation of space-borne telescopes and interferometry arrays offers hope that within a few decades we may be able to pick out Earthlike worlds orbiting other stars. Spectroscopic analysis might even make it possible to detect the presence of water oceans and oxygen in their atmospheres.

Now that we have seen other solar systems, we have gained a better perspective on our own. Planetary formation, it now seems, is a lot like any sort of birth—messy, chaotic and painful. Planets, like people, may be born in one place and wind up somewhere else entirely. Fortunately, Earth and its neighbors survived their awkward adolescence and now enjoy a stable and temperate middle age. The order we see today in our own solar system may be a rare exception, or it may turn out to be the cosmic norm, the unremarkable outcome of a few billion years worth of planetary jostling and jockeying for position. Within a few decades we may know for sure, one way or the other.

Chapter Six:
The Planetary Tour

Introduction

When the achievements of our generation are tallied in some future reckoning, one really exciting plus is that we have explored the neighborhood of the Sun with spacecraft. With the exception of Pluto, every planet, most of their moons, a sampling of asteroids and two comets have been visited.

We begin with Mercury, indeed the forgotten planet. Although it is relatively close to us, few astronomers have even seen this rocky world. Forever close to the Sun, Mercury can be seen occasionally in the evening after sunset, or in the morning before sunrise. Mercury was visited by the Mariner 10 spacecraft three times in 1974.

Venus, Earth and Mars appear to be three experiments on the possibilities for life. Two of those tries failed, one thanks to a runaway greenhouse effect, the other because of a lack of size, heat and air. Only on Earth did nature find just the right conditions to start the pageant of life of which we are a part. Since Mariner 2 visited Venus in December 1962, many spacecraft have flown by, especially Magellan, a spacecraft which successfully mapped Venus with radar. Mars has been a target of many missions since 1964, particularly Viking in 1976 and Mars Pathfinder in 1997.

The outer worlds are utterly different. Large bodies of gas, they have little in common with the terrestrial planets. Jupiter was visited by Pioneer 10 in 1973, Jupiter and Saturn by Pioneer 11. Voyager 2 completed a tour of the planets and moons of Jupiter, Saturn, Uranus, and Neptune. Only Pluto remains unexplored.

Mercury: The Forgotten Planet

Robert M. Nelson

The planet closest to the Sun, Mercury is a world of extremes. Of all the objects that condensed from the presolar nebula, it formed at the highest temperatures. The planet's dawn-to-dusk day, equal to 176 Earth days, is the longest in the solar system, longer in fact than its own year. When Mercury is at perihelion (the point in its orbit closest to the Sun), it moves so swiftly that, from the vantage of someone on the surface, the Sun would appear to stop in the sky and go backward—until the planet's rotation catches up and makes the Sun go forward again. During daytime, its ground temperature reaches 700 kelvins, the highest of any planetary surface (and more than enough to melt lead); at night, it plunges to a mere 100 kelvins (enough to freeze krypton).

Such oddities make Mercury exceptionally intriguing to astronomers. The planet, in fact, poses special challenges to scientific investigation. Its extreme properties make Mercury difficult to fit into any general scheme for the evolution of the solar system. In a sense, Mercury's unusual attributes provide an exacting and sensitive test for astronomers' theories. Yet even though Mercury ranks after Mars and Venus as one of Earth's nearest neighbors, distant Pluto is the only planet we know less about. Much about Mercury—its origins and evolution, its puzzling magnetic field, its tenuous atmosphere, its possibly liquid core and its remarkably high density—remains obscure.

Mercury shines brightly, but it is so far away that early astronomers could not discern any details of its terrain; they could map only its motion in the sky. As the innermost planet, Mercury (as seen from Earth) never wanders more than 27 degrees from the Sun. This angle is less than that made by the hands on a watch at one o'clock. It can thus be observed only during the day, but scattered sunlight makes it difficult to see, or shortly before sunrise and after sunset, with the Sun hanging just over the horizon. At dawn or dusk, however, Mercury is very low in the sky, and the light from it must pass through up to 10 times as much turbulent air as when it is directly overhead. The best Earth-based telescopes can see only those features on Mercury that are a few hundred kilometers across or wider—a resolution far worse than that for the moon seen with the unaided eye.

Despite these obstacles, terrestrial observation has yielded some interesting results. In 1955 astronomers were able to bounce radar waves off Mercury's surface. By measuring the so-called Doppler shift in the frequency of the reflections, they learned of

Mercury's 59-day rotational period. Until then, Mercury had been thought to have an 88-day period, identical to its year, so that one side of the planet always faced the Sun. The simple two-to-three ratio between the planet's day and year is striking. Mercury, which initially rotated much faster, probably dissipated energy through tidal flexing and slowed down, becoming locked into this ratio by an obscure process.

The new space-based observatories, such as the Hubble Space Telescope, are not limited by the problems of atmospheric distortion, and one might think them ideal tools for studying Mercury. Unfortunately, the Hubble, like many other sensors in space, cannot point at Mercury, because the rays of the nearby Sun might accidentally damage sensitive optical instruments on board.

The only other way to investigate Mercury is to send a spacecraft to examine it up close. Only once has a probe made the trip: Mariner 10 flew by in the 1970s as part of a larger mission to explore the inner solar system. Getting the spacecraft there was not a trivial task. Falling directly into the gravitational potential well of the Sun was impossible; the spacecraft had to ricochet around Venus to relinquish gravitational energy and thus slow down for a Mercury encounter. Mariner's orbit around the Sun provided three close flybys of Mercury: on March 29, 1974; September 21, 1974; and March 16, 1975. The spacecraft returned images of about 40 percent of Mercury, showing a heavily cratered surface that, at first glance, appeared similar to that of the Moon.

The pictures, sadly, led to the mistaken impression that Mercury differs very little from the Moon and just happens to occupy a different region of the solar system. As a result, Mercury has become the neglected planet of the American space program. There have been more than 40 missions to the Moon, 20 to Venus and more than 15 to Mars. By the end of the next decade, an armada of spacecraft will be in orbit about Venus, Mars, Jupiter and Saturn, returning detailed information about these planets and their environs for many years to come. But Mercury will remain largely unexplored.

THE IRON QUESTION

It was the Mariner mission that elevated scientific understanding of Mercury from almost nothing to most of what we currently know. The ensemble of instruments carried on that probe sent back about 2,000 images, with an effective resolution of about 1.5 kilometers, comparable to shots of the Moon taken from Earth through a large telescope. Yet those many pictures captured only one face of Mercury; the other side has never been seen.

By measuring the acceleration of Mariner in Mercury's surprisingly strong gravitational field, astronomers confirmed one of the planet's most unusual characteristics: its high density. The other terrestrial (that is, nongaseous) bodies—Venus, the Moon, Mars and Earth—exhibit a fairly linear relation between density and size. The largest, Earth and

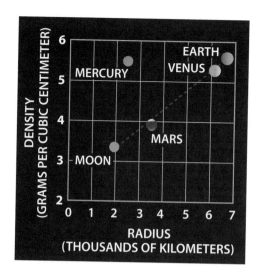

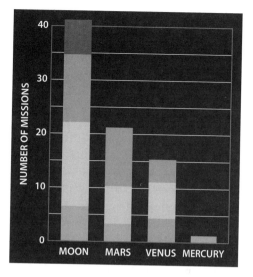

Venus, are quite dense, whereas the smaller ones, the Moon and Mars, have lower densities. Mercury is not much bigger than the Moon, but its density is typical of a far larger planet such as Earth.

This observation provides a fundamental clue about Mercury's interior. The outer layers of a terrestrial planet consist of lighter materials such as silicate rocks. With depth, the density increases, because of compression by the overlying rock layers and the different composition of the interior materials. The high-density cores of the terrestrial planets are probably made largely of iron.

Oddly enough, careful analysis of the Mariner findings, along with laborious spectroscopic observations from Earth, has failed to detect even trace amounts of iron in Mercury's crustal rocks. The apparent dearth of iron on the surface contrasts sharply with its presumed abundance in Mercury's interior. Iron occurs on Earth's crust and has been detected by spectroscopy on the rocks of the moon and Mars. So Mercury may be the only planet in the inner solar system with all its high-density iron concentrated in the interior and only low-density silicates in the crust. It may be that Mercury was molten for so long that the heavy substances settled at the center, just as iron drops below slag in a smelter.

Mariner 10 also found that Mercury has a relatively strong magnetic field—the most powerful of all the terrestrial planets except Earth. The magnetic field of Earth is generated by electrically conductive molten metals circulating in the core, through a process called the self-sustaining dynamo. If Mercury's magnetic field has a similar source, then that planet must have a liquid interior.

But there is a problem with this hypothesis. Small objects like Mercury have a high proportion of surface area compared with volume. Therefore, other factors being equal, smaller bodies radiate their energy to space faster. If Mercury has a purely iron core, as its large density and strong magnetic field imply, then the core should have cooled and solidified eons ago. But a solid core cannot support a self-sustaining magnetic dynamo.

Once a planetary surface solidifies sufficiently, it may bend when stress is applied steadily over long periods, or it may crack like a piece of glass on sudden impact. After Mercury was born four billion years ago, it was bombarded with huge meteorites that broke through its fragile outer skin and released torrents of lava. More recently, smaller collisions have caused lava to flow. These impacts must have either released enough energy to melt the surface or tapped deeper, liquid layers. Mercury's surface is stamped with events that occurred after its outer layer solidified.

Planetary geologists have tried to sketch Mercury's history using these features—and without accurate knowledge of the rocks that constitute its surface. The only way to determine absolute age is by radiometric dating of returned samples (which so far are lacking). But geologists have ingenious ways of assigning relative ages, mostly based on the principle of superposition: any feature that overlies or cuts across another is the younger. This principle is particularly helpful in establishing the relative ages of craters.

A FRACTURED HISTORY

Mercury has several large craters that are surrounded by multiple concentric rings of hills and valleys. The rings probably originated when a meteorite hit, causing shock waves to ripple outward like waves from a stone dropped into a pond, and then froze in place. Caloris, a behemoth 1,300 kilometers in diameter, is the largest of these craters. The impact that created it established a flat basin—wiping the slate clean, so to speak—on which a fresh record of smaller impacts has built up. Given an estimate of the rate at which projectiles hit the planet, the size distribution of these craters indicates that the Caloris impact probably occurred around 3.6 billion years ago; it serves as a reference point in time. The collision was so violent that it disrupted the surface on the opposite side of Mercury: the antipode of Caloris shows many cracks and faults.

Mercury's surface is also crosscut by linear features of unknown origin that are preferentially oriented north-south, northeast-southwest and northwest-southeast. These lineaments are called the Mercurian grid. One explanation for the checkered pattern is that the crust solidified when the planet was rotating much faster, perhaps with a day of only 20 hours. Because of its rapid spin, the planet would have had an equatorial bulge; after it slowed to its present period, gravity pulled it into a more spherical shape. The lineaments likely arose as the surface accommodated this change. The wrinkles do not cut across the Caloris crater, indicating that they were established before that impact occurred.

Caloris crater (left) was formed when a giant projectile hit Mercury 3.6 billion years ago. Shock waves radiated through the planet, creating hilly and lineated terrain on the opposite side. The rim of Caloris itself (below) consists of concentric waves that froze in place after the impact. The flattened bed of the crater has since been covered with smaller craters.

While Mercury's rotation was slowing, the planet was also cooling, so that the outer regions of the core solidified. The accompanying shrinkage probably reduced the planet's surface area by about a million square kilometers, producing a network of faults that are evident as a series of curved scarps, or cliffs, crisscrossing Mercury's surface.

Compared with Earth, where erosion has smoothed out most craters, Mercury, Mars and the Moon have heavily cratered surfaces. The craters on these three planets also show a similar distribution of sizes, except that Mercury's craters tend to be somewhat larger. The objects striking Mercury most likely had higher velocity than those hitting the other planets. Such a pattern is to be expected if the projectiles were in elliptical orbits about the Sun: they would have been moving faster in the region of Mercury's orbit than they were farther out. So these rocks may have been all from the same family, one that probably originated in the asteroid belt. In contrast, the moons of Jupiter have a different distribution of crater sizes, indicating that they collided with a different group of objects.

A TENUOUS ATMOSPHERE

Mercury's magnetic field is strong enough to trap charged particles, such as those blowing in with the solar wind (a stream of protons ejected from the Sun). The magnetic field forms a shield, or magnetosphere, that is a miniaturized version of the one surrounding Earth. Magnetospheres change constantly in response to the Sun's activity; Mercury's magnetic shield, because of its smaller size, can change much faster than Earth's. Thus, it responds quickly to the solar wind, which is 10 times denser at Mercury than at Earth.

The fierce solar wind steadily bombards Mercury on its illuminated side. The magnetic field is just strong enough to prevent the wind from reaching the planet's surface, except when the Sun is very active or when Mercury is at perihelion. At these times, the solar wind reaches all the way down to the surface, and its energetic protons knock material off the crust. The particles thus ejected can then get trapped by the magnetosphere.

Objects as hot as Mercury do not, however, retain appreciable atmospheres around them, because gas molecules tend to move faster than the escape velocity of the planet. Any significant amount of volatile material on Mercury should soon be lost to space. For this reason, it had long been thought that Mercury did not have an atmosphere. But the ultraviolet spectrometer on Mariner 10 detected small amounts of hydrogen, helium and oxygen, and subsequent Earth-based observations have found traces of sodium and potassium.

The source and ultimate fate of this atmospheric material is a subject of animated argument. Unlike Earth's gaseous cloak, Mercury's atmosphere is constantly evaporating and being replenished. Much of the atmosphere is probably created, directly or indirectly, by the solar wind. Some components of the thin atmosphere may come from the magnetosphere or from the direct infall of cometary material. And once an atom is "sputtered" off the surface by the solar wind, it also adds to the tenuous atmosphere. It is even possible that the planet is still outgassing the last remnants of its primordial inventory of volatile substances.

OBSTACLES TO EXPLORATION

Why has Mercury been left out of the efforts to explore the solar system for nearly a quarter century? One possibility, as mentioned, is the superficial similarity between Mercury and the Moon. Another, more subtle factor arises from the way planetary missions are devised. The members of peer-review panels for the National Aeronautics and Space Administration have generally been involved in NASA's most recent missions. The preponderance of missions has been to other planets, so that these planetary scientists have developed a highly specialized body of expertise and interests. In contrast to the planets thus favored, Mercury has a small advocacy group.

A mission to orbit Mercury poses a special technical hurdle. The spacecraft must be protected against the intense energy radiating from the Sun and even against the solar energy reflected off Mercury. Because the spacecraft will be close to the planet, at times "Mercury-light" can become a greater threat than the direct Sun itself. Despite all the challenges, NASA received one Discovery mission proposal for a Mercury orbiter in 1994 and two in 1996.

The 1994 proposal, called Hermes '94, employed a traditional hydrazine–nitrogen tetroxide propulsion system, requiring as much as 1,145 kilograms of propellants. Much of this fuel is needed to slow the spacecraft as it falls toward the Sun. The mission's planners, who include myself, could have reduced the fuel mass only by increasing the number of planetary encounters (to remove gravitational energy). Unfortunately, these maneuvers would have increased the time spent in space, where exposure to radiation limits the lifetime of critical solid-state components.

NASA has, however, selected one proposal for a Mercury orbiter for intensive consideration in the 1996 cycle of Discovery missions (Herme '96). This design, called Messenger, was developed by engineers at the Applied Physics Laboratory in Maryland. Like Hermes '94, it would rely on traditional chemical propulsion and carry similar sensors. Moreover, it would have two devices that could determine the proportions of the most abundant elements of the crustal rocks. Although these two instruments are scientifically attractive, their additional mass requires that the spacecraft swoop by Venus twice and Mercury three times before it goes into orbit. This trajectory will lengthen the journey to Mercury to more than four years (about twice that of Hermes '96). Messenger is also the most costly Discovery mission under consideration.

Global Climate Change on Venus

Mark A. Bullock and David H. Grinspoon

Emerging together from the presolar cauldron, Earth and Venus were endowed with nearly the same size and composition. Yet they have developed into radically different worlds. The surface temperature of Earth's sister planet is about 460 degrees Celsius—hot enough for rocks to glow visibly to any unfortunate carbon-based visitors. A deadly efficient greenhouse effect prevails, sustained by an atmosphere whose major constituent, carbon dioxide, is a powerful insulator. Liquid water is nonexistent. The air pressure at the surface is almost 100 times that on Earth; in many ways it is more an ocean than an atmosphere. A mélange of gaseous sulfur compounds, along with what little water vapor there is, provides chemical fodder for the globally encircling clouds of sulfuric acid.

This depiction of hell has been brought to us by an armada of 22 robotic spacecraft that have photographed, scanned, analyzed and landed on Venus over the past 37 years [since 1962]. Throughout most of that time, however, Venus's obscuring clouds hindered a full reconnaissance of its surface. Scientists' view of the planet remained static because they knew little of any dynamic processes, such as volcanism or tectonism, that might have occurred there. The Magellan spacecraft changed that perspective. From 1990 to 1994 it mapped the entire surface of the planet at high resolution by peering through the clouds with radar. It revealed a planet that has experienced massive volcanic eruptions in the past and is almost surely active today. Coupled with this probing of Venusian geologic history, detailed computer simulations have attempted to reconstruct the past billion years of the planet's climate history. The intense volcanism, researchers are realizing, has driven large-scale climate change. Like Earth but unlike any other planet astronomers know, Venus has a complex, evolving climate.

From our human vantage point next door in the solar system, it is sobering to ponder how forces similar to those on Earth have had such a dissimilar outcome on Venus. Studying that planet has broadened research on climate evolution beyond the single example of Earth and given scientists new approaches for answering pressing questions: How unique is Earth's climate? How stable is it? Humankind is engaged in a massive, uncontrolled experiment on the terrestrial climate brought on by the growing effluent from a technological society. Discerning the factors that affect the evolution of climate on other planets is crucial to understanding how natural and anthropogenic forces alter

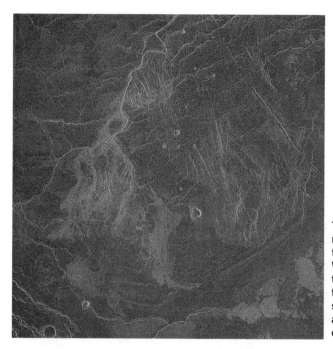

This delta exists at the terminus of a narrow channel that runs for 800 kilometers through Venus's northern volcanic plains. Water could not have carved it; Venus is too hot and dry. Instead it was probably the work of lavas rich in carbonate and sulfate salts—which implies that the average temperature used to be several tens of degrees higher than it is today.

the climate on Earth.

To cite one example, long before the ozone hole became a topic of household discussion, researchers were trying to come to grips with the exotic photochemistry of Venus's upper atmosphere. They found that chlorine reduced the levels of free oxygen above the planet's clouds. The elucidation of this process for Venus eventually shed light on an analogous one for Earth, whereby chlorine from artificial sources destroys ozone in the stratosphere.

CLIMATE AND GEOLOGY

The climate of Earth is variable partly because its atmosphere is a product of the ongoing shuffling of gases among the crust, the mantle, the oceans, the polar caps and outer space. The ultimate driver of geologic processes, geothermal energy, is also an impetus for the evolution of the atmosphere. Geothermal energy is a product primarily of the decay of radioactive elements in the interior, and a central problem in studying solid planets is understanding how they lose their heat. Two mechanisms are chiefly responsible: volcanism and plate tectonics.

The interior of Earth cools mainly by means of its plate tectonic conveyor-belt system, whose steady recycling of gases has exerted a stabilizing force on Earth's climate. Whereas volcanoes pump gases into the atmosphere, the subduction of lithospheric plates returns them to the interior. Most volcanoes are associated with plate tectonic

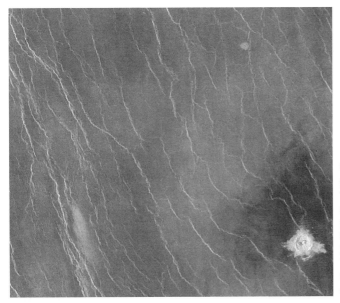

Wrinkle ridges are the most common feature on the volcanic plains of Venus. They are parallel and evenly spaced, suggesting that they formed when the plains as a whole were subjected to stress—perhaps induced by a dramatic, rapid change in surface temperature. This region, which is part of the equatorial plains known as Rusalka Planitia, is approximately 300 kilometers across.

activity, but some of the largest volcanic edifices on Earth (such as the Hawaiian Islands) have developed as "hot spots" independent of plate boundaries. Historically, the formation of immense volcanic provinces—regions of intense eruptions possibly caused by enormous buoyant plumes of magma within the underlying mantle—may have spewed large amounts of gases and led to periods of global warming.

What about Venus? Before the Magellan mission, much of the planet's geologic history remained speculative, relegated to comparisons with Earth and to extrapolations based on presumed similarities in composition and geothermal heat production. Now a global picture of the history of Venus's surface is emerging. Plate tectonics is not in evidence, except possibly on a limited scale. It appears that heat was transferred, at least in the relatively recent past, by the eruption of vast plains of basaltic lava and later by the volcanoes that grew on top of them. Understanding the effects of volcanoes is the starting point for any discussion of climate.

A striking feature of Magellan's global survey is the paucity of impact craters. Although Venus's thick atmosphere can shield the planet's surface from small impactors—it stops most meteoroids smaller than a kilometer in diameter, which would otherwise gouge craters up to 15 kilometers (nine miles) across—there is a shortage of larger craters as well. Observations of the number of asteroids and comets in the inner solar system, as well as crater counts on the Moon, give a rough idea of how quickly Venus should have collected impact scars: about 1.2 craters per million years. Magellan saw only, by the latest count, 963 craters spread randomly over its surface. Somehow impacts from the first 3.7 billion years of the planet's history have been eradicated.

A sparsity of craters is also evident on Earth, where old craters are eroded by wind and water. Terrestrial impact sites are found in a wide range of altered states, from the nearly pristine bowl of Meteor Crater in Arizona to the barely discernible outlines of buried Precambrian impacts in the oldest continental crust. Yet the surface of Venus is far too hot for liquid water to exist and surface winds are mild. In the absence of erosion, the chief processes altering and ultimately erasing impact craters should be volcanic and tectonic activity. That is the paradox. Most of the Venusian craters look fresh: only 6 percent of them have lava lapping their rims, and only 12 percent have been disrupted by folding and cracking of the crust. So where did all the old ones go, if most of those that remain are unaltered? If they have been covered up by lava, why do we not see more craters that are partially covered? And how have they been removed so that their initial random placement has been preserved?

To some researchers, the random distribution of the observed craters and the small number of partially modified ones imply that a geologic event of global proportions abruptly wiped out all the old craters some 800 million years ago. In this scenario, proposed in 1992 by Gerald G. Schaber of the U.S. Geological Survey (USGS) and Robert G. Strom of the University of Arizona, impacts have peppered the newly formed surface ever since.

But the idea of paving over an entire planet is unpalatable to many geologists. It has no real analogue on Earth. Roger J. Phillips of Washington University proposed an alternative model the same year, known as equilibrium resurfacing, which hypothesized that steady geologic processes continually eradicate craters in small patches, preserving an overall global distribution that appears random. A problem with this idea is that some geologic features on Venus are immense, suggesting that geologic activity would not wipe craters out cleanly and randomly everywhere.

CHOCOLATE-COVERED CARAMEL CRUST

Although there is no doubt that volcanism has been a major force in shaping Venus's surface, the interpretation of some enigmatic geologic features has until recently resisted integration into a coherent picture of the planet's evolution. Some of these features hint that the planet's climate may have changed drastically.

First, several striking lineaments resemble water-carved landforms. Up to 7,000 kilometers long, they are similar to meandering rivers and floodplains on Earth. Many end in outflow channels that look like river deltas. The extreme dryness of the environment makes it highly unlikely that water carved these features. So what did? Perhaps calcium carbonate, calcium sulfate and other salts are the culprit. The surface, which is in equilibrium with a hefty carbon dioxide atmosphere laced with sulfur gases, should be replete with these substances. Indeed, the Soviet Venera landers found that surface rocks are about 7 to 10 percent calcium minerals (almost certainly carbonates) and 1 to 5 percent sulfates.

Lavas laden with these salts melt at temperatures of a few tens to hundreds of degrees higher than Venusian surface temperatures today. Jeffrey S. Kargel of the USGS and his coworkers have hypothesized that vast reservoirs of molten carbonatite (salt-rich) magma, analogous to water aquifers on Earth, may exist a few hundred meters to several kilometers under the surface. Moderately higher surface temperatures in the past could have spilled salt-rich fluid lavas onto the surface, where they were stable enough to carve the features we see today.

Second, the mysterious tesserae—the oldest terrain on Venus—also hint at higher temperatures in the past. These intensely crinkled landscapes are located on continentlike crustal plateaus that rise several kilometers above the lowland lava plains. Analyses by Phillips and by Vicki L. Hansen of Southern Methodist University indicate that the plateaus were formed by extension of the lithosphere (the rigid exoskeleton of the planet, consisting of the crust and upper mantle). The process was something like stretching apart a chocolate-covered caramel that is gooey on the inside with a thin, brittle shell. Today the outer, brittle part of the lithosphere is too thick to behave this way. At the time of tessera formation, it must have been thinner, which implies that the surface was significantly hotter.

Finally, cracks and folds crisscross the planet. At least some of these patterns, particularly the so-called wrinkle ridges, may be related to temporal variations in climate. We and Sean C. Solomon of the Carnegie Institution of Washington have argued that the plains preserve globally coherent episodes of deformation that may have occurred over short intervals of geologic history. That is, the entire lithosphere seems to have been stretched or compressed all at the same time. It is hard to imagine a mechanism internal to the solid planet that could do that. But what about global climate change? Solomon calculated that stresses induced in the lithosphere by fluctuations in surface temperature of about 100 degrees C (210 degrees Fahrenheit) would have been as high as 1,000 bars—comparable to those that form mountain belts on Earth and sufficient to deform Venus's surface in the observed way.

If volcanoes really did repave the Venusian surface 800 million years ago, they should have also injected a great deal of greenhouse gases into the atmosphere in a relatively short time. A reasonable estimate is that enough lava erupted to cover the planet with a layer one to 10 kilometers thick. In that case, the amount of carbon dioxide in the atmosphere would have hardly changed—there is already so much of it. But the abundances of water vapor and sulfur dioxide would have increased 10- and 100-fold, respectively. Fascinated by the possible implications, we modeled the planet's climate as an interconnected system of processes, including volcanic outgassing, cloud formation, the loss of hydrogen from the top of the atmosphere and reactions of atmospheric gases with surface minerals.

The interaction of these processes can be subtle. Although carbon dioxide, water vapor and sulfur dioxide all warm the surface, the last two also have a countervailing effect: the production of clouds. Higher concentrations of water vapor and sulfur dioxide would not only enhance the greenhouse effect but also thicken the clouds, which reflect sunlight back into space and thereby cool the planet. Because of these competing effects, it is not obvious what the injection of the two gases did to the climate.

THE PLANETARY PERSPECTIVE

Because sulfur dioxide and water vapor are continuously lost, clouds require ongoing volcanism for their maintenance. We calculated that volcanism must have been active within the past 30 million years to support the thick clouds observed today. The interior processes that generate surface volcanism occur over periods longer than tens of millions of years, so volcanoes are probably still active. This finding accords with observations of varying amounts of sulfur dioxide on Venus. In 1984 Larry W. Esposito of the University of Colorado at Boulder noted that cloud-top concentrations of sulfur dioxide had declined by more than a factor of 10 in the first five years of the Pioneer Venus mission, from 1978 to 1983. He concluded that the variations in this gas and associated haze particles were a result of volcanism. Surface temperature fluctuations, precipitated by volcanism, are also a natural explanation for many of the enigmatic features found by Magellan.

Fortunately, Earth's climate has not experienced quite the same extremes in the geologically recent past. Although it is also affected by volcanism, the oxygen-rich atmosphere—provided by biota and plentiful water—readily removes sulfur gases. Therefore, water clouds are the key to the planet's heat balance. The amount of water vapor available to these clouds is determined by the evaporation of the oceans, which in turn depends on surface temperature. A slightly enhanced greenhouse effect on Earth puts more water into the atmosphere and results in more cloud cover. The higher reflectivity reduces the incoming solar energy and hence the temperature. This negative feedback acts as a thermostat, keeping the surface temperature moderate over short intervals (days to years). An analogous feedback, the carbonate-silicate cycle, also stabilizes the abundance of atmospheric carbon dioxide. Governed by the slow process of plate tectonics, this mechanism operates over timescales of about half a million years.

These remarkable cycles, intertwined with water and life, have saved Earth's climate from the wild excursions its sister planet has endured. Anthropogenic influences, however, operate on intermediate timescales. The abundance of carbon dioxide in Earth's atmosphere has risen by a quarter since 1860. Although nearly all researchers agree that global warming is occurring, debate continues on how much of it is caused by the burning of fossil fuels and how much stems from natural variations. Whether there is a critical amount of carbon dioxide that overwhelms Earth's climate regulation cycles is not known. But one thing is certain: the climates of Earthlike planets can undergo abrupt transitions because of interactions among planetary-scale processes. In the long term, Earth's fate is sealed. As the Sun ages, it brightens. In about a billion years, the oceans will begin to evaporate rapidly and the climate will succumb to a runaway greenhouse. Earth and Venus, having started as nearly identical twins and diverged, may one day look alike.

Studying Venus, however alien it may seem, is essential to the quest for the general principles of climate variation—and thus to understanding the frailty or robustness of our home world.

The Earth's Elements

Robert P. Kirshner

Matter in the universe was born in violence. Hydrogen and helium emerged from the intense heat of the big bang some 15 billion years ago. More elaborate atoms of carbon, oxygen, calcium and iron, out of which we are made, had their origins in the burning depths of stars. Heavy elements such as uranium were synthesized in the shock waves of supernova explosions. The nuclear processes that created these ingredients of life took place in the most inhospitable of environments.

Matter was created in a violent explosion, known as the big bang, some 15 billion years ago. Within a minute fraction of a second, newborn quarks coalesced into protons. These fused further into the nuclei of helium atoms. Gravitational forces amplified ripples in this primordial soup, pulling the densest regions together into a giant cosmic tapestry of galaxies and voids. Inside galaxies, thick clouds of gas spawned stars. Traces of those early ripples can be seen in the cosmic microwave radiation, which still bears traces of the structure in the infant universe.

The large-scale unfolding of the universe was accompanied by a parallel change in the microscopic structure of matter. Carbon and nitrogen and other elements essential to life on the Earth were synthesized in the interiors of stars now long deceased. Within the Milky Way Galaxy, in the familiar stars of the night sky, astronomers can study these processes of microscopic change. In the early 1900s, such studies led to the first of several paradoxes regarding the ages of planets and stars.

The study of natural radioactivity on the Earth provided clues about the ages of the elements. Geophysicists looking at the slow decay of uranium into lead computed an age for the Earth of a few billion years. But astrophysicists of the early 20th century, not knowing about nuclear processes, computed that a sun powered by chemical burning or gravitational shrinking could shine only for a few million years.

The discrepancy mattered. An age of billions of years for the Earth provides a much more plausible calendar for biological and geologic evolution, where humans often find that change is imperceptibly slow. Even though the rug in most astronomy departments is lumpy from all the discrepancies that have been swept under it, a factor of 1,000 demands attention.

Curiously, the key to the problem was found in the processes of nuclear physics that, in the form of radioactivity, had first posed it. If stars live for billions of years instead of millions, they must have a continuing source of energy 1,000 times larger than chemical energy. Ordinary chemical changes involve the electrical force rearranging electrons in the outer regions of atoms. Nuclear changes involve the strong force rearranging neutrons and protons within the nucleus of an atom. The products of the reaction sometimes have less mass than the ingredients; the excess mass is converted to energy according to the well-known formula $E=mc^2$.

In nuclear reactions the energy yield is extremely large, typically a million times the energy produced by chemical reactions. Even the terminology for nuclear weapons reflects this factor. The unit of nuclear energy is a megaton—the energy of a million tons of chemical explosive.

A star that burns hydrogen, such as the Sun, has an ample supply of energy for a lifetime of 10 billion years. The ashes of nuclear burning—the elements of the periodic table—are the materials out of which living things are made. Perhaps most important, nuclear fusion, occurring steadily over the lifetime of a star, ensures a continuous supply of energy for billions of years and allows time for life and intelligence to develop.

Stars, after all, are not such ordinary places in the universe. A star is a ball of gas neatly balanced between the inward pull of its own gravitation and the outward pressure of the hot gas within. The compressed hydrogen gas usually has the density of the water in Boston Harbor, some 10^{30} times higher than the norm in the universe. And in a universe with a typical temperature of three kelvins (–270 degrees Celsius), the center of a star is at 15 million kelvins.

At such extreme temperatures the hydrogen atoms are stripped of their electrons. The naked protons undergo frequent, jarring collisions as they buzz furiously in the star's dense interior. Near the center the temperature and density are highest. There the protons, despite the electrical repulsion between them, are pushed so close together that the strong and the weak nuclear forces can come into play.

In a series of nuclear reactions, hydrogen nuclei (protons) fuse into helium nuclei (two protons and two neutrons), emitting two positrons, two neutrinos and energy. If the elements synthesized were limited to helium (which is also made in the big bang) and if it stayed locked up in the cores of stars, this would not be quite such an interesting story—and we would not be here to discuss it. After a long and steady phase of hydrogen fusion, which leads to helium accumulating in the core, the star changes dramatically.

The core shrinks and heats as four nucleons are locked up in each helium nucleus. The temperature and density of the core increase to maintain the pressure balance. The star as a whole becomes less homogeneous. While the core becomes smaller, the outer layers swell up to 50 times their previous radius. A star the size of the Sun will swiftly transform into a cool, but luminous, red giant.

But interesting events take place inside red giants. As the core contracts, the central furnace grows denser and hotter. Then nuclear reactions that were previously impossible become the principal source of energy. For example, the helium that accumulates during hydrogen burning can now become a fuel. As the star ages and the core temperature rises, brief encounters between helium nuclei produce fusion events.

The collision of two helium nuclei leads initially to an evanescent form of beryllium having four neutrons and four protons. Amazingly enough, another helium nucleus collides with this short-lived target, leading to the formation of carbon. The process would seem about as likely as crossing a stream by stepping fleetingly on a log. A delicate match between the energies of helium, the unstable beryllium and the resulting carbon allows the last to be created. Without this process, we would not be here.

Carbon and oxygen, formed by fusing one more helium with carbon, are the most abundant elements formed in stars. The many collisions of protons with helium atoms do not give rise to significant fusion products. Lithium, beryllium and boron—the nuclei of which are smaller than those of carbon—are a million times less abundant than carbon. Thus, abundances of elements are determined by often obscure details of nuclear physics. A star of the Sun's mass endures as a red giant for only a few hundred million years. The last stages of burning are unstable: the star pushes off its outer layers to form a shell of gas called a planetary nebula. In some stars, carbon-rich matter from the core is dredged up by convection. The freshly synthesized matter then escapes, forming a sooty cocoon of graphite. Eventually fuel runs out, and the inner core of the red giant congeals into a white dwarf.

A white dwarf is protected from total gravitational collapse not by the kinetic pressure of gases; the carbon and oxygen in its interior are in an almost crystalline state. The star is held up by the quantum repulsion of its free electrons. Quantum mechanics forbids electrons from sharing the lowest energy state. This restriction forces most electrons to occupy higher energy states even though the gas is relatively cold. These electrons provide the pressure to support a white dwarf. There is no more generation of nuclear energy, and no new elements are synthesized.

Many white dwarfs in our galaxy come to this dull end, slowly cooling, dimming and slipping below the edge of detection. Sometimes a too generous neighboring star may supply gas that streams onto a white dwarf, provoking it into a Type I supernova and a sudden synthesis of new elements.

The most significant locations for the natural alchemy of fusion are, however, stars more massive than the Sun. Although rarer, a heavy star follows a shorter and more intense path to destruction. To support the weight of the star's massive outer layers, the temperature and pressure in its core have to be high. A star of 20 solar masses is more than 20,000 times as luminous as the Sun. Rushing through its hydrogen-fusion phase 1,000 times faster, it swells up to become a red giant in just 10 million years instead of the Sun's 10 billion.

The high central temperature leads as well to a more diverse set of nuclear reactions. A Sunlike star builds up carbon and oxygen that stays locked in the cooling ember of a white dwarf. Inside a massive star, carbon nuclei fuse further to make neon and magnesium. Fusion of oxygen yields silicon as well, along with sulfur. Silicon burns to make iron. Intermediate stages of fusion and decay make many different elements, all the way up to iron.

The iron nucleus occupies a special place in nuclear physics and, by extension, in the composition of the universe. Iron is the most tightly bound nucleus. Lighter nuclei, when fusing together, release energy. To make a nucleus heavier than iron, however, requires an expenditure of energy. This fact, established in terrestrial laboratories, is instrumental in the violent death of stars. Once a star has built an iron core, there is no way it can generate energy by fusion. The star, radiating energy at a prodigious rate, becomes like a teenager with a credit card. Using resources much faster than can be replenished, it is perched on the edge of disaster.

So what happens? For the star, at least, the disaster takes the form of a supernova explosion. The core collapses inward in just one second to become a neutron star or black hole. The material in the core is as dense as that within a nucleus. The core can be compressed no further. When even more material falls into this hard core, it rebounds like a train hitting a wall. A wave of intense pressure traveling faster than sound—a sonic boom—thunders across the extent of the star. When the shock wave reaches the surface, the star suddenly brightens and explodes. For a few weeks, the surface shines as brightly as a billion suns while the emitting surface expands at several thousand kilometers per second. The abrupt energy release is comparable to the total energy output of the Sun in its entire lifetime.

Such Type II supernova explosions play a special role in the chemical enrichment of the universe. First, unlike stars of low mass that lock up their products in white dwarfs, exploding stars eject their outer layers, which are unburned. They belch out the helium that was formed from hydrogen burning and launch the carbon, oxygen, sulfur and silicon that have accumulated from further burning into the gas in their neighborhood.

New elements are synthesized behind the outgoing shock wave. The intense heat enables nuclear reactions that cannot occur in steadily burning stars. Some of the nuclear products are radioactive, but stable elements heavier than iron can also be synthesized. Neutrons bombard iron nuclei, forging them into gold. Gold is transformed into lead (an alchemist's nightmare!), and lead is bombarded to make elements all the way up to uranium. Elements beyond iron in the periodic table are rare in the cosmos. For every 100 billion hydrogen atoms, there is one uranium atom—each made at special expense in an uncommon setting.

This theoretical picture of the creation of heavy elements in supernova explosions was thoroughly tested in February 1987. A supernova, SN 1987A, exploded in the nearby Large Magellanic Cloud. Sanduleak –69° 202 Celsius, which in 1986 was noted as a star of 20 solar masses, is no longer there. Together the star and the supernova give dramatic evidence that at least one massive star ended its life in a violent way.

The supernova has provided dramatic confirmation of elaborate theoretical models of the origin of elements. Successive cycles of star formation and destruction enrich the interstellar medium with heavy elements. We can identify the substances in interstellar gas: they absorb particular wavelengths of light from more distant sources, leaving a characteristic imprint. The absorption lines tell us as well the abundance of the element—its amount compared with that of hydrogen.

In a spiral galaxy like the Milky Way, interstellar gas is associated with the spiral arms. Optical studies of the galaxy are hampered by the accompanying dust, which absorbs much of the light passing through. But the dust also shields the hydrogen atoms from ultraviolet light, allowing them to combine chemically and form molecules (H_2). In these hidden backwaters of the galaxy, other molecules such as water (H_2O), carbon monoxide (CO) and ammonia (NH_3) all assemble. The chemical variety is quite surprising: more than 100 molecules have been found in interstellar clouds.

We can learn much about the materials from which the Earth was formed by the simple act of picking up a pen. Made of carbon compounds and metals, the pen—and indeed the Earth itself—is typical of the cosmic pattern of abundances. Except for hydrogen and helium, which easily slip the gravitational grip of a small planet, the elements of the Earth are the elements of the universe: formed by stars and dispersed throughout the galaxy. (The jury is still out on the question of whether ordinary matter, composed of known subatomic particles, is a small fraction of the total mass in the universe. If so, then we are truly made of uncommon stuff.)

Whereas the Sun is 99 percent hydrogen and helium, the 1 percent of more complex nuclei includes traces of iron and other heavy elements. Thus, the solar system must have formed from elements synthesized by previous generations of stars. Like silver candlesticks from your grandmother (but much more valuable), we have inherited the carbon and oxygen produced by ancestral stars.

Astronomers can begin to trace a family tree for the solar system by examining massive stars within the Milky Way. If the massive stars in a star cluster are just now becoming red giants, the cluster must be young. If the stars currently headed toward the red giant phase have the mass of the Sun, the cluster must be old enough for its Sunlike stars to begin that change: about 10 billion years. The oldest clusters in our galaxy are the globular clusters, which appear to have an age of 12 to 18 billion years when measured in this way.

We recognize the globular clusters as an early generation of stars. The oldest of these are significantly different from the Sun; the abundances of elements such as iron are often 100 or even 1,000 times lower. Yet even these ancient stars contain a pinch of heavy elements. Thus, they evince the presence of a completely unseen generation of stars, which has no members left.

Given that the universe itself is only about 15 billion years old, the initial chemical enrichment of the Milky Way must have been very rapid. (Even quasars, extragalactic beacons from a time when the universe was only a fifth of its current age, contain carbon and nitrogen.) There has been much less change in recent times. The present-day chemical abundances in interstellar gas are about the same as in the Sun, locked in five billion years ago. This is the raw material for future stars and planets.

In neighboring gas clouds such as the Orion Nebula, astronomers can study intimate scenes of stellar birth. New infrared detectors are lifting the shroud from these cradles. (Although it blocks visible light, interstellar dust is transparent to infrared or radio waves.) We can see infant stars as they condense, even before they ignite hydrogen fuel in their cores. In addition, large telescopes such as the eight-meter Gemini telescopes in Hawaii and Chile promise much more detail about the process by which stars condense.

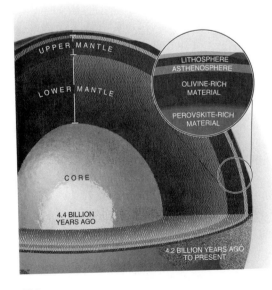

The differentiation of the planet took place quite quickly after the Earth was formed by the accretion of cosmic dust and meteorites. About 4.4 billion years ago the core—which, with the mantle, drives the geothermal cycle, including volcanism— appeared; gases emerging from the interior of the planet also gave rise to a nascent atmosphere. Somewhat later, although the issue has not been entirely resolved, it seems that continental crust formed as the various elements segregated into different depths.

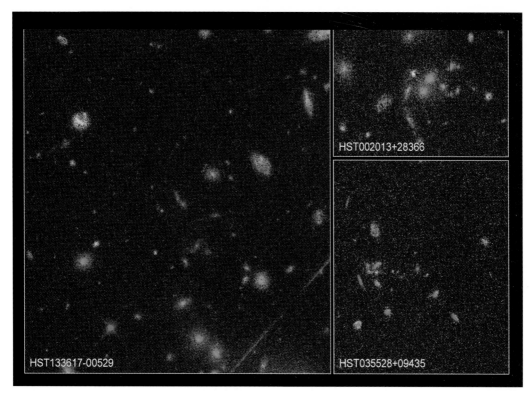

Above: A survey of galaxy clusters by NASA's Hubble Space Telescope in 1998 found some of the most distant clusters ever seen. About 10 to 20 of the furthest clusters may be over seven billion light years away, which means that the clusters were assembled early in the history of the universe.

Below: X-ray images of Coma (left) and Virgo (right) clusters show the hot intergalactic gas that dominates the luminous part of these structures. Both clusters are surrounded by infalling material.

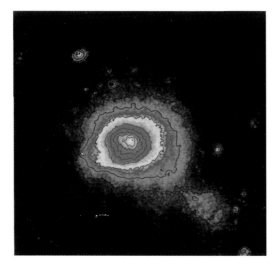
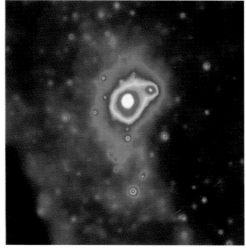

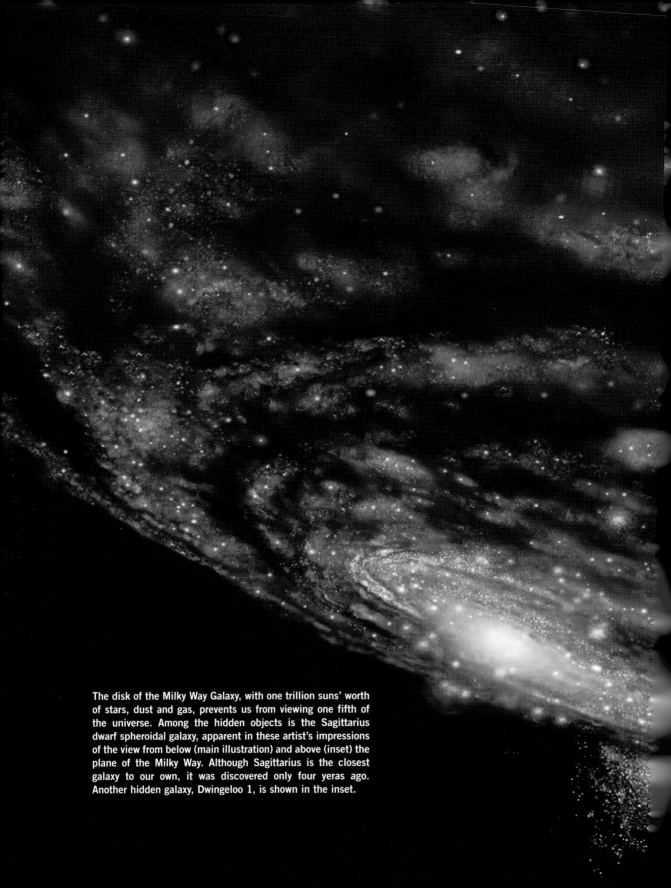

The disk of the Milky Way Galaxy, with one trillion suns' worth of stars, dust and gas, prevents us from viewing one fifth of the universe. Among the hidden objects is the Sagittarius dwarf spheroidal galaxy, apparent in these artist's impressions of the view from below (main illustration) and above (inset) the plane of the Milky Way. Although Sagittarius is the closest galaxy to our own, it was discovered only four yeras ago. Another hidden galaxy, Dwingeloo 1, is shown in the inset.

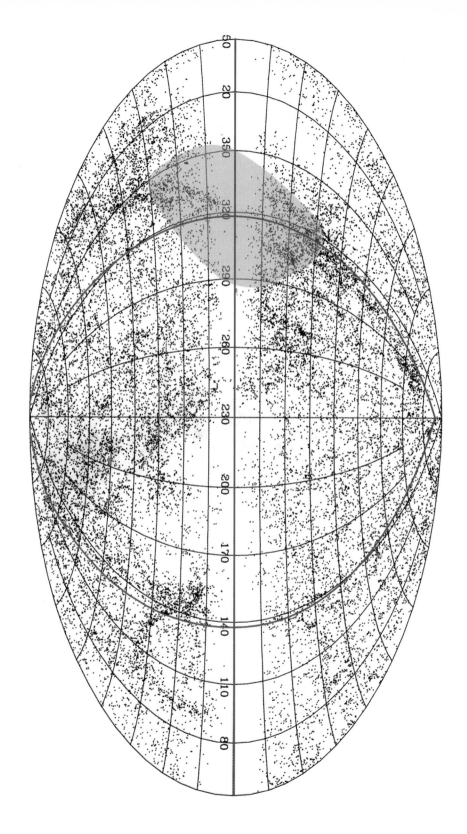

Thirty thousand galaxies, culled from three standard astronomical cata-logues, are shown as dots on this map. The galaxies appear all over the sky except in the so-called zone of avoidance, which corresponds to the plane of our Milky Way Galaxy (green horizontal center line). Outside the zone, the galaxies tend to clump near a line that traces out the Supergalactic Plane (purple line).

Above: The large Magellanic Cloud is one of the Milky Way's nearby satellite galaxies. As seen from Earth (or near-Earth orbit), dark objects in the cloud's halo gravitationally lens some of the cloud's stars, providing information—by inference—about the presence of dark matter in our own galaxy's halo.

Below: Galaxy M82, about 10 million light years from Earth, is distinguished by an outpouring of incandescent gas from the area around its core, causing dust and gas to rush into intergalactic space.

This Hubble Space Telescope image shows a pair of star clusters 166,000 light years away in the Large Magellanic Cloud. The field of view is 130 light years across. About 60 percent of the stars belong to the dominant yellow cluster called NGC 1850, which is estimated to be 50 million years old. Massive white stars scattered about the image are only about four million years old and represent about 20 percent of the stars in the image. Besides being much younger, the white stars are more loosely distributed than the yellow cluster. The difference between the two clusters suggests that they are separate start groups that lie along the same line of sight. The younger, more open cluster probably lies 200 light years beyond the older cluster. (If it were in the foreground, dust from the white cluster would obscure stars in the older yellow cluster.) To observe two well-defined star populations separated by such a small gap of space is unusual. This juxtaposition suggests that supernova explosions in the older cluster might have triggered the birth of the younger cluster.

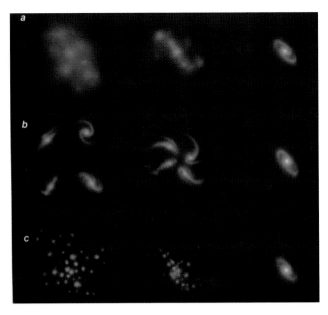

Above: Models of galaxy formation fall into three general categories. In the Eggen-Lynden-Bel-Sandage model (a), the Milky Way formed by the rapid collapse of a single gaseous proto-galaxy. In the Toomre model (b), several large aggregates of gas merged. The Searle-Zinn picture (c) is similar to the Toomre model except that the ancestral fragments consisted of much smaller but more numerous pieces.

Below: Hubble image of nearby galaxy Centaurus A shows "galactic cannibalism"—a massive black hole hidden at its center is feeding on a smaller galaxy in a spectacular collision. Such fireworks were common in the early universe as galaxies formed and evolved.

Interacting binary star system, consisting of a hot, dense white dwarf star and its cool, less massive red companion, quasiperiodically brightens as much as 100-fold. The white dwarf's gravity distorts the red star into a teardrop shape and pulls a stream of gas through the point of the teardrop.

Because of its orbital motion, the stream trails behind the white dwarf. Viscosity causes the stream to spread into a disk; gas in the disk falls inward and ultimately accretes onto the white star.

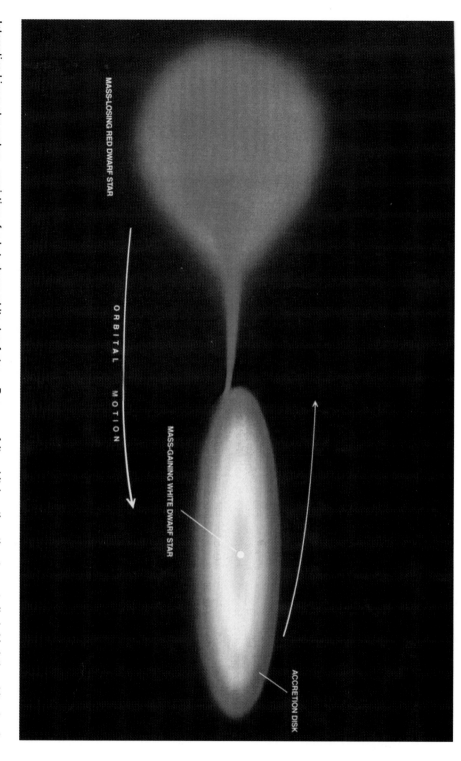

MASS-LOSING RED DWARF STAR

ORBITAL MOTION

MASS-GAINING WHITE DWARF STAR

ACCRETION DISK

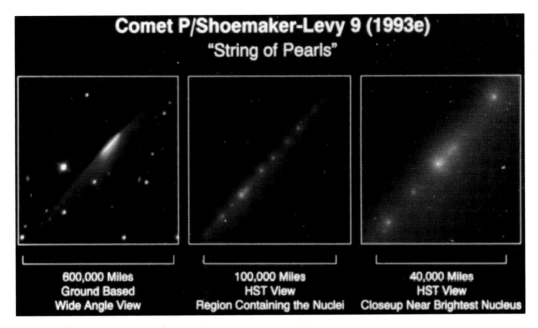

Comet P/Shoemaker-Levy 9 (1993e)
"String of Pearls"

600,000 Miles
Ground Based
Wide Angle View

100,000 Miles
HST View
Region Containing the Nuclei

40,000 Miles
HST View
Closeup Near Brightest Nucleus

Above: Comet Shoemaker-Levy 9 broke into more than a dozen pieces before Jupiter's gravity inexorably pulled them down to the planet's atmospherre. Dubbed the "String of Pearls," they are shown here in both ground-based and orbital photos.

Below: The Hubble Space Telescope zeroes in on the first visible blemish in Jupiter's upper atmosphere, a dark spot where one piece of the comet has exploded upon impact.

Jupiter · July 16, 1994 · 19:00 UT

Above: A limb of Mercury. This image is a photo-mosaic taken by the Mariner 10 spacecraft in 1974, showing the innermost planet's inhospitable, impact-cratered surface

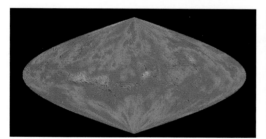

The surface of Venus, seen topographically, spans a wide range of elevations, about 13 kilometers from low (blue) to high (yellow). But 3/5s of the surface lies within 500 meters of the average elevation. In contrast, topography on Earth clusters around two distinct elevations, which correspond to continents and ocean floors.

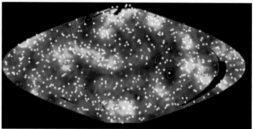

Impact craters are randomly scattered all over Venus. Most are pristine (white dots). Those modified by lava (red dots) or by faults (triangles) are concentrated in places such as Aphrodite Terra. Areas with a low density of craters (blue background) are often located in highlands. Higher crater densities (yellow background) are usually found in the lowland plains.

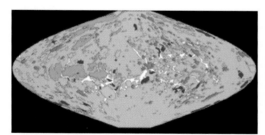

The terrain of Venus consists predominately of volcanic plains (blue). Within the plains are deformed areas such as tesserae (pink) and rift zones (white), as well as volcanic features such as coronae (peach), lava floods (red) and volcanoes of various sizes (orange). Volcanoes are not concentrated in chains as they are on Earth, indicating that plate tectonics do not operate here.

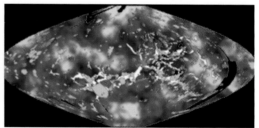

This geologic map shows the different terrains and their relative ages, as inferred from the crater density. Volcanoes and coronae tend to clump along equatorial rift zones, which are younger (blue) than the rest of the planet's surface. The tesserae, ridges and plains are older (yellow). In general, however, the surface lacks the extreme variation in age that is found on Earth and Mars.

Above: Earth seen from space has changed dramatically. One hundred million years after it had formed—some 4.35 billion years ago—the planet was probably undergoing meteor bombardment (left). At this time, it may have been studded with volcanic islands and shrouded by an atmosphere laden with carbon dioxide and heavy with clouds. Three billion years ago its face may have been obscured by an orange haze of methane, the product of early organisms (center). Today clouds, oceans and continents are clearly discernible (right).

Below: the icy outer surface of Europa has been continually tortured and fractured by Jupiter's gravitational forces. This computer-enhanced image of Europa's surface was taken by the *Galileo* space probe.

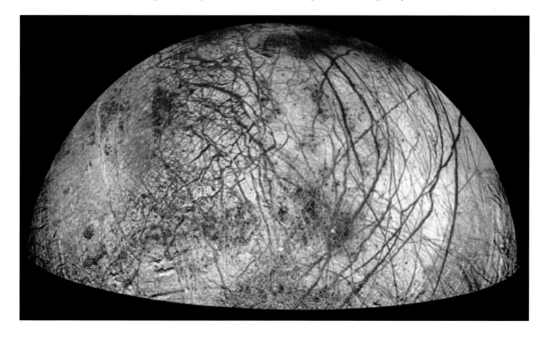

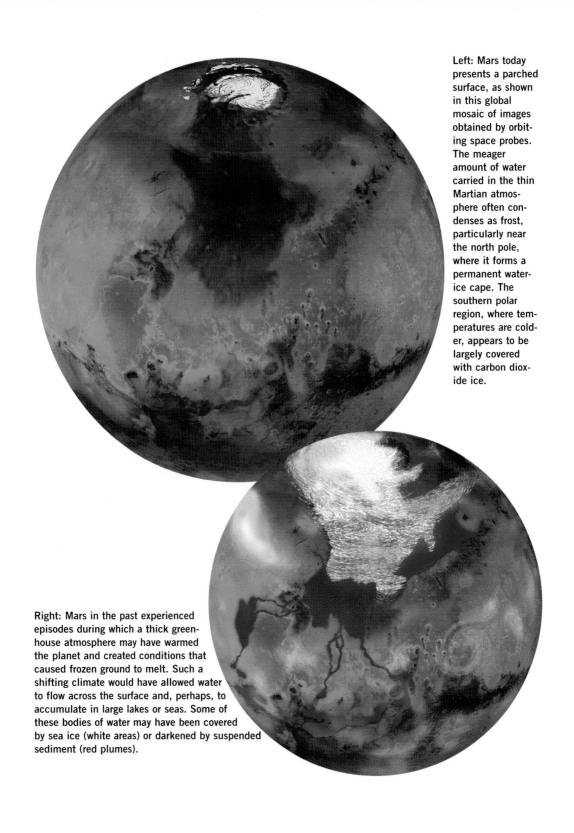

Left: Mars today presents a parched surface, as shown in this global mosaic of images obtained by orbiting space probes. The meager amount of water carried in the thin Martian atmosphere often condenses as frost, particularly near the north pole, where it forms a permanent water-ice cape. The southern polar region, where temperatures are colder, appears to be largely covered with carbon dioxide ice.

Right: Mars in the past experienced episodes during which a thick greenhouse atmosphere may have warmed the planet and created conditions that caused frozen ground to melt. Such a shifting climate would have allowed water to flow across the surface and, perhaps, to accumulate in large lakes or seas. Some of these bodies of water may have been covered by sea ice (white areas) or darkened by suspended sediment (red plumes).

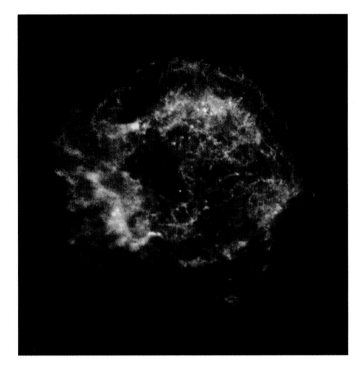

Above: Spectacular first-light image from the new Chandra x-ray orbital telescope shows the supernova remnant known as Cassiopeia A, which exploded about 300 years ago. Shock waves expanding at 10 million mph have heated this 10 light-year diameter bubble of stellar debris.

Below: A 3,700 light-year diameter dust disk encircles a 300 million solar-mass black hole in the center of the elliptical galaxy NGC 7052. The disk, possibly the remains of an ancient galaxy collision, will be swallowed up completely by the black hole in several billion years.

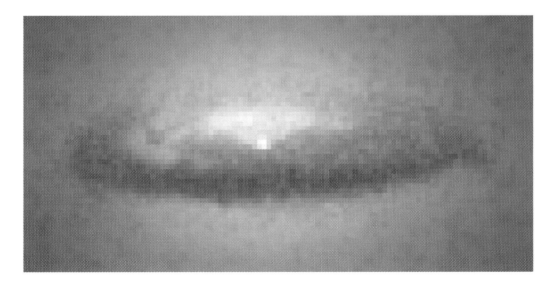

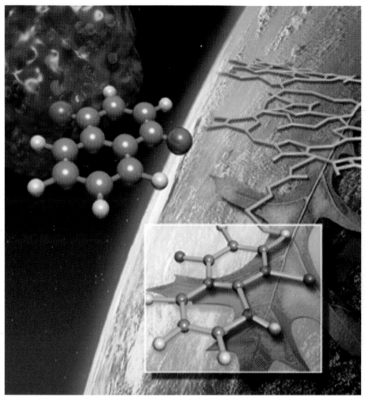

Left: Quinones from space have structures nearly identical to those that help chlorophyll molecules transfer light energy from one part of a plant cell to another.

Below: Interstellar ice begins to form when molecules such as water, methanol and hydrocarbon freeze to sandlike granules of silicate drifting in dense interstellar clouds (left). Ultraviolet radiation from nearby stars breaks some of the chemical bonds of the frozen compounds as the ice grain grows to no bigger than about one ten-thousandth of a millimeter across (center). Broken molecules recombine into structures such as quinones, which would never form if the fragments were free to float away (right).

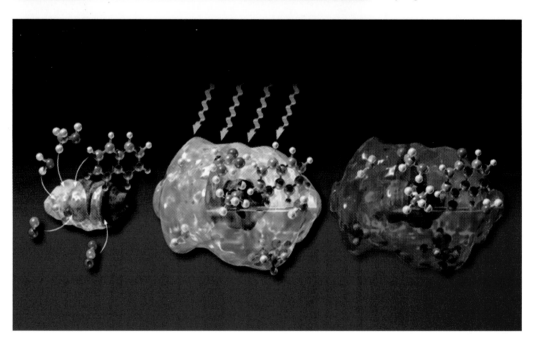

Left: Martian rock, also known as ALH-84001, is shown here at actual size. The meteorite consists mostly of orthopyroxense, a silicate mineral.

Below: The rock was cut, exposing a cross section. The crack slightly above the center of the cut face is a fracture through which fluid flowed and deposited globules of carbonate minerals. The findings suggest that the fluid that flowed through the fracture contained the decay products of living organisms, which were trapped by the forming globules.

The majestic spiral galaxy NGC 4414, imaged by the Hubble Space Telescope in 1995 and 1999, as part of a project to accurately measure distances to other nearby galaxies. NGC 4414 was measured to be about 60 million light years distant. This measurement, along with others in the project, contributes to astronomers' overall knowledge of the rate of expansion of the universe.

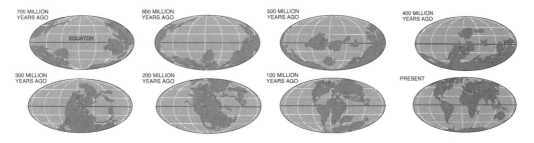

Continental shift has altered the face of the planet for nearly a billion years, as can be seen in the differences between the positions of the continents that we know today and those of 700 million years ago. Pangaea, the super aggregate of early continents, came together about 200 million years ago and then promptly, in geologic terms, broke apart.

As gas coalesces into a star, it first forms a rotating disk of gas and dust. While the star condenses, the dust aggregates into rocky planets, such as the Earth. Residual gas accumulates to make large gas planets, such as Jupiter. Disks, observed with infrared and radio techniques and, occasionally, glimpsed with optical methods, are common.

The composition of the Earth is the natural by-product of energy generation in stars and successive waves of stellar birth and death in our galaxy. We do not know if other stars have Earthlike planets where complex atoms, formed in stellar cauldrons, have organized themselves into intelligent systems. But understanding the history of matter and searching for its most interesting forms, such as galaxies, stars, planets and life, seem a suitable use for our intelligence.

The Evolution of the Earth

Claude J. Allègre and Stephen H. Schneider

Like the lapis lazuli gem it resembles, the blue, cloud-enveloped planet that we recognize immediately from satellite pictures seems remarkably stable. Continents and oceans, encircled by an oxygen-rich atmosphere, support familiar life-forms. Yet this constancy is an illusion produced by the human experience of time. The Earth and its atmosphere are continuously altered. Plate tectonics shift the continents, raise mountains and move the ocean floor while processes that no one fully comprehends alter the climate.

Such constant change has characterized the Earth since its beginning some 4.5 billion years ago. From the outset, heat and gravity shaped the evolution of the planet. These forces were gradually joined by the global effects of the emergence of life. Exploring this past offers us the only possibility of understanding the origin of life and, perhaps, its future.

Scientists used to believe the rocky planets, including the Earth, Mercury, Venus and Mars, were created by the rapid gravitational collapse of a dust cloud, a deflation giving rise to a dense orb. In the 1960s the Apollo space program changed this view. Studies of moon craters revealed that these gouges were caused by the impact of objects that were in great abundance about 4.5 billion years ago. Thereafter, the number of impacts appeared to have quickly decreased. This observation rejuvenated the theory of accretion postulated by Otto Schmidt. The Russian geophysicist had suggested in 1944 that planets grew in size gradually, step by step.

According to Schmidt, cosmic dust lumped together to form particulates, particulates became gravel, gravel became small balls, then big balls, then tiny planets, or planetesimals, and, finally, dust became the size of the moon. As the planetesimals became larger, their numbers decreased. Consequently, the number of collisions between planetesimals, or meteorites, decreased. Fewer items available for accretion meant that it took a long time to build up a large planet. A calculation made by George W. Wetherill of the Carnegie Institution of Washington suggests that about 100 million years could pass between the formation of an object measuring 10 kilometers in diameter and an object the size of the Earth.

The process of accretion had significant thermal consequences for the Earth, consequences that have forcefully directed its evolution. Large bodies slamming into the planet produced immense heat in the interior, melting the cosmic dust found there. The resulting furnace—situated some 200 to 400 kilometers underground and called a magma ocean—was active for millions of years, giving rise to volcanic eruptions. When the Earth was young, heat at the surface caused by volcanism and lava flows from the interior was supplemented by the constant bombardment of huge planetesimals, some of them perhaps the size of the Moon or even Mars. No life was possible during this period.

Beyond clarifying that the Earth had formed through accretion, the Apollo program compelled scientists to try to reconstruct the subsequent temporal and physical development of the early Earth. This undertaking had been considered impossible by founders of geology, including Charles Lyell, to whom the following phrase is attributed: No vestige of a beginning, no prospect for an end. This statement conveys the idea that the young Earth could not be re-created, because its remnants were destroyed by its very activity. But the development of isotope geology in the 1960s had rendered this view obsolete. Their imaginations fired by Apollo and the Moon findings, geochemists began to apply this technique to understand the evolution of the Earth.

Dating rocks using so-called radioactive clocks allows geologists to work on old terrains that do not contain fossils. The hands of a radioactive clock are isotopes—atoms of the same element that have different atomic weights—and geologic time is measured by the rate of decay of one isotope into another. Among the many clocks, those based on the decay of uranium 238 into lead 206 and of uranium 235 into lead 207 are special. Geochronologists can determine the age of samples by analyzing only the daughter product—in this case, lead—of the radioactive parent, uranium.

Isotope geology has permitted geologists to determine that the accretion of the Earth culminated in the differentiation of the planet: the creation of the core—the source of the Earth's magnetic field—and the beginning of the atmosphere. In 1953 the classic work of Claire C. Patterson of the California Institute of Technology used the uranium-lead clock to establish an age of 4.55 billion years for the Earth and many of the meteorites that formed it. Recent work [in the early 1980s] by one of us (Allègre) on lead isotopes, however, led to a somewhat new interpretation. As Patterson argued, some meteorites were indeed formed about 4.56 billion years ago, and their debris constituted the Earth. But the Earth continued to grow through the bombardment of planetesimals until some 120 to 150 million years later. At that time—4.44 to 4.41 billion years ago—the Earth began to retain its atmosphere and create its core. This possibility had already been suggested by Bruce R. Doe and Robert E. Zartman of the U.S. Geological Survey in Denver a decade ago [about 1984] and is in agreement with Wetherill's estimates.

The emergence of the continents came somewhat later. According to the theory of plate tectonics, these landmasses are the only part of the Earth's crust that is not recycled and, consequently, destroyed during the geothermal cycle driven by the convection in the mantle. Continents thus provide a form of memory because the record of early life can be read in their rocks. The testimony, however, is not extensive. Geologic activity, including plate tectonics, erosion and metamorphism, has destroyed almost all the ancient rocks. Very few fragments have survived this geologic machine.

One of the most important aspects of the Earth's evolution is the formation of the atmosphere, because it is this assemblage of gases that allowed life to crawl out of the oceans and to be sustained. Researchers have hypothesized since the 1950s that the terrestrial atmosphere was created by gases emerging from the interior of the planet. When a volcano spews gases, it is an example of the continuous outgassing, as it is called, of the Earth. But scientists have questioned whether this process occurred suddenly about 4.4 billion years ago when the core differentiated or whether it took place gradually over time.

The major challenge facing an investigator who wants to measure such ratios of decay is to obtain high concentrations of rare gases in mantle rocks because they are extremely limited. Fortunately, a natural phenomenon occurs at mid-oceanic ridges during which volcanic lava transfers some silicates from the mantle to the surface. The small amounts of gases trapped in mantle minerals rise with the melt to the surface and are concentrated in small vesicles in the outer glassy margin of lava flows. This process serves to concentrate the amounts of mantle gases by a factor of 104 or 105. Collecting these rocks by dredging the seafloor and then crushing them under vacuum in a sensitive mass spectrometer allow geochemists to determine the ratios of the isotopes in the mantle. The results are quite surprising. Calculations of the ratios indicate that between 80 and 85 percent of the atmosphere was outgassed in the first one million years; the rest was released slowly but constantly during the next 4.4 billion years.

The composition of this primitive atmosphere was most certainly dominated by carbon dioxide, with nitrogen as the second most abundant gas. Trace amounts of methane, ammonia, sulfur dioxide and hydrochloric acid were also present, but there was no oxygen. Except for the presence of abundant water, the atmosphere was similar to that of Venus or Mars. The details of the evolution of the original atmosphere are debated, particularly because we do not know how strong the Sun was at that time. Some facts, however, are not disputed. It is evident that carbon dioxide played a crucial role. In addition, many scientists believe the evolving atmosphere contained sufficient quantities of gases like ammonia and methane to give rise to organic matter.

Still, the problem of the Sun remains unresolved. One hypothesis holds that during the Archean era, which lasted from about 4.5 to 2.5 billion years ago, the Sun's power was only 75 percent of what it is today. This possibility raises a dilemma: How could life have survived in the relatively cold climate that should accompany a weaker Sun? A solution to the faint early Sun paradox, as it is called, was offered by Carl Sagan and George Mullen of Cornell University in 1970. The two scientists suggested that methane and ammonia, which are very effective at trapping infrared radiation, were quite abundant. These gases could have created a super-greenhouse effect. The idea was criticized on the basis that such gases were highly reactive and have short lifetimes in the atmosphere.

The rapid outgassing of the planet liberated voluminous quantities of water from the mantle, creating the oceans and the hydrologic cycle. The acids that were probably present in the atmosphere eroded rocks, forming carbonate-rich rocks. The relative importance of such a mechanism is, however, debated. Heinrich D. Holland of Harvard University believes the amount of carbon dioxide in the atmosphere rapidly decreased during the Archean and stayed at a low level.

Understanding the carbon dioxide content of the early atmosphere is pivotal to understanding the mechanisms of climatic control. Two sometimes conflicting camps have put forth ideas on how this process works. The first group holds that global temperatures and carbon dioxide were controlled by inorganic geochemical feedbacks; the second asserts that they were controlled by biological removal.

James C. G. Walker, James F. Kasting and Paul B. Hays, then at the University of Michigan, proposed the inorganic model in 1981. They postulated that levels of the gas were high at the outset of the Archean and did not fall precipitously. The trio suggested that as the climate warmed, more water evaporated, and the hydrologic cycle became more vigorous, increasing precipitation and runoff. The carbon dioxide in the atmosphere mixed with rainwater to create carbonic acid runoff, exposing minerals at the surface to weathering. Silicate minerals combined with carbon that had been in the atmosphere, sequestering it in sedimentary rocks. Less carbon dioxide in the atmosphere meant, in turn, less of a greenhouse effect. The inorganic negative feedback process offset the increase in solar energy.

This solution contrasts with a second paradigm: biological removal. One theory advanced by James E. Lovelock, an originator of the Gaia hypothesis, assumed that photosynthesizing microorganisms, such as phytoplankton, would be very productive in a high–carbon dioxide environment. These creatures slowly removed carbon dioxide from the air and oceans, converting it into calcium carbonate sediments. Critics retorted that phytoplankton had not even evolved for most of the time that the Earth has had life. (The Gaia hypothesis holds that life on the Earth has the capacity to regulate temperature and the composition of the Earth's surface and to keep it comfortable for living organisms.)

The issue of carbon remains critical to the story of how life influenced the atmosphere. Carbon burial is a key to the vital process of building up atmospheric oxygen concentrations—a prerequisite for the development of certain life-forms. In addition, global warming may be taking place now as a result of humans releasing this carbon. For one or two billion years, algae in the oceans produced oxygen. But because this gas is highly reactive and because there were many reduced minerals in the ancient oceans—iron, for example, is easily oxidized—much of the oxygen produced by living creatures simply got used up before it could reach the atmosphere, where it would have encountered gases that would react with it.

Even if evolutionary processes had given rise to more complicated life-forms during this anaerobic era, they would have had no oxygen. Furthermore, unfiltered ultraviolet sunlight would have likely killed them if they left the ocean. Researchers such as Walker and Preston Cloud, then at the University of California at Santa Barbara, have suggested that only about two billion years ago, after most of the reduced minerals in the sea were oxidized, did atmospheric oxygen accumulate. Between one and two billion years ago oxygen reached current levels, creating a niche for evolving life.

By examining the stability of certain minerals, such as iron oxide or uranium oxide, Holland has shown that the oxygen content of the Archean atmosphere was low, before two billion years ago. It is largely agreed that the present-day oxygen content of 20 percent is the result of photosynthetic activity.

The presence of oxygen in the atmosphere had another major benefit for an organism trying to live at or above the surface: it filtered ultraviolet radiation. Ultraviolet radiation breaks down many molecules—from DNA and oxygen to the chlorofluorocarbons that are implicated in stratospheric ozone depletion. Such energy splits oxygen into the highly unstable atomic form O, which can combine back into O_2 and into the very special molecule O_3, or ozone. Ozone, in turn, absorbs ultraviolet radiation. It was not until oxygen was abundant enough in the atmosphere to allow the formation of ozone that life even had a chance to get a root-hold or a foothold on land. It is not a coincidence that the rapid evolution of life from prokaryotes (single-celled organisms with no nucleus) to eukaryotes (single-celled organisms with a nucleus) to metazoa (multicelled organisms) took place in the billion-year-long era of oxygen and ozone.

Although the atmosphere was reaching a fairly stable level of oxygen during this period, the climate was hardly uniform. There were long stages of relative warmth or coolness during the transition to modern geologic time. The composition of fossil plankton shells that lived near the ocean floor provides a measure of bottom water temperatures. The record suggests that over the past 100 million years bottom waters cooled by nearly 15 degrees Celsius. Sea levels dropped by hundreds of meters, and continents drifted apart. Inland seas mostly disappeared, and the climate cooled an average of 10 to 15 degrees C. Roughly 20 million years ago permanent ice appears to have built up on Antarctica.

About two to three million years ago the paleoclimatic record starts to show significant expansions and contractions of warm and cold periods on 40,000-year or so cycles. This periodicity is interesting because it corresponds to the time it takes the Earth to complete an oscillation of the tilt of its axis of rotation. It has long been speculated, and recently calculated, that known changes in orbital geometry could alter the amount of sunlight coming in between winter and summer by about 10 percent or so and could be responsible for initiating or ending ice ages.

Most interesting and perplexing is the discovery that between 600,000 and 800,000 years ago the dominant cycle switched from 40,000-year periods to 100,000-year intervals with very large fluctuations. The last major phase of glaciation ended about 10,000 years ago. At its height 20,000 years ago, ice sheets a mile thick covered much of northern Europe and North America. Glaciers expanded in high plateaus and mountains throughout the world. Enough ice was locked up on land to cause sea levels to drop more than 100 meters below where they are today. Massive ice sheets scoured the land and revamped the ecological face of the Earth, which was five degrees C cooler on average than it is currently.

If we humans consider ourselves part of life—that is, part of the natural system—then it could be argued that our collective impact on the Earth means we may have a significant coevolutionary role in the future of the planet. The current trends of population growth, the demands for increased standards of living and the use of technology and organizations to attain these growth-oriented goals all contribute to pollution. When the price of polluting is low and the atmosphere is used as a free sewer, carbon dioxide, methane, chlorofluorocarbons, nitrous oxides, sulfur oxides and other toxics can build up.

The theory of heat trapping—codified in mathematical models of the climate—suggests that if carbon dioxide levels double sometime in the middle of the next century, the world will warm between one and five degrees C. The mild end of that range entails warming at the rate of one degree per 100 years—a factor of 10 faster than the one degree per 1,000 years that has historically been the average rate of natural change on a global scale. Should the higher end of the range occur, then we could see rates of climatic change 50 times faster than natural average conditions. Change at this rate would almost certainly force many species to attempt to move their ranges, just as they did from the ice age–interglacial transition between 10,000 and 15,000 years ago. Not only would species have to respond to climatic change at rates 10 to 50 times faster, but few would have undisturbed, open migration routes as they did at the end of the ice age and the onset of the interglacial era. It is for these reasons that it is essential to understand whether doubling carbon dioxide will warm the Earth by one degree or five.

To make the critical projections of future climatic change needed to understand the fate of ecosystems on this Earth, we must dig through land, sea and ice to uncover as much of the geologic, paleoclimatic and paleoecological records as we can. These records provide the backdrop against which to calibrate the crude instruments we must use to peer into a shadowy environmental future, a future increasingly influenced by us.

Global Climatic Change on Mars

Jeffrey S. Kargel and Robert G. Strom

To those of us who have spent a good part of our lives studying Mars, the newly discovered evidence that extraterrestrial microbes may have once lived in a rock cast off from that planet stirs feelings of awe. But the recent claim also evokes thoughts of Percival Lowell, a well-known American astronomer of the early 20th century, who turned his telescope toward Mars and saw a vast network of canals bordered by vegetation. His suggestion that Mars harbored such lushness had many people believing that the surface of the planet enjoyed conditions not so different from those on Earth. But in the 1960s three Mariner spacecraft flew by Mars and revealed the true harshness of its environment.

Observations from those unmanned probes indicated that Mars has an atmosphere that is thin, cold and dry. This tenuous shroud, composed almost entirely of carbon dioxide, provides less than 1 percent of the surface pressure found at sea level on Earth. The images radioed back during those first fleeting encounters three decades ago were fuzzy and few in number, but they were decidedly more accurate than Lowell's telescopic views. The Mariner cameras showed no canals, no water and no vegetation. They presented only a moonlike surface covered with craters. Sober scientists quickly dismissed any notion that the climate on Mars was sufficiently warm or wet to sustain life.

With its distant orbit—50 percent farther from the Sun than Earth—and slim atmospheric blanket, Mars experiences frigid weather conditions. Surface temperatures typically average about -60 degrees Celsius (-76 degrees Fahrenheit) at the equator and can dip to -123 degrees C near the poles. Only the midday Sun at tropical latitudes is warm enough to thaw ice on occasion, but any liquid water formed in this way would evaporate almost instantly because of the low atmospheric pressure.

Although the atmosphere holds a small amount of water, and water-ice clouds sometimes develop, most Martian weather involves blowing dust or carbon dioxide. Each winter, for example, a blizzard of frozen carbon dioxide rages over one pole, and a few meters of this dry-ice snow accumulate as previously frozen carbon dioxide evaporates from the opposite polar cap. Yet even on the summer pole, where the Sun remains in the sky all day long, temperatures never warm enough to melt frozen water.

Despite the abundant evidence for cold, dry conditions, the impression of Mars as a perpetually freeze-dried world has been steadily giving way since the Mariner probes first reported their findings. Planetary scientists, who continue to examine the voluminous data from both the Mariner and the later Viking missions of the 1970s, now realize that Mars has had a complex climatic history—one that was perhaps punctuated with many relatively warm episodes. At certain times, huge volumes of water flowed freely across the surface of the planet. Before considering what this astonishing fact means for the possibility of life evolving on Mars or the strategy for the next round of Martian exploration, it is instructive to review how this reversal in the way Mars is perceived came about.

MUDDY RECOLLECTIONS

Scrutinizing the Mariner and Viking images obtained from orbit, planetary scientists soon noticed that most old Martian craters (unlike lunar ones) are eroded and that features resembling mudflows occur around almost every large, young crater on Mars. Such muddy "ejecta" probably represent the frozen remnants of a cataclysmic moment in the past when an asteroid or comet collided with the Martian surface, melting a patch of icy permafrost (where water-saturated ground had been frozen) and excavating a large hole that tapped a zone containing liquid water deep underground. By the late 1970s planetologists concluded that a considerable amount of underground ice and water has been present below the Martian surface throughout much of the history of the planet.

Researchers have since found other indications that a thick substratum of frozen ground exists on Mars. They have also identified evidence that ice once formed on the surface, where it appears to have created characteristic glacial landscapes. These features include bouldery ridges of sediment left by melting glaciers at their margins and meandering lines of sand and gravel deposited beneath glaciers by streams running under the ice (so-called eskers).

Many telltale landforms on Mars resemble frosty sites on Earth. For example, the pitted terrain on Mars corresponds to an earthly equivalent called thermokarst, which forms when the ice contained at shallow levels melts and the ground collapses. The apron-shaped lobes of rocky debris seen on the flanks of some Martian mountains might be rubble-covered glaciers. Or, more likely, they represent "rock glaciers," like the ones that form within the Alaska Range and in the Antarctic Dry Valleys on Earth. These distinctive sloping surfaces result after thousands of freeze-thaw cycles cause the top meter or so of water-soaked ground to creep slowly downhill.

Glacial features and muddy ejecta around craters are not the only examples of water shaping the Martian surface. In some places, sinuous valleys one kilometer wide and many hundreds of kilometers long form large branching networks. Carl Sagan of Cornell University, Victor R. Baker of the University of Arizona and their colleagues

Right: Teardrop-shaped hills dot the floor of outflow channels on Mars, commonly forming behind raised crater rims.

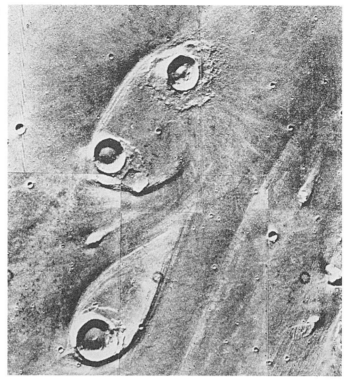

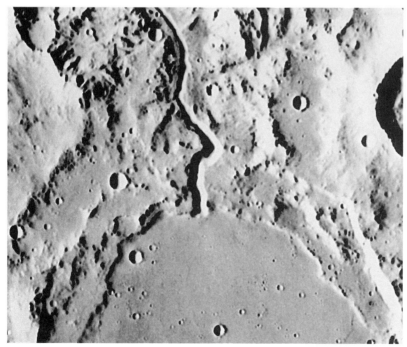

Left: Crater lakes can be found on Earth in the depressions left after an impact of an asteroid or comet. A similar body may have once occupied a smooth-floored crater of the Martian highlands, which shows both a terraced inlet and a deeply incised outlet channel.

suggested in the 1970s that such troughs were created by running water. Other Martian valleys have blunt starting points and short tributaries, characteristics that are typical of erosion by groundwater "sapping." That process, common on Earth, results from the seepage of water from underground springs, which causes the overlying rock and soil to wash away.

Images of Mars also reveal enormous outflow channels etched on the surface. Some of these structures are more than 200 kilometers wide and can stretch for 2,000 kilometers or more. These channels emanate from what is called chaotic terrain, regions of fractured, jumbled rocks that apparently collapsed when groundwater suddenly surged outward. The ensuing floods carved the vast channels, leaving streamlined islands more than 100 kilometers long and gouging cavernous potholes several hundred meters deep. Baker compared the Martian outflow channels to similar, albeit smaller, flood features found on Earth in parts of Oregon and Washington State. Those so-called channeled scablands of the Pacific Northwest formed after a glacier that had dammed a large lake broke open suddenly and caused a catastrophic flood.

The geometry of the Martian outflow channels indicates that water could have flowed along the surface as rapidly as 75 meters per second (170 miles per hour). Michael H. Carr of the U.S. Geological Survey estimates that the vast quantity of water necessary to create these many enormous channels would have been enough to fill a global Martian ocean that was 500 meters deep, although not all this liquid flowed at one time. One source for that great quantity of water may have been a large reservoir under ice-impregnated permafrost that had been warmed by heat from the interior of the planet.

Why should such an underground accumulation of water suddenly inundate the surface? Scientists are unsure of the exact cause, but this groundwater might have started to flow after the icy permafrost capping it thinned and weakened, perhaps because of a sudden climate warming, volcanism or tectonic uplift. Once water broke through to the surface, carbon dioxide from saturated groundwater—a Martian seltzer of sorts—may have erupted in tremendous geysers, further undermining the stability of the saturated underground layers. The result was to produce chaotic terrain and to unleash floods and mudflows of a magnitude that has rarely, if ever, been matched by any earthly deluge.

AN OCEAN AWAY

Some highland areas on Mars contain extensive systems of valleys that drained into sediment-floored depressions. These lowlands were at one time full of water. The largest of these Martian lakes filled two gigantic impact basins called Hellas and Argyre.

But these lakes may not have been the largest bodies of water on the planet. Research groups led by David H. Scott and Kenneth L. Tanaka of the U.S. Geological Survey and by Jeffrey M. Moore of the National Aeronautics and Space Administration Ames

Research Center independently concluded that repeated floods from the outflow channels emptied to the north and formed a succession of transient lakes and seas. We have interpreted many features bordering these ancient basins as marking where glaciers once emptied into these deep bodies of water. Tanaka and Moore believe that thick layers of sediment deposited in these seas now stretch across much of the extensive northern plains. According to several estimates, one of the larger of the northern seas on Mars could have displaced the combined volume of the Gulf of Mexico and the Mediterranean Sea.

Yet even that great body of water may not have been the supreme example: there may have been a Mars ocean. As early as 1973 the late Henry Faul of the University of Pennsylvania raised this intriguing possibility. During the past decade, other researchers, working with information acquired during the Viking missions, have revived Faul's idea.

For instance, in 1989 Timothy J. Parker and his colleagues at the Jet Propulsion Laboratory in Pasadena, Calif., again proposed a northern ocean (arguing that many features in the northern plains looked as if they had resulted from coastal erosion). Motivated in part by such work, Baker and several colleagues (including us) named this hypothetical northern ocean Oceanus Borealis. We calculated that it was possibly four times as large as the Arctic Ocean on Earth, and we proposed a scenario for the actions of the water cycle on Mars that could have accounted for it.

Whereas most planetary scientists now agree that large bodies of water formed repeatedly in the northern plains on Mars, many do not accept that there was ever a true ocean there. Some envision that only a vast, muddy slurry, or mud ocean, existed. In any case, it is clear that huge amounts of water once flowed over the surface of Mars. Yet the fate of that water remains unknown.

TRUST THE OLD SALTS

Although images of the landforms left by ancient glaciers, river valleys, lakes and seas are strong testament that Mars was once rich in water, evidence comes from other sources as well. Earth-based spectroscopic measurements of Mars reveal the presence of clay minerals. Even more directly, the two landers that set down on the surface during the Viking program analyzed Martian soil and found that it probably contains 10 to 20 percent salts. Martian rocks, like those on Earth, react to form salt and clay minerals when exposed to water. But such chemical weathering probably cannot occur under the cold and dry conditions that now reign on Mars.

Some scientists have also studied Martian rocks found here on Earth. These rare samples of the Martian surface were blasted into space by the impact of an asteroid or comet and later fell to Earth as meteorites. Allan H. Treiman of the Lunar and Planetary

Institute in Houston and James L. Gooding of the NASA Johnson Space Center have shown in the past several years that minerals in some of these so-called SNC meteorites were chemically altered by cool, salty water, whereas others were affected by warmer hydrothermal solutions. Their conclusions imply that Mars once had a relatively warm, wet climate and may have had hot springs. Just perhaps, conditions were right for life.

That possibility inspired David S. McKay of the NASA Johnson Space Center and his colleagues to examine an SNC meteorite for signs of ancient Martian life. Although their conclusion that fossil microbes are present is open to debate (and a vigorous one is indeed going on), the composition of the rock they studied—with fractures filled by minerals that probably precipitated from an aqueous solution—indicated that conditions on Mars a few billion years ago would have been compatible with the existence of life.

In agreement with this assessment, many atmospheric physicists had already concluded that Mars has lost immense quantities of water vapor to space over time. Their theoretical calculations are in good accord with measurements made by various Soviet space probes that showed oxygen and hydrogen atoms (derived from breakdown of atmospheric water exposed to sunlight) streaming away from Mars. The continuous loss of these elements implies that Mars must once have had all the water needed to fill an Oceanus Borealis.

But water was not the only substance lost. Recently David M. Kass and Yuk L. Yung of the California Institute of Technology examined the evolution of carbon dioxide—a potent greenhouse gas—in the atmosphere of Mars. They found that over time an enormous quantity of carbon dioxide has escaped to space. That amount of gaseous carbon dioxide would have constituted a thick Martian atmosphere with three times the pressure found at the surface of Earth. The greenhouse effect from that gas would have been sufficient to warm most of the surface of Mars above the freezing point of water. Thus, from this perspective, too, it seems quite plausible that the climate on Mars once was much warmer and wetter than it is today.

The constant loss of water and carbon dioxide from the atmosphere suggests that early epochs on Mars (that is, billions of years ago) may have been especially warm and wet. But some balmy periods may also have been relatively recent: Timothy D. Swindle of the University of Arizona and his colleagues studied minerals in an SNC meteorite created by aqueous alteration and determined that they formed 300 million years ago—a long time by human standards but only a few percent of the age of the 4.6-billion-year-old solar system. Their result was, however, accompanied by a considerable degree of uncertainty.

The duration of the wet periods on Mars is also difficult to gauge exactly. If the eroded Martian landscapes formed under conditions typical of terrestrial glacial environ-

Eskers are sinuous ridges made up of sand and gravel deposited by streams that formerly ran underneath a sheet of ice. Eskers appear to exist on the floor of Argyre basin on Mars, indicating that melting glaciers once covered the area.

ments, more than a few thousand but less than about a million years of warm, wet climate were required. Had these conditions endured substantially longer, erosion would have presumably erased all but traces of a few impact craters, just as it does on Earth.

This limitation does not apply to the earliest history of the planet, billions of years ago, before the craters now visible had formed. A young Mars may well have had vigorous erosion smoothing its face. But eventually, as the planet slipped toward middle age, its visage became cold, dry and pockmarked. Only scattered intervals of warmth have since rejuvenated the surface of the planet in certain regions. Yet the mechanism that causes Mars to switch between mild and frigid regimes remains largely mysterious.

Scientists can now venture only crude explanations for how these climate changes might have occurred.

TURNING ON THE HEAT

One hypothesis involves shifts in obliquity, the tilt of the spin axis from its ideal position, perpendicular to the orbital plane. Mars, like Earth, is now canted by about 24 degrees, and that tilt changes regularly over time. Jihad Touma and Jack L. Wisdom of the Massachusetts Institute of Technology discovered in 1993 that, for Mars, the tilt can also change abruptly. Excursions of the tilt axis through a range of as much as 60 degrees may recur sporadically every 10 million years or so. In addition, the orientation of the tilt axis and the shape of the orbit that Mars follows both change cyclically with time.

These celestial machinations, particularly the tendency of the spin axis to tilt far over, can cause seasonal temperature extremes. Even with a thin atmosphere such as the one that exists today, summer temperatures at middle and high Martian latitudes during periods with large obliquity could have climbed above freezing for weeks on end, and Martian winters would have been even harsher than they are currently.

But with sufficient summer warming of one pole, the atmosphere may have changed drastically. Releases of gas from the warmed polar cap, from seltzer groundwater or from carbon dioxide–rich permafrost may have thickened the atmosphere sufficiently to create a temporary greenhouse climate. Water could then have existed on the surface. Aqueous chemical reactions during such warm periods would in turn form salts and carbonate rocks. That process would slowly draw carbon dioxide from the atmosphere, thereby reducing the greenhouse effect. A return to moderate levels of obliquity might further cool the planet and precipitate dry-ice snow, thinning the atmosphere even more and returning Mars to its normal, frigid state.

By 300 million years ago on Earth, amphibians that evolved from fish had crawled out of the sea and inhabited swampy coastlines. Might other complex creatures have flourished simultaneously along Martian shores? The basic conditions for life may have existed for a million years late in Martian history—perhaps much longer during an earlier period. Were these intervals conducive for organisms to evolve into forms that could survive the dramatic changes in climate? Could Martian organisms still survive today in underground hot springs? The next decade of concentrated exploration may provide the definitive answers, which, if positive, would mark an intellectual leap as great as any in human history.

The Outer Planets

Mark Washburn

Jupiter and Saturn, known to the ancients, and Uranus, Neptune and Pluto—discovered in the 18th, 19th, and 20th centuries, respectively—are so unlike the familiar worlds of the inner solar system that if planets were animals, we would say that each belongs to a different species. Yet all nine planets share a common, if somewhat murky, origin, and their individual differences provide insights into topics as diverse as the structure of spiral galaxies and the origin of life.

Although they are collectively known as the gas giants, Jupiter, Saturn, Uranus and Neptune are not exactly a matched set. Each is a rapidly rotating sphere of gas surrounding a hot, dense core, but they differ from one another in matters as fundamental as size, temperature, composition, internal heat and magnetic fields. Indeed, each of them seems to serve as a fiendishly designed control experiment for the other three. Even their ring systems are individual and unique.

Jupiter, with a diameter of 143,000 km, is the largest planet and was likely the first of the outer worlds to begin forming in the gas and dust of the solar nebula, although the process was probably much the same for the others. Asteroid-sized planetesimals within the cloud formed dense clumps that served as nuclei for the planets-to-be. In the cold of the outer solar nebula, unlike the warmer inner region where Earth formed, both rock and ice were available for planet-building. When the accreting bodies reached a hypothetical limit of some 10 to 20 times the mass of the Earth, their gravitational pull drew in gas (almost all of it hydrogen and helium) from the surrounding nebula. As the gas accumulated, Jupiter eventually reached a total of 318 Earth masses, and Saturn, 95. But out in the suburbs of the solar nebula, where Uranus and Neptune formed, there was less material to work with, and time was running out. The solar nebula began to dissipate as the protosun turned on, and the planet-building era ended before Uranus and Neptune could achieve the huge dimensions of their sister worlds. Consequently, Uranus weighs in at 14.5 Earth masses and Neptune at 17.2.

The weight of all that infalling gas caused the interior pressure and temperature of the young planets to skyrocket. Gravitational energy was converted to kinetic energy as the gas giants slowly contracted, and their interiors became hot. Near Jupiter's cloudtops, at an atmospheric pressure of one bar, the temperature is just 165 degrees K, but 10,000 km beneath the cloudtops, the pressure is one million bars and the temperature is 6,000 degrees K, or about the same as on the surface of the Sun. But Jupiter and the other gas giants are far too small to generate the enormous pressure and temperature necessary to touch off nuclear fusion in their cores, as happens in stars.

Under the extreme conditions in gas giant interiors, familiar elements can take on strange and wonderful properties. Deep inside Jupiter and Saturn, beyond the one megabar level, hydrogen becomes ionized and behaves as if it were molten metal. This layer of electrically conducting liquid metallic hydrogen is probably responsible for generating the powerful magnetic fields of Jupiter and Saturn. Uranus and Neptune grabbed much less gas from the nebula and are only about 15 percent hydrogen by mass; they are mostly core material, with just one or two Earth masses of atmospheric gases. The pressure on the hydrogen does not exceed about 100,000 bars, so it remains in its normal molecular form.

Three of the four gas giants slowly dissipate their primordial heat by radiating it away into space. Uranus, for some reason, does not. Infrared studies of the heat flow from Jupiter, Saturn and Neptune have given us a window into their interiors and offer possible mechanisms to explain the structure and dynamics of their upper atmospheres. However, theories of internal pressure and temperature tend to be much more neat and tidy than the actual observed behavior of the gas giants' cloudtops.

Centuries of telescopic observation had led to a picture of a Jovian atmosphere composed of orderly bands of high, bright "zones" and dark, descending "belts." But the view from close up proved to be far more complicated. A week before Voyager 1's closest approach in 1979, the Imaging Team Leader Bradford A. Smith told the press, "the existing atmospheric circulation models have all been shot to hell by Voyager." Images from the spacecraft had revealed a churning cauldron of towering, multicolored clouds, where oppositely directed zonal jets blow at more than 100 meters per second. Energy from the jets seems to maintain long-lived atmospheric "storms," the largest of which is the famous Great Red Spot, which has been observed for more than 300 years.

Atmospheric circulation models are in somewhat better shape now, but they are far from definitive. According to some computer models, Jupiter's circulation may be simply a "surface" phenomenon, limited to a thin outer layer of the atmosphere. Other models suggest that the visible belts and zones may be the outer manifestation of huge, nested, rotating cylinders that extend straight through the planet.

Scientists have recently enjoyed two opportunities to get data from beneath the cloudtops. The first came unexpectedly in 1994, when fragments of Comet Shoemaker-Levy 9 crashed into Jupiter, throwing up plumes of material from the interior. The second was the entry of the Galileo probe in 1995; the spacecraft relayed data for nearly an hour before it was vaporized by the heat and pressure of the gas giant. Data from both events confirmed the basic "solar composition" model of Jupiter, although the Galileo probe seems to have entered an atypically dry region of the atmosphere—a sort of Jovian desert.

Saturn is smaller, less dense and somewhat better-behaved than Jupiter. Although Saturn's zonal jets whistle through the upper atmosphere at speeds of up to 500 meters per second, the sixth planet is more muted in appearance and harbors fewer of the stormlike atmospheric ovals that are so common on Jupiter. Still, Saturn is far from placid. Every thirty years (that is, once each Saturnian year), large atmospheric disturbances have been observed. The latest, in 1990, spread through 25 degrees of latitude before fading away after a few months.

Saturn used to be known as "the one with the rings," but since the 1970s, we have learned that the other gas giants also have their own unique ring systems. But Saturn's are still the biggest, brightest and, by far, the most beautiful. The Voyagers revealed that the broad, bright rings seen from Earth are actually composed of thousands of discrete "ringlets," whose structure seems to be controlled by the presence of small moons that gravitationally "shepherd" the icy ring particles into preferred orbits. The particles also seem to be influenced by spiral density waves, which apparently also control the structure of spiral galaxies and the frequency of traffic jams on the Santa Monica Freeway. Basically, ring particles, like SUV's, clump together and form knots, with the disturbances propagating across the ring system, creating other traffic jams.

Jupiter's ring, discovered by Voyager 1 in 1979, is narrow and dark and composed of particles averaging just 1 to 2 microns across. Particles so small are influenced by forces such as drag and electromagnetism, which should dissipate the ring in a thousand years or less. Either the Jovian ring is a very short-lived phenomenon, or it is being replenished constantly, possibly by material from Io's volcanoes and the small inner moon, Amalthea. The nine rings of Uranus, discovered by observations from the Kuiper Airborne Observatory in 1977, are also dark and narrow, and must also be replenished from some yet to be identified source. But the Uranian rings also resemble the Saturnian system in their sharply defined boundaries, implying the presence of unseen embedded moonlets that gravitationally control the ring structure.

Neptune's five dark rings were also discovered from Earth, but the initial observations were puzzling because the rings appeared to be incomplete "ring arcs." Theorists breathed a sigh of relief when Voyager 2 found that the rings are actually continuous, but the "arcs" are real, too. Ring material is being confined to preferred locations, creating dense, clumpy segments covering from one to 10 degrees of arc; small moons and "disrupted moons" that never fully formed are probably controlling the strange structure and providing ring material.

The big surprise in ring studies has been the realization that the lifetime of a typical ring particle is quite short. With a variety of forces working to erode and disrupt the rings, it seems doubtful that they can be primordial objects, formed at the same time as the planets themselves. The rings we see today are unlikely to be more than a few hundred million years old, a circumstance which raises a number of serious questions about where rings come from and how they evolve and survive. Theorists are still grappling with this problem.

In any comparison of the four gas giants, Uranus is usually the odd man out. It has a good excuse, however—in its youth, Uranus was soundly whacked by another body and knocked over on its side. The collision left Uranus tilted more than 90 degrees, with respect to the plane of the ecliptic. Consequently, during parts of its 84-year orbit around the Sun, one pole sees continuous sunshine while the other is left in darkness for 20 years at a time. This uneven input of solar energy should disrupt the thermal equilibrium of the Uranian atmosphere, but observations have been unable to detect any obvious mechanisms for thermal transport from poles to equator.

Visual observations, in fact, have revealed relatively little about the seventh planet. Even in images from Voyager 2's 1986 encounter, Uranus is a bland, bluish orb with few distinguishing characteristics. Although there is some evidence of horizontal banding, as on Jupiter, details of the Uranian atmosphere remain obscured by a high haze of methane clouds.

Neither Uranus nor Neptune is massive enough to possess a deep layer of liquid metallic hydrogen. Instead, beneath their hydrogen atmospheres, both worlds probably possess global oceans composed of water, methane and ammonia and other chemical components derived from them in the high temperature and pressure of the interior. Convection currents within these hot, briny seas may be responsible for generating the magnetic fields of each planet, which are not as strong as those of Jupiter and Saturn.

The magnetic fields of both inner gas giants are closely aligned with the planets' axes of rotation, as is the case on Earth. But the Uranian magnetic field is offset from the rotation axis by 58.6 degrees. It is tempting to believe that this offset is somehow related to the tilt of the entire planet, but Voyager 2 found that Neptune's field is also offset by nearly 50 degrees. Both fields are displaced from the centers of the planets. In consequence, the fields of both Uranus and Neptune wobble and wander all over the place, making for strange and complex interactions with the solar wind.

Neptune is a deeper blue than Uranus, probably due to differing depths of cloud layers. But unlike Uranus, Neptune revealed considerable atmospheric activity to the prying eyes of Voyager 2. Most notably, the spacecraft returned images of a huge, dark atmospheric oval that was immediately dubbed the Great Dark Spot (GDS). Unlike the Great

Red Spot on Jupiter, which is long-lived and remains at the same latitude, the GDS was drifting slowly toward the equator when first seen in 1989. Within a few years, it could no longer be seen in images from the Hubble Space Telescope, and it may well have dissipated entirely.

Surprisingly, considering Neptune's great distance from the Sun, the eighth planet boasts atmospheric zonal jets that are more energetic than those of Jupiter. Internal factors, rather than solar input, must be primarily responsible for supplying energy to maintain the flow. Yet Neptune seems to lack the small-scale turbulence that is so prevalent on Jupiter. That, in turn, again raises the fundamental question about the gas giants that have kept theorists scratching their heads: How deep does the flow go? Competing models have offered no firm answers—which is perhaps not surprising, considering that we still have much to learn about our own swirling atmosphere.

In contrast to the gaudy gas giant worlds, the ninth planet, Pluto, is tiny, solid, and yet to be visited by a spacecraft. We know so little about it that some observers question whether it even deserves to be called a planet. Discovered in 1930 by Clyde Tombaugh, working at the Lowell Observatory in Flagstaff, Arizona, Pluto is only about 2,300 km in diameter, making it slightly smaller than Neptune's moon, Triton, and barely half the size of our own Moon. Its mean distance from the Sun is about 6 billion kilometers and it takes 248 terrestrial years to complete one orbit. But Pluto's orbit is tilted to the plane of the ecliptic and so highly elliptical that it actually spends about 20 years inside the orbit of Neptune during its circuit around the Sun. The last such Plutonian summer began in 1979 and came to an end about 1999—it's Labor Day on Pluto.

In 1978, the astronomer James Christy discovered that Pluto has a companion on its lonely journey. The moon, dubbed Charon, is more than half the size of Pluto, making it the largest moon in the solar system relative to the size of its planet. The two bodies are in tidal lock; Pluto rotates and Charon revolves around it once every 6.4 days. Interestingly, Charon orbits in a north-south plane, relative to the ecliptic, so it is likely that the rotation axis of Pluto is tipped over to nearly the same extent as that of Uranus.

Pluto's surface is probably very similar to Triton's, composed of a mixture of water, methane and ammonia ices. However, carbon dioxide, which has been found on Triton, has not been observed on Pluto. Nevertheless, it is likely that Pluto and Triton shared a similar origin, somewhere in the cold and dark of the Kuiper belt. Both bodies may be aggregates of icy planetesimals, like those that formed the cores of the gas giant worlds.

Pluto and Charon, in fact, belong to a family of icy Kuiper Belt objects that share similar inclinations to the ecliptic and make two laps around the Sun for every three orbits by Neptune. These bodies are known as "Plutinos," and may consist of shards of an ancient collision that excavated Charon from the original proto-Pluto. About 50 Plutinos have been identified, and astronomers are searching for more.

Is Pluto a planet at all? Its similarity to Triton and the Kuiper belt objects make it suspect in some minds, and it is certainly a maverick compared with the other eight worlds of our solar system. But *every* planet is unique, one way or another, and after 70 years of recognizing Pluto as a planet, it would be churlish of us to kick it out of the club at this late date. Perhaps a final determination must await a closer look—which could be provided by the Pluto Express mission, which is in the planning stages. This fast flyby mission would take about 16 years just to reach Pluto, so there is no need for a rush to rewrite textbooks. For now, at least, Pluto is a planet.

The Outer Moons

Mark Washburn

"It shouldn't look like this at all!"

The bemused and amazed protest of the geologist Laurence Soderblom as he got his first good look at the surface of Jupiter's moon Io could well serve as a motto for the exploration of all the moons of the outer solar system. Voyager 1's close flyby of Io on March 4, 1979, provided a shocking but entirely appropriate introduction to the bizarre gallery of satellites orbiting the four gas giant worlds—Jupiter, Saturn, Uranus and Neptune. By the time Voyager 2 sailed past Neptune's moon Triton in 1989, scientists had seen some forty of these small bodies, an experience that revolutionized their understanding of the outer solar system.

Before the first Voyager encounters, little was known about the moons of the giant planets. Only the four large Galilean satellites of Jupiter—Io, Europa, Ganymede and Callisto—and Saturn's cloud-shrouded moon, Titan, had been studied in any detail, but even these planet-sized bodies retained their air of mystery. The dozens of smaller moons could scarcely be studied at all, appearing as little more than pinpoints of light in even the best telescopic views.

Scientists generally expected that the outer moons would turn out to be rather dull, frozen, crater-pocked bodies of rock and ice. It was believed that in the cold of the outer solar system, such small bodies could not retain enough energy to drive an active geology. Their surfaces should have been ancient, virtually unchanged in the four and a half billion years since their formation. Instead, Soderblom and the other Voyager scientists were stunned to find that their spacecraft's first port of call, Io, boasted the *youngest* surface and the most active geology of any body in the solar system.

Even before the Voyager encounter, there had been hints that something strange was going on at Io. Astronomers had seen evidence of a large toroidal (doughnut-shaped) cloud of plasma in Io's orbital path around Jupiter, and there were signs of energetic electromagnetic links between the moon and its parent world. And just weeks before Voyager 1's arrival, a remarkably well-timed paper by Stanton Peale, Patrick Cassen and Raymond Reynolds proposed that Io, caught in a gravitational tug-of-war between Jupiter and the other Galilean satellites, was being internally flexed by the tidal interactions, possibly producing enough heat to drive volcanism. Still, Voyager's discovery of no less than eight active volcanoes on Io's surface was a profound shock.

Io's geology is not merely active, but hyperactive. Infrared measurements of the moon's heat flow show that it is releasing an average of about 2.5 watts of energy for every square meter of surface area—compared with a paltry 0.06 watts per square meter for Earth. Io possesses more than 200 volcanic calderas larger than 20 km in diameter (Earth has just 15 calderas that large) and lakes of molten lava as hot as 1,800 degrees K. The lava itself seems to be mainly silicate material from melted rocks (as on Earth), but sulfur also plays a major role on Io. The sulfur eruptions, which may be more similar to geysers than volcanoes, inject material into the plasma torus and are responsible for painting the moon's surface with a frost of sulfur dioxide. They also maintain Io's tenuous sulfur dioxide atmosphere. The total absence of impact craters on Io suggests that the moon is being resurfaced at a rate averaging at least a millimeter per year, and perhaps as much as a centimeter per year. In just four months, between the encounters of Voyager 1 and Voyager 2, distinct changes were observed in some areas of the surface, and many more were catalogued when Galileo went into orbit around Jupiter in 1995.

If Io defied scientists' expectations of cold, dead moons, Callisto, at first glance, seemed to confirm them. As seen by Voyager's cameras, Callisto's surface appeared to be saturated with impact craters. One huge, ringed structure, dubbed Valhalla, seemed to be the site of a colossal ancient impact whose ripples created a bulls-eye pattern covering an entire hemisphere of the moon. However, the keener gaze of Galileo revealed a startling absence of small impact craters. On presumably dead Callisto, some process was reworking the surface material.

Callisto can best be understood by comparing it with its near-twin, Ganymede. The two moons are both about the size of the planet Mercury, and have densities compatible with a composition of 60 percent rock and 40 percent ice. But gravity data acquired by measuring minute Doppler shifts in Galileo's radio signal as it flew past the two moons indicated that they have very different internal structures. Ganymede, 1,070,000 km from Jupiter, formed in a warmer environment than Callisto (1,833,000 km) and may, at one time, have been subject to tidal interactions similar to (but much weaker than) those afflicting Io. In addition, Ganymede may have received a larger inventory of heat-producing radioactive elements than its sister moon. As a result, Ganymede's interior was sufficiently hot at some point for it to undergo differentiation; dense, rocky material sank to form an iron core while ice melted, rose to the surface and refroze. But Callisto's interior was never hot enough to sustain differentiation, and the moon is simply an ice-rock mix all the way through, with a thin, ice-rich outer layer.

Ice is what makes outer solar system moons so fundamentally different from the rocky bodies of the inner solar system. Familiar processes such as impact cratering, volcanism and erosion take on a new twist in the icy environment beyond the asteroid belt. As one Voyager scientist put it, "When you take a sledgehammer to a car windshield, it doesn't melt." But ice-worlds do melt when hit hard enough, and ice can pinch-hit for rock in a number of geological processes. The paucity of small impact craters on Callisto is probably attributable to the high ice content of the surface layer. High-resolution Galileo images show bright, eroded crater rims; the exposed ice may simply sublime into space over time, causing darker surface material to fill in the ancient impact sites. The process may still be active today.

Ganymede's greater reservoir of internal heat transformed that moon's surface in much more dramatic ways. Ganymede is a patchwork satellite, where dark, heavily cratered ancient terrain intersects with and is overlapped by brighter, younger, "grooved" terrain. As differentiation proceeded in the interior, melted ice flooded the surface, refroze and expanded. The expansion produced extensional faulting and fresh ice rose to fill in the resulting cracks. Under pressure from below, water or warm, mushy ice may have erupted onto the surface in a display of cryovolcanism. Most of this activity probably occurred some four billion years ago, before Ganymede's interior cooled; however, Galileo data offers intriguing hints that Ganymede may have been active much more recently.

Galileo discovered that Ganymede possesses an intrinsic magnetic field, implying that it has an iron core that is still at least partially molten. Callisto has no *intrinsic* magnetic field produced by a core dynamo, but it does appear to have a slight *induced* magnetic field, which may be created by electrical currents flowing through a global, subsurface layer of salt water some 10 km thick and 100 km deep. Ganymede may also possess such a layer of brine, although the intrinsic magnetic field would make it harder to detect. The possible presence of liquid water within these two moons raises some very interesting questions—as it does for their sister moon, Europa.

The smallest of the Galilean satellites, Europa, was not seen from close range by the Voyagers, but images did reveal a bright, icy cueball of a world, with little surface relief and few obvious impact craters. The surface was crisscrossed by strange triple ridge systems extending for hundreds of kilometers. Like Io, Europa is subject to internal flexing because of tidal interactions with Jupiter and the other large moons. Thus, Europa probably underwent differentiation, resulting in a suspected iron-rich core, a silicate mantle and an outer layer of water ice some 150 km thick. From Voyager images, it was not possible to determine whether Europa's icy surface was ancient or had been resurfaced through ongoing processes such as cryovolcanism.

Early on, scientists began to wonder if there might be a global ocean of liquid water lurking beneath that bright, icy crust. The speculation was intriguing because it implied conditions that might support the existence of life on, or within, Europa. Comet impacts should have delivered the necessary organic matter to Europa, and deep thermal vents, such as those found on terrestrial oceanic ridges, could supply the energy needed to maintain some form of biological activity.

Galileo provided a much better view of Europa, imaging the entire world to a resolution of 2 km or less, with details as small as a few meters visible in some areas. Much of the surface consists of bright, icy plains, with large patches of mottled terrain where darker material has intruded. Some form of tectonic activity seems to have disrupted the mottled terrain, moving and twisting immense blocks of ice. Numerous models have been proposed to account for the surface deformation and the morphology of the triple ridges, but no clear favorite has yet emerged. As on Ganymede, cryovolcanism may play an important role.

A better understanding of Europa may have to wait for future spacecraft that are equipped to peer beneath the crustal ice. A Europa orbiter with a radar system could gauge the depth of the ice and determine whether the suspected ocean of liquid water actually exists. If it does, we might eventually send landers to melt their way through the ice and release robotic submarines to cruise through the hidden waters in search of extraterrestrial life; as matters now stand, Europa may be the most likely place in the solar system to find it.

If not Europa, then perhaps Titan. Speculations about conditions on Saturn's large moon (which is slightly smaller than Ganymede, the solar system's largest satellite) have been rife ever since the 1940s, when Gerard P. Kuiper discovered evidence of a methane atmosphere. Titan is the only moon with a significant atmosphere, and the presence of methane implied that a complex chemical cycle was at work. If Titan's atmosphere was dense enough, a greenhouse effect could provide Earthlike temperatures at the surface, making even more complex chemistry possible—perhaps even biochemistry of some sort.

When Voyager 1 passed within 4000 km of Titan in the fall of 1980, a flood of new data resolved some questions about Titan, but raised many more. Methane turned out to be but one of many complex hydrocarbons in the Titanian atmosphere, but its principal constituent is nitrogen. With a surface pressure of about 1.5 bars, greenhouse heating is not much of a factor, and the surface temperature is a chilly 94 degrees K. Terrestrial-style biology seemed to be ruled out, but "prebiological" processes may well be occurring on Titan.

It is difficult to tell exactly what is happening on Titan because the surface is totally obscured by a dense, photochemical smog. Voyager images revealed only a bland, orange fuzzball, but the color itself is intriguing. The smog is actually a haze of aerosols produced by the condensation of hydrocarbon molecules, but those particles are normally white or gray in color. The orange tint may be caused by the creation of ever-bigger long-chain polymers such as hydrogen cyanide and acetylene.

As they grow larger, the condensates inevitably fall toward the surface, possibly coating it with deep drifts of hydrocarbon snow. But the surface temperature is just right for ethane to arrive there in liquid form, and there should be enough of it to submerge the entire surface in an ethane sea several kilometers deep. However, radar and infrared studies indicate that continent-sized structures exist on Titan's surface, so there can be no more than large lakes or seas of ethane. What happens to the excess ethane is a mystery that will only be solved when we gain a better understanding of the complex chemical cycles at work beneath the clouds.

That day may arrive in 2004, when NASA's Cassini spacecraft delivers the European Space Agency's Huygens probe to Titan. During the probe's three-hour parachute descent to the surface, it will make direct measurements of the atmospheric gases and aerosols. A camera should give us a clear view of Titan's ethane seas and hydrocarbon snowdrifts—and whatever new surprises Titan may have in store.

Titan is Saturn's only large satellite, but unlike Jupiter—whose non-Galilean satellites are mostly tiny lumps of rock that are probably captured asteroids—Saturn is home to an array of mid-size satellites, ranging from 400 to 1500 km in diameter. These bodies, composed mainly of water ice, should have fit the "dull, cold, dead, crater-pocked" stereotype of pre-Voyager expectations. But Voyager showed that each of these moons has had its own complex history, and no two are exactly alike.

Tiny Mimas (398 km diameter) was seen in Voyager images to have been the site of a mammoth impact event that must have come close to shattering the small world. Too small to have generated significant internal heat, Mimas preserves on its surface the record of four billion years worth of outer-solar-system violence. Yet Enceladus, just 100 km larger, stunned Voyager scientists with its broad swathes of relatively crater-free terrain broken by ridges and faults. Bright, fresh ice has somehow paved over much of the ancient cratering record. Tethys and Dione, twice as large as Enceladus, also display regions of younger terrain and evidence of geological activity.

The fresh ice on Enceladus is so completely free of craters that it seems certain that the resurfacing process is still at work today. Gravitational resonances with other moons probably provide the energy to drive this process through tidal heating similar to that seen on Europa. But Tethys and Dione have no strong resonances, so other factors must be invoked to explain the signs of past activity on those moons. The likeliest candidate is ice volcanism. If ice contains a small amount of ammonia, it melts not at 273 degrees K, but at just 176 degrees K. Thus, even on these cold Saturnian satellites, ice-ammo-

Jupiter's satellite Io, the most volcanically active body yet found in the solar system. Impact craters are absent because volcanism continually resurfaces the moon.

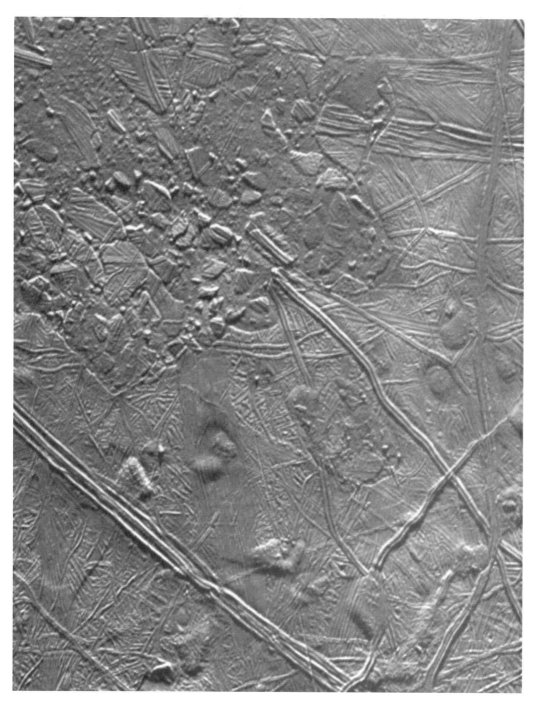

This view of Jupiter's moon Europa shows a portion of its icy surface that has been highly disrupted by fractures and ridges. Although some impact craters are visible, their general absence indicates a youthful surface.

nia mixtures could percolate upward and spread across the surface like a flood of molasses. Voyager was unable to identify ammonia, but Cassini should be able to verify its presence and map deposits of ammonia-ice on the surfaces of these small moons.

Cassini will also pay particular attention to Saturn's strange, two-toned moon, Iapetus. The leading hemisphere of Iapetus, as it orbits Saturn, is as dark as coal; its trailing hemisphere is as bright as snow. Scientists suspect that Iapetus is sweeping up dust from the small, retrograde moon, Phoebe, which is preferentially deposited on Iapetus's leading hemisphere. However, the dark material on Iapetus does not spectroscopically match the composition of Phoebe, so something more complicated may be happening.

Uranus, like Saturn, boasts a suite of icy, mid-sized moons, where many of the same processes seem to have been at work. On average, the Uranian satellites are darker and somewhat rockier than the Saturnian moons. Titania and Ariel, in particular, show evidence of large-scale geological activity and resurfacing of ancient terrain. Ariel has the brightest surface of any of the Uranian satellites, evidence of extensive ice volcanism. Ariel's energy source may have been a transitory tidal coupling with Umbriel.

During its 1986 encounter with Uranus, Voyager 2 saw only one moon from close range, but it was a shocker. Five-hundred-km Miranda displayed a crazy quilt of bizarre terrains such as had never been seen before anywhere in the solar system. Huge, trapezoidal features (labeled coronae) dominated the surface, looking like misfit pieces in some gigantic jigsaw puzzle. To some Voyager scientists, notably the late Eugene Shoemaker, it looked as if Miranda had been completely shattered—perhaps several times—by ancient impacts. The pieces finally reassembled in an apparently random mix, resulting in such uniquely Mirandan features as a sheer cliff face, 20 km high.

Subsequent analysis led other scientists to a less apocalyptic scenario. The coronae, cliffs, and other odd features, they proposed, were the result of extensional faulting, combined with episodes of cryovolcanism. But tiny Miranda's internal heat apparently depended on a short-lived resonance with the moons Umbriel and Ariel. When orbits shifted and the resonance went away, so did Miranda's energy source. Miranda froze in mid-transformation, leaving it a geological mess.

Neptune has just one large moon, Triton, which Voyager 2 determined to have a diameter of 2710 km, making it slightly larger than the planet Pluto. In fact, Triton may be a twin to Pluto, and the 1989 Voyager encounter may have given us a sneak preview of conditions on the ninth planet. Triton's retrograde orbit strongly implies that it formed elsewhere in the solar nebula—perhaps in the Kuiper belt—and was captured by Neptune. Triton shows relatively few impact craters, making its surface one of the youngest in the outer solar system. The gravitational energy unleashed by the capture of Triton probably melted its interior and completely resurfaced the body. Its most striking feature is terrain that bears an unmistakable resemblance to the skin of a cantaloupe.

Methane, nitrogen, carbon monoxide, carbon dioxide and water all exist as ice on the surface of Triton, where the mean temperature is a frosty 38 degrees K. But Triton, due to its odd orbit, experiences extreme seasonal variations, causing a slow migration of the methane, CO and nitrogen ices, while the water ice serves as solid bedrock.

On this icy body, even a tiny change in temperature can produce some spectacular consequences. An increase of just 2 degrees K could be enough to cause nitrogen ice to sublime, leading to a buildup of gas pressure beneath the surface—enough pressure to cause eruptions to burst through the crust and jet several kilometers into the wispy nitrogen atmosphere. That, at least, is the favored explanation for Voyager 2's stunning observation of dark plumes erupting from Triton's surface. These nitrogen geysers reach an altitude of 8 km before trailing off in the prevailing atmospheric currents. Downwind, they spread dark streaks extending a hundred or so km across the icy terrain.

Before the epic journey of the Voyager spacecraft, the notion of nitrogen geysers, sulfur volcanoes, and global oceans would have seemed fantastic. But the outer moons, in their variety and complexity, have led scientists to a new level of understanding about the basic nature of our solar system. And they turned out to be anything but dull.

What's New in the Solar System

Peter Jedicke

From the cosmic perspective, the solar system is our own back yard. The thrust to explore this back yard, and to understand what is going on there, has led to a wide variety of scientific and technical advances. These, in turn, rely on all the theoretical and technological progress made by the past generation of researchers and engineers. In particular, space probes have made possible close up and even *in situ* investigations of the bodies in the solar system, and powerful radar systems have enabled observations of the nearer bodies directly from Earth.

The availability of detailed photographs has been of enormous importance in planetary science. It is now known that craters are the dominant surface features of many planets, moons and even asteroids. The impacts that formed the craters we see today were the final stages in the origin of the solar system. Although the worst of the impact episodes took place within the first billion years after the Sun became a star, the process is by no means ended. At a slow, but cosmically significant, pace, the solar system is still "under construction."

Scientists once viewed collisions among solar system bodies and volcanic processes within those bodies as competing theories to explain the craters and other surface features, but now it is recognized that both mechanisms have played key roles in the development of the solar system. Some of the heat required for the volcanic processes was provided by the kinetic energy of impacting objects. The relative significance of this source of energy and other sources is nowhere more vivid than on the surface of Jupiter's moon Io, where an active volcanic surface is maintained by the energy of tidal forces. Solid outer crusts of planets like Earth and Venus still betray a relationship with the heat within; on Earth, vast tectonic plates ride on softer, warmer materials below, while the radar images taken by the Magellan spacecraft around Venus showed numerous ridges and cracks that formed because the surface crinkled as it cooled.

The crash of Comet Shoemaker-Levy 9 on Jupiter in the summer of 1994 was among the most significant natural events ever observed by scientists, and it tied together a number of themes in planetary science. In the clouds of Jupiter, the fragments of Shoemaker-Levy 9 left dark gray blemishes large enough to be seen by small telescopes on Earth, and experts saw sulfur compounds and ammonia in Jupiter's atmosphere. Auroral activity on Jupiter indicated massive disruptions in Jupiter's magnetic field that were caused by the impact, and the gradual fading of the impact patterns demonstrated the effects of high-altitude winds in Jupiter's atmosphere. But the data were not quite sufficient to resolve the debate over whether the comet's fragments were kilometers across or merely hundreds of meters.

The significance of collisions was also emphasized by the growing consensus that an impact 65 million years ago in the Yucatan peninsula was a major factor in the extinction of the dinosaurs, and that earlier impacts were probably responsible for previous extinction events in the history of life on Earth. The theory that comets brought a significant fraction of Earth's water supply from space was in serious doubt even though water was among the molecules that were identified in comets such as Shoemaker-Levy 9, C/1995O1 (Hale-Bopp) and C/1996B2 (Hyakutake), because the ratio of deuterium to normal hydrogen in the water molecules is very different from the same ratio in the oceans of the Earth. In a related development, microscopic water droplets were also found bound to salt crystals in a meteorite that fell in Texas in 1998. At the very least, it seems that water is not uncommon in the solar system.

Water is the subject of much scrutiny in planetary science. The space probes Clementine and Lunar Prospector found strong evidence that there are reservoirs of ice in shadowed craters near the poles of the Moon. Surface features on Mars, even at the local scale seen by the Pathfinder lander, look like they were formed by running water during catastrophic floods, although there has been no hint of where all that water might be now. But perhaps the most fascinating prospect is that there are almost certainly watery oceans beneath the ice that forms the surface of Jupiter's moon Europa. Tidal heating may provide enough energy in those oceans to foster the kind of chemical reactions that are the basis of simple life, and the discovery of tenacious life forms in deep sea vents on Earth raises interest in what could be happening on Europa.

Meanwhile, the presence of polyaromatic hydrocarbon chemicals, as well as microscopic features that just might be fossilized bacteria, in a meteorite that fell to the Earth thousands of years ago after being kicked off the surface of Mars millions of years before has given a new currency to the idea that something like life previously existed on Mars. Indeed, there has been renewed discussion of what the definition of life is in the first place. Much has been learned about the surface and atmospheric conditions on Mars. Pathfinder, which was renamed Sagan Memorial Station after the late Carl Sagan, reported a warmest temperature of only -10 degrees Celsius and a coldest temperature of -76 degrees. Dust devils and thin clouds made of water ice have also been observed there. Pictures of the pinkish rocks show effects that were likely created by liquid water a long time ago, and by wind erosion over the past few billion years. The robot rover Sojourner poked around on the surface of Mars for more than 12 weeks. It has been suggested that Mars's gravity and its orbit are more friendly to meteorites,

The asteroid designated as 243 Ida is now known to have a tiny orbiting companion, named Dactyl. This image was taken by the Galileo spacecraft.

and that there may be significant numbers of them among the loose surface gravel, waiting to be identified and analysed.

Controversy continues over the classification of small bodies in the solar system, most notably the status of Pluto. The discovery of asteroids beyond the orbit of Neptune, beginning with 2060 Chiron, has accelerated and there are now almost 200 such objects known. They are collectively called Edgeworth-Kuiper Objects (EKOs) after the two scientists who hypothesized their existence, or simply Trans-Neptunian Objects (TNOs). The question is whether Pluto deserves to be among the EKOs or deserves to be considered the ninth planet. On another level, scientists wonder whether asteroids and comets are really so different, after all. Asteroid 7968 Elst Pizarro may be a comet instead; 3200 Phaethon is almost certainly a dormant comet; and asteroid 4015 is the same object as Comet 107P/Wilson-Harrington. Perhaps objects should be considered to have more or less rocky material and more or less icy material without being distinguished simply as asteroids or comets.

Images taken by the Galileo spacecraft highlighted the irregular shapes of asteroids 951 Gaspra and 243 Ida, and also revealed that 243 Ida has a tiny companion, now named Dactyl, as it moves in its orbit. The NEAR spacecraft added 253 Mathilde and 433 Eros to the list of targets, and other asteroids have been studied by radar beamed out from Earth. The potato-like shapes have provided a challenge for geographers who make maps and models of such objects. Peanut-shaped asteroids such as 4769 Castalia and 1999 JM8 hint that the distinction between asteroids with odd shapes and asteroids with companions may also be blurred, and such objects are now called contact binaries.

Radar and studies of the variation of reflected light have shown that larger asteroids and even most asteroids smaller than 30 kilometers across rotate slowly. But a handful of smaller asteroids have been observed to turn around in less than four hours, and asteroid 1998 KY26 spins in about 12 minutes. Therefore 1998 KY26 must be a single solid piece of rock, not a "rock pile." Also, it is thought that these fast rotating objects were involved in collisions, because it is easy to understand how a glancing collision would make a small object spin so quickly. The early history of all asteroids, which are divided into families according to common orbital characteristics, is now thought to have involved a break-up of at least one primordial body. The breakup may have been initiated by tidal forces from Jupiter. After the breakup, a long period of collisions, and the continuous action of tidal forces from Jupiter and the other planets, rearranged the asteroid objects in their orbits. In fact, the incredible power of modern computers has made it possible for theoreticians to develop highly sophisticated models of the interactions of gravitational forces among the planets and asteroids. From these models, an improved understanding of the stability of orbits in the solar system has been cultivated.

There have been many successful programs to discover new objects in the inner solar system, as well. Following the success of the Spacewatch project, similar search schemes have been implemented, such as the Catalina Sky Survey, the Lowell Observatory's Near-Earth Object Search (LONEOS), and most notably LINEAR, the Lincoln Laboratory's near-Earth asteroid search, which discovered 173 near-Earth asteroids in just over a year after becoming fully operational. The known population of objects whose orbits lie just outside the Earth's orbit (Amor objects), or cross the Earth's orbit (Apollo objects), or lie entirely inside the Earth's orbit (Aten objects), has grown dramatically. Statistical analysis has hinted that there may be as many as one thousand objects larger than one kilometer in size that may collide with Earth, but only a fraction of these have actually been seen and plotted. This has brought heightened awareness of and even concern over the risk of such an object being on a collision course with Earth—so much so that there has been public alarm in dramatic cases such as 1997 XF11 and 1999 AN10, which of course proved to be no threat at all. Future search programs hope to discover at least 90 percent of the potentially hazardous objects.

Chapter Seven:
Life on Earth . . . and Elsewhere

Introduction

If we're alone in our galaxy, thought Carl Sagan, it's an awfully big waste of space, and if the Nobel-prize winning scientist George Wald is right, the Universe is designed for life. Three of his lines of evidence are: ice floats, the night sky is dark, and the proton and electron are precisely opposite.

What if ice didn't float? As water cooled below freezing, the top layer of ice would sink, allowing the next layer to freeze more quickly and sink. This layering effect would continue until all the water had frozen solid. And, as Earth's early ponds froze solid, life would have had a hard time making it past the first winter.

The darkness of the night sky is something we take for granted. But two centuries ago, a German physician named Heinrich Olbers looked up at the sky and wondered why it was dark. With stars and galaxies increasing exponentially as we go farther out into space, there should be so much radiation that the whole sky should be as bright as the Sun. If the sky were that bright, the amount of radiation pouring onto every planet would make life impossible anywhere. The answer to this paradox—the light is there yet it is not there—was learned not very long ago and has to do with conditions that were set at the very start of the universe.

If the universe were hundreds of billions of years old, we might very well see all the energy that ever existed within it—radiation from all the stars, galaxies and superclusters of galaxies might indeed blind us out. But the universe is not nearly as old as it is big; and light from its most distant objects has not reached us yet. Also, since the universe has been expanding since its genesis, radiation from distant objects is weakened, making it less intense.

What about protons and electrons? If their charges were not precisely equal, the electromagnetic force, and not gravity, would have determined the large-scale structure of the universe. The universe would consist of vast clouds with lots of lightning. There would be no galaxies, no stars, no planets and no life.

The question of life is clearly one that involves more than biology. It concerns the basic structure of the atom, the chemical behavior of water as it cools to freezing, and the physics of vast distances. The articles in this chapter examine this question from several angles, in particular the classic articles "The Structure of Hereditary Material" by Francis Crick, codiscoverer of the structure of DNA, and "The Search for Extraterrestrial Intelligence" by Carl Sagan and Frank Drake. One looks at the quesion from biology, the other from the point of view of likelihood. The subject is utterly fascinating, and complicated, and yet the question is so simple as we look up at the stars at night: are we alone?

The Structure of Hereditary Material

Francis H. C. Crick

Viewed under a microscope, the process of mitosis, by which one cell divides and becomes two, is one of the most fascinating spectacles in the whole of biology. No one who watches the event unfold in speeded-up motion pictures can fail to be excited and awed. As a demonstration of the powers of dynamic organization possessed by living matter, the act of division is impressive enough, but even more stirring is the appearance of two identical sets of chromosomes where only one existed before. Here lies biology's greatest challenge: How are these fundamental bodies duplicated?

One approach is the study of the nature and behavior of whole living cells; another is the investigation of substances extracted from them. This essay discusses only the second approach, but both are indispensable if we are ever to solve the problem; indeed, some of the most exciting results are being obtained by what might loosely be described as a combination of the two methods.

Chromosomes consist mainly of three kinds of chemical: protein, deoxyribonucleic acid (DNA) and ribonucleic acid (RNA). (Since RNA is only a minor component, we shall not consider it in detail here.) The nucleic acids and the proteins have several features in common. They are all giant molecules, and each type has the general structure of a main backbone with side groups attached. The proteins have about 20 different kinds of side groups; the nucleic acids usually only four (and of a different type). The smallness of these numbers itself is striking, for there is no obvious chemical reason why many more types of side groups should not occur. Another interesting feature is that no protein or nucleic acid occurs in more than one optical form; there is never an optical isomer, or mirror-image molecule. This shows that the shape of the molecules must be important.

These generalizations (with minor exceptions) hold over the entire range of living organisms, from viruses and bacteria to plants and animals. The impression is inescapable that we are dealing with a very basic aspect of living matter, and one having far more simplicity than we would have dared to hope. It encourages us to look for simple explanations for the formation of these giant molecules.

The most important role of proteins is that of the enzymes—the machine tools of the living cell. An enzyme is specific, often highly specific, for the reaction which it catalyzes. Moreover, chemical and x-ray studies suggest that the structure of each enzyme

257

is itself rigidly determined. The side groups of a given enzyme are probably arranged in a fixed order along the polypeptide backbone.

We believe that this order is controlled by the chromosomes. DNA is found in all chromosomes—and only in the chromosomes (with minor exceptions). The amount of DNA per chromosome set is in many cases a fixed quantity for a given species. The sperm, having half the chromosomes of the normal cell, has about half the amount of DNA, and tetraploid cells in the liver, having twice the normal chromosome complement, seem to have twice the amount of DNA. This constancy of the amount of DNA is what one might expect if it is truly the material that determines the hereditary pattern.

DNA can be extracted from cells by mild chemical methods, and much experimental work has been carried out to discover its chemical nature. This work has been conspicuously successful. It is now known that DNA consists of a very long chain made up of alternate sugar and phosphate groups. The sugar is always the same sugar, known as deoxyribose. And it is always joined onto the phosphate in the same way, so that the long chain is perfectly regular, repeating the same phosphate-sugar sequence over and over again.

But while the phosphate-sugar chain is perfectly regular, the molecule as a whole is not, because each sugar has a "base" attached to it and the base is not always the same. Four different types of base are commonly found: two of them are purines, called adenine and guanine, and two are pyrimidines, known as thymine and cytosine. So far as is known the order in which they follow one another along the chain is irregular and probably varies from one piece of DNA to another. In fact, we suspect that the order of the bases is what confers specificity on a given DNA. Because the sequence of the bases is not known, one can only say that the *general* formula for DNA is established. Nevertheless, this formula should be reckoned one of the major achievements of biochemistry, and it is the foundation for all the ideas described in the rest of this essay.

Although we know from the chemical formula of DNA that it is a chain, this does not in itself tell us the shape of the molecule, for the chain, having many single bonds around which it may rotate, might coil up in all sorts of shapes. However, we know from physical-chemical measurements and electron microscope pictures that the molecule usually is long, thin and fairly straight, rather like a stiff bit of cord. It is only about 20 angstroms thick (one angstrom equals one 100 millionth of a centimeter). This is very small indeed, in fact not much more than a dozen atoms thick. The length of the DNA seems to depend somewhat on the method of preparation. A good sample may reach a length of 30,000 angstroms, so that the structure is more than 1,000 times as long as it is thick. The length inside the cell may be much greater than this, because there is always the chance that the extraction process may break it up somewhat.

This floor-to-ceiling model of the DNA double helix molecule was built for an exhibit on the "Origin and Structure of Life" at the New York Museum of Natural History in 1964.

None of these methods tells us anything about the detailed arrangement in space of the atoms inside the molecule. For this, it is necessary to use x-ray diffraction. The average distance between bonded atoms in an organic molecule is about $1^{1}/_{2}$ angstroms; between unbonded atoms, three to four angstroms. X rays have a small enough wavelength (1 1/2 angstroms) to resolve the atoms, but unfortunately, an x-ray diffraction photograph is not a picture in the ordinary sense of the word. We cannot focus x rays as we can ordinary light; hence, a picture can be obtained only by roundabout methods. Moreover, it can show clearly only the periodic, or regularly repeated, parts of the structure.

With patience and skill, several English workers obtained good diffraction pictures of DNA extracted from cells and drawn into long fibers. The first studies, even before details emerged, produced two surprises. First, they revealed that the DNA structure could take two forms. In relatively low humidity, when the water content of the fibers was about 40 percent, the DNA molecules gave a crystalline pattern, showing that they were aligned regularly in all three dimensions. When the humidity was raised and the fibers took up more water, they increased in length by about 30 percent, and the pattern tended to become "paracrystalline," which means that the molecules were packed side by side in a less regular manner, as if the long molecules could slide over one another somewhat. The second surprising result was that DNA from different species appeared to give identical x-ray patterns, despite the fact that the amounts of the four bases present varied. This was particularly odd because of the existence of the crystalline form just mentioned. How could the structure appear so regular when the bases varied? It seemed that the broad arrangement of the molecule must be independent of the exact sequence of the bases, and it was therefore thought that the bases play no part in holding the structure together. As we shall see, this turned out to be wrong.

The early x-ray pictures showed a third intriguing fact: namely, the repeats in the crystallographic pattern came at much longer intervals than the chemical repeat units in the molecule. The distance from one phosphate to the next cannot be more than about seven angstroms, yet the crystallographic repeat came at intervals of 28 angstroms in the crystalline form and 34 angstroms in the paracrystalline form; that is, the chemical unit repeated several times before the structure repeated crystallographically.

J. D. Watson and I, working in the Medical Research Council Unit at the Cavendish Laboratory in Cambridge, were convinced that we could get somewhere near the DNA structure by building scale models based on the x-ray patterns obtained by M. H. F. Wilkins, Rosalind Franklin and their coworkers at King's College, London. A great deal is known about the exact distances between bonded atoms in molecules, about the angles between the bonds and about the size of atoms—the so-called van der Waals distance between adjacent nonbonded atoms. This information is easy to embody in scale models. The problem is rather like a three-dimensional jigsaw puzzle with curious pieces joined together by rotatable joints (single bonds between atoms).

To get anywhere at all, we had to make some assumptions. The most important one had to do with the fact that the crystallographic repeat did not coincide with the repetition of chemical units in the chain but came at much longer intervals. A possible explanation was that all the links in the chain were the same, but the x rays were seeing every tenth link, say, from the same angle and the others from different angles. What sort of chain might produce this pattern? The answer was easy: the chain might be coiled in a helix. (A helix is often loosely called a spiral; the distinction is that a helix winds not around a cone but around a cylinder, as a winding staircase usually does.) The distance between crystallographic repeats would then correspond to the distance in the chain between one turn of the helix and the next.

We had some difficulty at first because we ignored the bases and tried to work only with the phosphate-sugar backbone. Eventually we realized that we had to take the bases into account, and this led us quickly to a structure which we now believe to be correct in its broad outlines.

This particular model contains a pair of DNA chains wound around a common axis. The two chains are linked together by their bases. A base on one chain is joined by very weak bonds to a base at the same level on the other chain, and all the bases are paired off in this way right along the structure. Paradoxically, in order to make the structure as symmetric as possible, we had to have the two chains run in opposite directions; that is, the sequence of the atoms goes one way in one chain and the opposite way in the other. Thus, the figure looks exactly the same whichever end is turned up.

Now we found that we could not arrange the bases any way we pleased; the four bases would fit into the structure only in certain pairs. In any pair there must always be one big one (purine) and one little one (pyrimidine). A pair of pyrimidines is too short to bridge the gap between the two chains, and a pair of purines is too big to fit into the space.

At this point we made an additional assumption. The bases can theoretically exist in a number of forms depending on where the hydrogen atoms are attached. We assumed that for each base one form was much more probable than all the others. The hydrogen atoms can be thought of as little knobs attached to the bases, and the way the bases fit together depends crucially on where these knobs are. With this assumption the only possible pairs that will fit in are adenine with thymine and guanine with cytosine.

Hydrogen bonds hold the two bases of a pair together. They are very weak bonds; their energy is not many times greater than the energy of thermal vibration at room temperature. (Hydrogen bonds are the main forces holding different water molecules together, and it is because of them that water is a liquid at room temperatures and not a gas.)

Adenine must always be paired with thymine, and guanine with cytosine; it is impossible to fit the bases together in any other combination in our model. (This pairing is likely to be so fundamental for biology that I could not help wondering whether some day an enthusiastic scientist will christen his newborn twins Adenine and Thymine!) The model places no restriction, however, on the sequence of pairs along the structure. Any specified pair can follow any other. This is because a pair of bases is flat, and since in this model they are stacked roughly like a pile of coins, it does not matter which pair goes above which.

It is important to realize that the specific pairing of the bases is the direct result of the assumption that both phosphate-sugar chains are helical. This regularity implies that the distance from a sugar group on one chain to that on the other at the same level is always the same, no matter where one is along the chain. It follows that the bases linked to the sugars always have the same amount of space in which to fit. It is the regularity of the phosphate-sugar chains, therefore, that is at the root of the specific pairing.

Many of the physical and chemical properties of DNA can now be understood in terms of this model. For example, the comparative stiffness of the structure explains rather naturally why DNA keeps a long, fiberlike shape in solution. The hydrogen bonds of the bases account for the behavior of DNA in response to changes in pH. Most striking of all is the fact that in every kind of DNA so far examined the amount of adenine is about equal to the amount of thymine and the guanine equal to the cytosine, while the cross-ratios (between, say, adenine and guanine) can vary considerably from species to species. This remarkable fact, first pointed out by Erwin Chargaff, is exactly what one would expect according to our model, which requires that every adenine be paired with a thymine and every guanine with a cytosine.

It may legitimately be asked whether the artificially prepared fibers of extracted DNA, on which our model is based, are really representative of intact DNA in the cell. There is every indication that they are. It is difficult to see how the very characteristic features of the model could be produced as artifacts by the extraction process. Moreover, Wilkins has shown that intact biological material, such as sperm heads and bacteriophages, gives x-ray patterns very similar to those of the extracted fibers.

The present position, therefore, is that in all likelihood this statement about DNA can safely be made: its structure consists of two helical chains wound around a common axis and held together by hydrogen bonds between specific pairs of bases.

Now the exciting thing about a model of this type is that it immediately suggests how the DNA might produce an exact copy of itself. The model consists of two parts, each of which is the complement of the other. Thus, either chain may act as a sort of mold on which a complementary chain can be synthesized. The two chains of a DNA, let us say, unwind and separate. Each begins to build a new complement onto itself. When the process is completed, there are two pairs of chains where we had only one. Moreover, because of the specific pairing of the bases the sequence of the pairs of bases will have been duplicated exactly; in other words, the mold has not only assembled the building blocks but also has put them together in just the right order.

Let us imagine that we have a single helical chain of DNA and that floating around it inside the cell is a supply of precursors of the four sorts of building blocks needed to make a new chain. In any case, from time to time a loose unit will attach itself by its base to one of the bases of the single DNA chain. Another loose unit may attach itself to an adjoining base on the chain. Now if one or both of the two newly attached units is not the correct mate for the one it has joined on the chain, the two newcomers will be unable to link together, because they are not the right distance apart. One or both will soon drift away, to be replaced by other units. When, however, two adjacent newcomers are the correct partners for their opposite numbers on the chain, they will be in just the right position to be linked together and begin to form a new chain. Thus, only the unit with the proper base will gain a permanent hold at any given position, and eventually the right partners will fill in the vacancies all along the forming chain. While this is going on, the other single chain of the original pair also will be forming a new chain complementary to itself.

A more fundamental difficulty is to explain how the two chains of DNA are unwound in the first place. There would have to be a lot of untwisting, for the total length of all the DNA in a single chromosome is something like four centimeters (400 million angstroms). This means that there must be more than 10 million turns in all, although the DNA may not be all in one piece.

The duplicating process can be made to appear more plausible by assuming that the synthesis of the two new chains begins as soon as the two original chains start to unwind, so that only a short stretch of the chain is ever really single. In fact, we may postulate that it is the growth of the two new chains that unwinds the original pair. This is likely in terms of energy because, for every hydrogen bond that has to be broken, two new ones will be forming. Moreover, plausibility is added to the idea by the fact that the paired chain forms a rather stiff structure, so that the growing chain would tend to unwind the old pair.

The difficulty of untwisting the two chains is a topological one and is caused by the fact that they are intertwined. There would be no difficulty in "unwinding" a single helical chain, because there are so many single bonds in the chain about which rotation is possible. If in the twin structure one chain should break, the other one could easily spin around. This might relieve accumulated strain, and then the two ends of the broken chain, still being in close proximity, might be joined together again.

There remains the fundamental puzzle as to how DNA exerts its hereditary influence. A genetic material must carry out two jobs: duplicate itself and control the development of the rest of the cell in a specific way. We have seen how it might do the first of these, but the structure gives no obvious clue concerning how it may carry out the second. We suspect that the sequence of the bases acts as a kind of genetic code. Such an arrangement can carry an enormous amount of information. If we imagine that the pairs of bases correspond to the dots and dashes of the Morse code, there is enough DNA in a single cell of the human body to encode about 1,000 large textbooks. As we have seen, the three key components of living matter—protein, RNA and DNA—are probably all based on the same general plan. Their backbones are regular, and the variety comes from the sequence of the side groups. It is therefore natural to suggest that the sequence of the bases of the DNA is in some way a code for the sequence of the amino acids in the polypeptide chains of the proteins that the cell must produce.

What, then, one may reasonably ask, are the virtues of the proposed model, if any? The prime virtue is that the configuration suggested is not vague but can be described in terms acceptable to a chemist. The pairing of the bases can be described rather exactly. Then the structure brings together two striking pieces of evidence which at first sight seem to be unrelated—the analytical data, showing the one-to-one ratios for adenine-thymine and guanine-cytosine, and the helical nature of the x-ray pattern. These can now be seen to be two facets of the same thing. Finally, is it not perhaps a remarkable coincidence, to say the least, to find in this key material a structure of exactly the type one would need to carry out a specific replication process: namely, one showing both variety and complementarity?

The model is also attractive in its simplicity. While it is obvious that whole chromosomes have a fairly complicated structure, it is not unreasonable to hope that the molecular basis underlying them may prove to be rather simple. Be that as it may, we now have for the first time a well-defined model for DNA and for a possible replication process, and this in itself should make it easier to devise the crucial experiments.

Life's Far-Flung Raw Materials

Max P. Bernstein, Scott A. Sandford and Louis J. Allamandola

For centuries, comets have imprinted disaster on the human mind. By 400 B.C. Chinese astronomers had sketched 29 varieties of comets, many foretelling calamity. Aristotle's assumption that comets were a warning from the gods gripped Western civilization for two millennia after the heyday of the ancient Greeks. Even at the close of the 20th century, comets and meteors play starring roles in cinematic tales of doom and destruction. The comet threat, it turns out, is not merely mythological. Modern science has revealed that a giant collision probably did in the dinosaurs, and in 1994 human beings nervously watched Comet Shoemaker-Levy 9 smash into Jupiter.

In light of their ominous reputation, it is ironic to consider that such far-flung space debris might be responsible for making Earth the pleasant, life-covered planet it is today. Since the early 1960s, space scientists have speculated that comets and other remnants of solar system formation hauled in gas and water molecules and that these components provided the atmosphere and oceans that made the planet habitable. A growing number of investigators, including our team at the Astrochemistry Laboratory at the National Aeronautics and Space Administration Ames Research Center, now believe that some important raw materials needed to build life also hitched a ride from space. Some of these extraterrestrial organic molecules formed leaky capsules that could have housed the first cellular processes. Other molecules could have absorbed part of the Sun's ultraviolet radiation, thereby sheltering less hardy molecules, and could have helped convert that light energy into chemical food.

In this scenario, the stage for life was set more than four billion years ago when a cold, dark interstellar cloud collapsed into the swirling disk of fiery gas and dust that spawned our solar system. Earth coalesced not long after the Sun, about 4.5 billion years ago, and was long thought to have retained water and the ingredients for life since then. Many scientists today, however, suspect that its earliest days were hot, dry and sterile. It is now clear that space debris bombarded the young planet, creating cataclysms equivalent to the detonation of countless atomic bombs. In fact, the Moon may be a chunk of Earth that was blown off in a collision with an object the size of Mars (see "The Scientific Legacy of Apollo," by G. Jeffrey Taylor; *Scientific American*, July 1994). Impacts of this kind, common until about 4 billion years ago, surely aborted any fledgling life struggling to exist before that time.

As new research is pushing forward the day the planet became habitable, other discoveries are pushing back the first signs of life. Microfossils found in ancient rocks from Australia and South Africa demonstrate that terrestrial life was certainly flourishing by 3.5 billion years ago. Even older rocks from Greenland, 3.9 billion years old, contain isotopic fingerprints of carbon that could have belonged only to a living organism. In other words, only 100 million years or so after the earliest possible point when Earth could have safely supported life, organisms were already well enough established that evidence of them remains today. This narrowing window of time for life to have emerged implies that the process might have required help from space molecules.

ORIGINS OF ORIGINS

The planet's first single-celled organisms presumably owe their primeval debut to a series of chemical steps that led up to carbon-rich molecules such as amino acids. Under the right conditions, the amino acids linked into chainlike proteins, the building blocks of life. One of the first researchers to show how these jump-starter amino acids might have originated was Stanley L. Miller, a graduate student in Harold C. Urey's University of Chicago laboratory in the early 1950s. Miller, now at the University of California at San Diego, sent sparks akin to lightning through a primitive "atmosphere" of simple hydrogen-rich molecules enclosed in a glass flask. Over a few weeks' time, the reaction yielded an array of organic molecules—among them amino acids—in a second flask simulating ocean water below.

New evidence has drawn the components of Miller's atmosphere into question, but his primordial soup theory for how life's ingredients were spawned in a warm pond or ocean on the planet's surface still has a strong following. Some scientists have recently moved the soup pot to the seafloor, where they say murky clouds of minerals spewing from hot springs may have generated life's precursor molecules. But a growing group of other researchers are looking at an altogether different source for life-giving molecules: space.

Juan Oró of the University of Houston suggested extraterrestrial input in 1961, and Sherwood Chang at NASA Ames revived the theory in 1979. Since 1990 Christopher R. Chyba of the Search for Extraterrestrial Intelligence (SETI) Institute in Mountain View, California, has been the premier advocate of the idea that small comets, meteorites and interplanetary dust particles transported the planet's water and atmospheric gases from space.

Not all scientists agree about how Earth got its oceans, but most concur that space debris contributed. Hundreds of tons of dust alone are estimated to drift down to the planet's surface every day. These tiny flecks—the largest no bigger than a grain of sand—litter the inner solar system and sometimes streak across the night sky as shooting stars. Growing evidence now argues that in addition to hauling in the gases and water that made the planet habitable, comets and their cousins peppered the primordial soup with ready-made organic molecules of the kind seen in living systems today.

Recent observations of comet celebrities Halley, Hale-Bopp and Hyakutake revealed that these icy visitors are rife with organic compounds. In 1986 cameras on board the Giotto and Vega spacecrafts captured images of dark material on Halley's surface that resembles the coallike kerogen in some meteorites, and mass spectrometers caught glimpses of carbon-rich molecules. More recently, ground-based telescopes inspecting the coma and tail of comets Hyakutake and Hale-Bopp distinguished a number of specific organic compounds, including methane and ethane. Several space probes will explore other comets during the next 20 years.

When a comet passes through the warm inner solar system, part of it boils away as gas and dust, some of which is later swept up by Earth's gravitational pull. NASA scientists snag comet particles in the upper atmosphere using ER2 aircraft that fly twice as high as a typical commercial jetliner. At altitudes of 62,000 feet, the space dust sticks to oil-coated plastic plates inside collectors under the plane's wings. One of us (Sandford), among other researchers who analyzed these microscopic particles, found that some contain as much as 50 percent organic carbon, more than any other known extraterrestrial object. Even composed of only 10 percent carbon on average, space dust brings about 30 tons of organic material to Earth every day.

Better understood than distant comets and microscopic dust are the large chunks of asteroids that actually smack into Earth as meteorites. Made up mostly of metal and rock, some meteorites also bear compounds such as nucleobases, ketones, quinones, carboxylic acids, amines and amides. Of the slew of complex organics extracted from meteorites, the 70 varieties of amino acids have attracted the most attention. Only eight of these amino acids are part of the group of 20 employed by living cells to build proteins, but those of extraterrestrial origin embody a trait intrinsic to earthly life.

Amino acids exist in mirror-image pairs, a molecular quality called chirality. Just as a person's hands look alike when pressed palm to palm but different when placed palm to knuckles, individual amino acids are either left-handed or right-handed. For little-known reasons and with rare exceptions, amino acids in living organisms are left-handed. One criticism of Miller-type experiments is that they produce equal numbers of both forms. This is where extraterrestrial amino acids come out ahead. Since his first report in 1993, John R. Cronin of Arizona State University has demonstrated a slight surplus of left-handedness in several amino acids extracted from two different meteorites. Some researchers believe life's left-handedness is by chance, but extraterrestrial starting ingredients may have predetermined this molecular peculiarity.

Amino acids may be the most biologically relevant carbon molecules in meteorites, but they are not the most abundant. Most of the carbon is tied up in kerogen, a material composed partly of polycyclic aromatic hydrocarbons, compounds perhaps best known as carcinogenic pollutants on Earth. A product of combustion found in soot, grilled hamburgers and automobile exhaust, these special hydrocarbons also caused a stir when they were detected in the controversial Martian meteorite ALH 84001, which some scientists think harbors evidence of fossilized Martian microbes.

ICEBOX OR FIRESTORM?

Although it is clear that comets, meteorites and dust carry interesting molecules to Earth, finding out where these molecules originated has been tougher to determine. Some scientists have suggested that reactions in liquid water trickling through the parent comets or through asteroids of some meteorites are partly responsible for their rich organic chemistry. But these reactions could hardly account for the carbon molecules frozen in dark interstellar clouds.

Scientists increasingly believe that comet ice is a remnant of the dark cloud that collapsed into the fiery solar nebula, the swirling disk of gas and dust that gave birth to the Sun and planets. The ice has remained unchanged because it stayed protected in the deep freeze at the system's fringe. Other scientists still assert an older claim that extraterrestrial organic molecules were born within the nebula. According to this theory, ice from the mother cloud boiled off, and molecules broke apart and were rearranged in the violence of planet formation.

Molecules tortured in the solar nebula, and only later frozen into comets, should bear the isotopic signatures common to planets and other objects in the inner solar system. On the contrary, most comet dust is enriched in rare elements such as deuterium (an isotope of hydrogen with one extra neutron). Deuterium enrichment is a characteristic of chemical reactions in the low-temperature environment of interstellar space. Out where temperatures hover just above absolute zero, there is enough energy to shake apart only a few of the molecules made from the heavier isotopes, so they tend to build up over time.

The true origin of most comets and meteorites almost certainly combines the pure interstellar icebox and the nebular firestorm. This duality is manifest in space dust comprising materials that have been altered by great heat right next to others that have not. Still, a barrage of evidence during the two years since the observations of Comets Hale-Bopp and Hyakutake has bolstered the case for comets' interstellar heritage. For example, dozens of researchers have detected striking similarity between specific molecules and deuterium enrichments in comets and those commonly observed in interstellar ice grains. In addition, the spin state of hydrogen atoms—a measure of the conditions the ice has experienced—in water from Comet Hale-Bopp confirms that the ice formed at, and was never warmed above, approximately 25 kelvins (-400 degrees Fahrenheit).

If comet ice came from an interstellar cloud, it is easy to believe that organic molecules did, too. Astronomers see signatures of a range of organic compounds throughout the universe, especially among the clouds. For example, a decade of research conducted by one of us (Allamandola) and others has revealed that polycyclic aromatic hydrocarbons are the most abundant class of carbon-bearing compounds in the universe, trapping as much as 20 percent of the total galactic carbon in their molecular lattices.

Deducing the composition of microscopic particles of dust and ice hundreds of light-years away is possible in part through astronomical observations of clouds such as the Eagle Nebula. Dark clouds absorb some of the infrared radiation from nearby stars. When the remaining radiation reaches detectors on Earth and is spread out into a spectrum, light missing at certain wavelengths corresponds to particular chemical bonds with the capacity to absorb light.

CLOUDS IN THE LAB

By comparing the infrared spectra of clouds in space with similar measurements of interstellar ice analogues made in the laboratory, our group at NASA Ames and several other teams around the world determined that the ice grains in the dark clouds are frozen on cores of silicate or carbon. The ice is composed primarily of water but often contains up to 10 percent simple molecules such as carbon dioxide, carbon monoxide, methane, methanol and ammonia.

We wanted to understand how these very simple and abundant interstellar molecules undergo reactions in the ice that transform them into the more complicated compounds seen in meteorites. Allamandola, who had trained as a cryogenics chemist, decided to build an interstellar cloud in the laboratory.

Refrigerators and pumps generate a frigid vacuum of space inside a metal chamber about 20 centimeters (about 8 inches) on a side. A mist of simple gas molecules sprayed from a copper tube freezes onto a lollipop-size disk of aluminum or cesium iodide, which plays the role of the space grain's core. To make the environment of the interstellar cloud complete, a small ultraviolet lamp projects starlike radiation into the chamber.

Our experiments reveal that even at the extremely low temperatures and pressures of space, the ultraviolet radiation breaks chemical bonds just as it does in Earth's atmosphere. There the radiation is infamous for breaking apart chemicals such as chlorofluorocarbons, whose newly freed atoms attack the protective ozone molecules that keep the radiation from baking the planet down below.

In space, when atoms are locked in ice, this bond-breaking process can make molecular fragments recombine into unusually complex structures that would not be possible if these segments were free to drift apart. Everywhere in space where these ice grains are seen, complex compounds are forming—especially in the ultraviolet-rich regions around young stars. In our cloud chamber, we bathe the growing ice grain in radiation equal to what a space grain would endure in thousands of years.

When one of us (Bernstein) started with a simple ice of frozen water, methanol and ammonia—in the same proportions seen in space ice—the experiment yielded complex compounds such as the ketones, nitriles, ethers and alcohols found in carbon-rich meteorites. We also created hexamethylenetetramine, or HMT, a six-carbon molecule known to produce amino acids in warm, acidic water. Molecules with as many as 15 carbon bonds also showed up in the mix.

Some of these compounds display a curious tendency that may have housed the activities of early life. David W. Deamer, a chemist at the University of California at Santa Cruz, found that some of the molecules in the cloud-chamber ice grains form capsule-like droplets in water. These capsules are strikingly similar to those that he produced in the late 1980s using extracts of the meteorite from Murchison, Australia. When Deamer mixed organic compounds from the meteorite with water, they spontaneously assembled into spherical structures similar to cell membranes. Our colleague Jason Dworkin has shown that these capsules are made up of a host of complex organic molecules.

For this self-organization to occur, the molecules usually have a dozen carbon atoms or more, and they must be amphiphilic. That means that their hydrophilic, or water-loving, heads line up facing the water, while their hydrophobic tails stay tucked away inside the membrane. Bubbles in both the meteorite and cloud-chamber extracts also fluoresce, indicating that additional organic material is trapped inside.

Of the compounds we produce, those of perhaps the greatest biological significance are made when we start with water ices embedded with the polycyclic aromatic hydrocarbons known to be abundant in the clouds. Under interstellar conditions, the hydrocarbons convert to many of the components of carbon-rich meteorites, including more complex alcohols, ethers and, perhaps most significantly, quinones. Ubiquitous in living systems today, quinones can stabilize unpaired electrons, an ability living cells need for various energy-transfer activities. For example, the active ingredients in aloe and henna are quinones.

The electron-transport ability of these versatile molecules plays an essential role in converting light into chemical energy in modern photosynthesis. This ability proves more intriguing in the early-Earth scenario when coupled with the quinones' ability to absorb ultraviolet radiation—a grave danger to fragile molecules such as amino acids. Extraterrestrial quinones may have acted as ultraviolet shields before Earth's protective ozone layer developed. In addition, they may have been the molecules that the planet's first life-forms used to trap light for the primitive precursor of photosynthesis.

FROM MOLECULES TO LIFE

We know from laboratory experiments and astronomical observations that the seemingly barren conditions of deep space generate complex organic compounds that meteorites and dust bring to us even today. Reconsidering the emergence of life in this light, we can see that the arrival of amino acids, quinones, amphiphilic molecules and other extraterrestrial organics may well have made it possible for life to flourish or at least may have facilitated its development. Perhaps extraterrestrial amino acids built the first proteins, and perhaps amphiphilic molecules housed the light-harnessing capacity of the quinones, but the exact roles these organic compounds played is not clear. Extraterrestrial organics may have been nothing more than starting materials for chemical reactions that produced other molecules entirely.

One can imagine that a molecule, literally dropped from the sky, could have jump-started or accelerated a simple chemical reaction key to early life. If life's precursor molecules really linked up in a primordial soup, amino acids from space may have provided the crucial quantities to make those steps possible. Likewise, life-building events taking place on the seafloor might have incorporated components of extraterrestrial compounds that were raining into the oceans. Being able to carry out this chemistry more efficiently could have conferred an evolutionary advantage. In time, that simple reaction would become deeply embedded in what is now a biochemical reaction regulated by a protein.

Of course, a huge gap still yawns between even the most complex organic compounds and the genetic code, metabolism and self-replication that are crucial to the definition of life. But given their omnipresence, if organic molecules from space had something to do with life here, that means they were—and always are—available to help with the development of life elsewhere.

Hints of life-friendly conditions on Mars and under the icy surface of Jupiter's moon Europa suggest that other places in our solar system may have benefited from extraterrestrial input. The ubiquity of complex organic molecules across space, combined with the recent discoveries of planets around other stars, also makes it more likely that the conditions conducive to life, if not life itself, have developed in other solar systems as well.

The Evolution of Life on the Earth

Stephen Jay Gould

Some creators announce their inventions with grand éclat. God proclaimed, "*Fiat lux*," and then flooded his new universe with brightness. Others bring forth great discoveries in a modest guise, as did Charles Darwin in defining his new mechanism of evolutionary causality in 1859: "I have called this principle, by which each slight variation, if useful, is preserved, by the term Natural Selection."

Natural selection is an immensely powerful yet beautifully simple theory that has held up remarkably well, under intense and unrelenting scrutiny and testing, for 135 years. In essence, natural selection locates the mechanism of evolutionary change in a "struggle" among organisms for reproductive success, leading to improved fit of populations to changing environments. (Struggle is often a metaphorical description and need not be viewed as overt combat, guns blazing. Tactics for reproductive success include a variety of nonmartial activities such as earlier and more frequent mating or better cooperation with partners in raising offspring.) Natural selection is therefore a principle of local adaptation, not of general advance or progress.

Yet powerful though the principle may be, natural selection is not the only cause of evolutionary change (and may, in many cases, be overshadowed by other forces). This point needs emphasis because the standard misapplication of evolutionary theory assumes that biological explanation may be equated with devising accounts, often speculative and conjectural in practice, about the adaptive value of any given feature in its original environment (human aggression as good for hunting, music and religion as good for tribal cohesion, for example). Darwin himself strongly emphasized the multifactorial nature of evolutionary change and warned against too exclusive a reliance on natural selection.

Natural selection is not fully sufficient to explain evolutionary change for two major reasons. First, many other causes are powerful, particularly at levels of biological organization both above and below the traditional Darwinian focus on organisms and their struggles for reproductive success. At the lowest level of substitution in individual base pairs of DNA, change is often effectively neutral and therefore random. At higher levels, involving entire species or faunas, punctuated equilibrium can produce evolutionary trends by selection of species based on their rates of origin and extirpation, whereas mass extinctions wipe out substantial parts of biotas for reasons unrelated to adaptive struggles of constituent species in "normal" times between such events.

Second, and the focus of this essay, no matter how adequate our general theory of evolutionary change, we also yearn to document and understand the actual pathway of life's history. Theory, of course, is relevant to explaining the pathway (nothing about the pathway can be inconsistent with good theory, and theory can predict certain general aspects of life's geologic pattern). But the actual pathway is strongly *underdetermined* by our general theory of life's evolution. This point needs some belaboring as a central yet widely misunderstood aspect of the world's complexity. Webs and chains of historical events are so intricate, so imbued with random and chaotic elements, so unrepeatable in encompassing such a multitude of unique (and uniquely interacting) objects, that standard models of simple prediction and replication do not apply.

History can be explained, with satisfying rigor if evidence be adequate, after a sequence of events unfolds, but it cannot be predicted with any precision beforehand. History includes too much chaos, or extremely sensitive dependence on minute and unmeasurable differences in initial conditions, leading to massively divergent outcomes based on tiny and unknowable disparities in starting points. And history includes too much contingency, or shaping of present results by long chains of unpredictable antecedent states, rather than immediate determination by timeless laws of nature.

Homo sapiens did not appear on the Earth, just a geologic second ago, because evolutionary theory predicts such an outcome based on themes of progress and increasing neural complexity. Humans arose, rather, as a fortuitous and contingent outcome of thousands of linked events, any one of which could have occurred differently and sent history on an alternative pathway that would not have led to consciousness.

Therefore, to understand the events and generalities of life's pathway, we must go beyond principles of evolutionary theory to a paleontological examination of the contingent pattern of life's history on our planet—the single actualized version among millions of plausible alternatives that happened not to occur. Such a view of life's history is highly contrary both to conventional deterministic models of Western science and to the deepest social traditions and psychological hopes of Western culture for a history culminating in humans as life's highest expression and intended planetary steward.

Science can, and does, strive to grasp nature's factuality, but all science is socially embedded, and all scientists record prevailing "certainties," however hard they may be aiming for pure objectivity. Darwin himself, in the closing lines of *The Origin of Species,* expressed Victorian social preference more than nature's record in writing: "As natural selection works solely by and for the good of each being, all corporeal and mental endowments will tend to progress towards perfection."

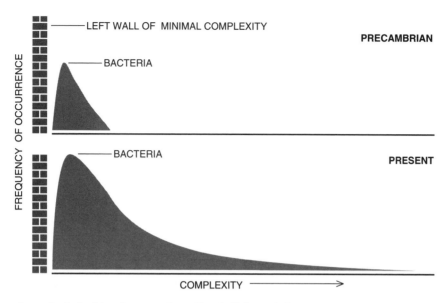

Progress does not rule (and is not even a primary thrust of) the evolutionary process. For reasons of chemistry and physics, life arises next to the "left wall" of its simplest conceivable and preservable complexity. This style of life (bacterial) has remained most common and most successful. A few creatures occasionally move to the right, thus extending the right tail in the distribution of complexity. Many always move to the left, but they are absorbed within space already occupied. Note that the bacterial mode has never changed in position but just grown higher.

Life's pathway certainly includes many features predictable from laws of nature, but these aspects are too broad and general to provide the "rightness" that we seek for validating evolution's particular results—roses, mushrooms, people and so forth. Organisms adapt to, and are constrained by, physical principles.

Predictable ecological rules govern the structuring of communities by principles of energy flow and thermodynamics (more biomass in prey than in predators, for example). Evolutionary trends, once started, may have local predictability ("arms races," in which both predators and prey hone their defenses and weapons, for example—a pattern that Geerat J. Vermeij of the University of California at Davis has called "escalation" and documented in increasing strength of both crab claws and shells of their gastropod prey through time). But laws of nature do not tell us why we have crabs and snails at all, why insects rule the multicellular world and why vertebrates rather than persistent algal mats exist as the most complex forms of life on the Earth.

Relative to the conventional view of life's history as an at least broadly predictable process of gradually advancing complexity through time, three features of the paleontological record stand out in opposition and shall therefore serve as organizing themes for the rest of this essay: the constancy of modal complexity throughout life's history; the concentration of major events in short bursts interspersed with long periods of relative stability; and the role of external impositions, primarily mass extinctions, in disrupting patterns of "normal" times. These three features, combined with more general themes of chaos and contingency, require a new framework for conceptualizing and drawing life's history, and this essay therefore closes with suggestions for a different iconography of evolution.

The primary paleontological fact about life's beginnings points to predictability for the onset and very little for the particular pathways thereafter. The Earth is 4.6 billion years old, but the oldest rocks date to about 3.9 billion years because the Earth's surface became molten early in its history, a result of bombardment by large amounts of cosmic debris during the solar system's coalescence, and of heat generated by radioactive decay of short-lived isotopes. These oldest rocks are too metamorphosed by subsequent heat and pressure to preserve fossils (though some scientists interpret the proportions of carbon isotopes in these rocks as signs of organic production). The oldest rocks sufficiently unaltered to retain cellular fossils—African and Australian sediments dated to 3.5 billion years old—do preserve prokaryotic cells (bacteria and cyanophytes) and stromatolites (mats of sediment trapped and bound by these cells in shallow marine waters). Thus, life on the Earth evolved quickly and is as old as it could be. This fact alone seems to indicate an inevitability, or at least a predictability, for life's origin from the original chemical constituents of atmosphere and ocean.

No one can doubt that more complex creatures arose sequentially after this prokaryotic beginning—first eukaryotic cells, perhaps about two billion years ago, then multicellular animals about 600 million years ago, with a relay of highest complexity among animals passing from invertebrates, to marine vertebrates and, finally (if we wish, albeit parochially, to honor neural architecture as a primary criterion), to reptiles, mammals and humans.

I do not deny the facts of the preceding paragraph but wish to argue that our conventional desire to view history as progressive, and to see humans as predictably dominant, has grossly distorted our interpretation of life's pathway by falsely placing in the center of things a relatively minor phenomenon that arises only as a side consequence of a physically constrained starting point. The most salient feature of life has been the stability of its bacterial mode from the beginning of the fossil record until today and, with little doubt, into all future time so long as the Earth endures.

For reasons related to the chemistry of life's origin and the physics of self-organization, the first living things arose at the lower limit of life's conceivable, preservable complexity. Call this lower limit the "left wall" for an architecture of complexity. Since so little space exists between the left wall and life's initial bacterial mode in the fossil record this move toward greater complexity is probably only a bias inspired by parochial focus on ourselves, and consequent overattention to complexifying creatures, while we ignore just as many lineages adapting equally well by becoming simpler in form. The morphologically degenerate parasite, safe within its host, has just as much prospect for evolutionary success as its gorgeously elaborate relative coping with the slings and arrows of outrageous fortune in a tough external world.

Even if complexity is only a drift away from a constraining left wall, we might view trends in this direction as more predictable and characteristic of life's pathway as a whole if increments of complexity accrued in a persistent and gradually accumulating manner through time. But nothing about life's history is more peculiar with respect to this common (and false) expectation than the actual pattern of extended stability and rapid episodic movement, as revealed by the fossil record.

Life remained almost exclusively unicellular for the first five sixths of its history—from the first recorded fossils at 3.5 billion years to the first well-documented multicellular animals less than 600 million years ago. (Some simple multicellular algae evolved more than a billion years ago, but these organisms belong to the plant kingdom and have no genealogical connection with animals.) This long period of unicellular life does include, to be sure, the vitally important transition from simple prokaryotic cells without organelles to eukaryotic cells with nuclei, mitochondria and other complexities of intracellular architecture—but no recorded attainment of multicellular animal organization for a full three billion years. If complexity is such a good thing, and multicellularity represents its initial phase in our usual view, then life certainly took its time in making this crucial step. Such delays speak strongly against general progress as the major theme of life's history, even if they can be plausibly explained by lack of sufficient atmospheric oxygen for most of Precambrian time or record, only one direction for future increment exists—toward greater complexity at the right.

Thus, every once in a while, a more complex creature evolves and extends the range of life's diversity in the only available direction. In technical terms, the distribution of complexity becomes more strongly right-skewed through these occasional additions. But the additions are rare and episodic. They do not even constitute an evolutionary series but form a motley sequence of distantly related taxa, usually depicted as eukaryotic cell, jellyfish, trilobite, nautiloid, eurypterid (a large relative of horseshoe crabs), fish, an amphibian such as Eryops, a dinosaur, a mammal and a human being. This sequence cannot be construed as the major thrust or trend of life's history. Think rather of an occasional creature tumbling into the empty right region of complexity's space. Throughout this entire time, the bacterial mode has grown in height and remained constant in position. Bacteria represent the great success story of life's pathway. They occupy a wider domain of environments and span a broader range of biochemistries than any other group. They are adaptable, indestructible and astoundingly diverse.

One might grant that complexification for life as a whole represents a pseudotrend based on constraint at the left wall but still hold that evolution within particular groups differentially favors complexity when the founding lineage begins far enough from the left wall to permit movement in both directions. Empirical tests of this interesting hypothesis are just beginning (as concern for the subject mounts among paleontologists), and we do not yet have enough cases to advance a generality. But the first two studies—by Daniel W. McShea of the University of Michigan on mammalian vertebrae and by George F. Boyajian of the University of Pennsylvania on ammonite suture lines—show no evolutionary tendencies to favor increased complexity.

More curiously, all major stages in organizing animal life's multicellular architecture then occurred in a short period beginning less than 600 million years ago and ending by about 530 million years ago—and the steps within this sequence are also discontinuous and episodic, not gradually accumulative.

The Cambrian then began with an assemblage of bits and pieces, frustratingly difficult to interpret, called the "small shelly fauna." The subsequent main pulse, starting about 530 million years ago, constitutes the famous Cambrian explosion, during which all but one modern phylum of animal life made a first appearance in the fossil record. (Geologists had previously allowed up to 40 million years for this event, but an elegant study, published in 1993, clearly restricts this period of phyletic flowering to a mere five million years.) The Bryozoa, a group of sessile and colonial marine organisms, do not arise until the beginning of the subsequent, Ordovician period, but this apparent delay may be an artifact of failure to discover Cambrian representatives.

Although interesting and portentous events have occurred since, from the flowering of dinosaurs to the origin of human consciousness, we do not exaggerate greatly in stating that the subsequent history of animal life amounts to little more than variations on anatomical themes established during the Cambrian explosion within five million years. Three billion years of unicellularity, followed by five million years of intense creativity and then capped by more than 500 million years of variation on set anatomical themes can scarcely be read as a predictable, inexorable or continuous trend toward progress or increasing complexity.

We do not know why the Cambrian explosion could establish all major anatomical designs so quickly. In any case, this initial period of both internal and external flexibility yielded a range of invertebrate anatomies that may have exceeded (in just a few million years of production) the full scope of animal form in all the Earth's environments today (after more than 500 million years of additional time for further expansion). Scientists are divided on this question. Cambrian diversity at least equaled the modern range—so even the most cautious opinion holds that 500 million subsequent years of opportunity have not expanded the Cambrian range, achieved in just five million years. The Cambrian explosion was the most remarkable and puzzling event in the history of life.

Moreover, we do not know why most of the early experiments died, while a few survived to become our modern phyla. It is tempting to say that the victors won by virtue of greater anatomical complexity, better ecological fit or some other predictable feature of conventional Darwinian struggle. But no recognized traits unite the victors, and each surviving lineage, including our own phylum of vertebrates, inhabits the Earth today more by the luck of the draw than by any predictable struggle for existence. The history of multicellular animal life may be more a story of great reduction in initial possibilities, with stabilization of lucky survivors, than a conventional tale of steady ecological expansion and morphological progress in complexity.

Finally, this pattern of long stasis, with change concentrated in rapid episodes that establish new equilibria, may be quite general at several scales of time and magnitude, forming a kind of fractal pattern in self-similarity. According to the punctuated equilibrium model of speciation, trends within lineages occur by accumulated episodes of geologically instantaneous speciation, rather than by gradual change within continuous populations (like climbing a staircase rather than rolling a ball up an inclined plane).

Even if evolutionary theory implied a potential internal direction for life's pathway (although previous facts and arguments in this essay cast doubt on such a claim), the occasional imposition of a rapid and substantial, perhaps even truly catastrophic, change in environment would have intervened to stymie the pattern. These environmental changes trigger mass extinction of a high percentage of the Earth's species and may so derail any internal direction and so reset the pathway that the net pattern of life's history looks more capricious and concentrated in episodes than steady and directional. Mass extinctions have been recognized since the dawn of paleontology; the major divisions of the geologic timescale were established at boundaries marked by such events. But until the revival of interest that began in the late 1970s, most paleontologists treated mass extinctions only as intensifications of ordinary events, leading (at most) to a speeding up of tendencies that pervaded normal times. In this gradualistic theory of mass extinction, these events really took a few million years to unfold (with the appearance of suddenness interpreted as an artifact of an imperfect fossil record), and they only made the ordinary occur faster (more intense Darwinian competition in tough times, for example, leading to even more efficient replacement of less adapted by superior forms).

The reinterpretation of mass extinctions as central to life's pathway and radically different in effect began with the presentation of data by Luis and Walter Alvarez in 1979, indicating that the impact of a large extraterrestrial object (they suggested an asteroid seven to 10 kilometers in diameter) set off the last great extinction at the Cretaceous-Tertiary boundary 65 million years ago.

This reawakening of interest also inspired paleontologists to tabulate the data of mass extinction more rigorously. Work by David M. Raup, J. J. Sepkoski, Jr., and David Jablonski of the University of Chicago has established that multicellular animal life experienced five major (end of Ordovician, late Devonian, end of Permian, end of Triassic and end of Cretaceous) and many minor mass extinctions during its 530-million-year history. We have no clear evidence that any but the last of these events was triggered by catastrophic impact, but such careful study leads to the general conclusion that mass extinctions were more frequent, more rapid, more extensive in magnitude and more different in effect than paleontologists had previously realized. These four properties encompass the radical implications of mass extinction for understanding life's pathway as more contingent and chancy than predictable and directional.

Mass extinctions are not random in their impact on life. Some lineages succumb and others survive as sensible outcomes based on presence or absence of evolved features. But especially if the triggering cause of extinction be sudden and catastrophic, the reasons for life or death may be random with respect to the original value of key features when first evolved in Darwinian struggles of normal times. This "different rules" model of mass extinction imparts a quirky and unpredictable character to life's pathway based on the evident claim that lineages cannot anticipate future contingencies of such magnitude and different operation.

We all know that dinosaurs perished in the end Cretaceous event and that mammals therefore rule the vertebrate world today. Most people assume that mammals prevailed in these tough times for some reason of general superiority over dinosaurs. But such a conclusion seems most unlikely. Mammals and dinosaurs had coexisted for 100 million years and mammals had remained rat-sized or smaller, making no evolutionary "move" to oust dinosaurs. No good argument for mammalian prevalence by general superiority has ever been advanced, and fortuity seems far more likely. As one plausible argument, mammals may have survived partly as a result of their small size (with much larger, and therefore extinction-resistant, populations as a consequence, and less ecological specialization with more places to hide, so to speak). Small size may not have been a positive mammalian adaptaton at all, but more a sign of inability ever to penetrate the dominant domain of dinosaurs. Yet this "negative" feature of normal times may be the key reason for mammalian survival and a prerequisite to my writing and your reading this essay today.

The Darwinian revolution remains woefully incomplete because, even though thinking humanity accepts the fact of evolution, most of us are still unwilling to abandon the comforting view that evolution means (or at least embodies a central principle of) progress defined to render the appearance of something like human consciousness either virtually inevitable or at least predictable. The pedestal is not smashed until we abandon progress or complexification as a central principle and come to entertain the strong possibility that *H. sapiens* is but a tiny, late-arising twig on life's enormously arborescent bush—a small bud that would almost surely not appear a second time if we could replant the bush from seed and let it grow again.

Primates are visual animals, and the pictures we draw betray our deepest convictions and display our current conceptual limitations. Artists have always painted the history of fossil life as a sequence from invertebrates, to fishes, to early terrestrial amphibians and reptiles, to dinosaurs, to mammals and, finally, to humans. There are no exceptions; all sequences painted since the inception of this genre in the 1850s follow the convention.

Yet we never stop to recognize the almost absurd biases coded into this universal mode. No scene ever shows another invertebrate after fishes evolved, but invertebrates did not go away or stop evolving! After terrestrial reptiles emerge, no subsequent scene ever shows a fish (later oceanic tableaus depict only such returning reptiles as ichthyosaurs and plesiosaurs). But fishes did not stop evolving after one small lineage managed to invade the land. In fact, the major event in the evolution of fishes, the origin and rise to dominance of the teleosts, or modern bony fishes, occurred during the time of the dinosaurs and is therefore never shown at all in any of these sequences—even though teleosts include more than half of all species of vertebrates. Why should humans appear at the end of all sequences? Our order of primates is ancient among mammals, and many other successful lineages arose later than we did.

We will complete Darwin's revolution until we find, grasp and accept another way of drawing life's history. J. B. S. Haldane proclaimed nature "queerer than we can suppose," but these limits may only be socially imposed conceptual locks rather than inherent restrictions of our neurology. New icons might break the locks. Trees—or rather copiously and luxuriously branching bushes—rather than ladders and sequences hold the key to this conceptual transition.

We must learn to depict the full range of variation, not just our parochial perception of the tiny right tail of most complex creatures. We must recognize that this tree may have contained a maximal number of branches near the beginning of multicellular life and that subsequent history is for the most part a process of elimination and lucky survivorship of a few, rather than continuous flowering, progress and expansion of a growing multitude. We must understand that little twigs are contingent nubbins, not predictable goals of the massive bush beneath. We must remember the greatest of all biblical statements about wisdom: "She is a tree of life to them that lay hold upon her; and happy is every one that retaineth her."

The Case for Relic Life on Mars

Everett K. Gibson, Jr., David S. McKay, Kathie Thomas-Keptra and Christopher S. Romanek

If all the scientific subjects that have seized the public psyche, few have held on as tightly as the idea of life on Mars. Starting not long after the invention of the telescope and continuing for a good part of the past three centuries, the subject has inspired innumerable studies, ranging from the scientific to the speculative. But common to them all was recognition of the fact that in our solar system, if a planet other than Earth harbors life, it is almost certainly Mars.

Interest in Martian life has tended to coincide with new discoveries about the mysterious red world. Historically, these discoveries have often occurred after one of the periodic close approaches between the two planets. Every 15 years, Mars comes within about 56 million kilometers of Earth (the next approach will occur in the summer of 2003).

It was after one of the close approaches in the late 19th century that Italian astronomer Giovanni V. Schiaparelli announced that he had seen great lines stretching across the planet's surface, which he called *canali*. At the turn of the century, U.S. astronomer Percival Lowell insisted that the features were canals constructed by an advanced civilization. In the 1960s and 1970s, however, any lingering theories about the lines and elaborate civilizations were put to rest after the United States and the Soviet Union sent the first space probes to the planet. The orbiters showed that there were in fact no canals, although there were long, huge canyons. Within a decade, landers found no evidence of life, let alone intelligent life and civilization.

Although the debate about intelligent life was essentially over, the discussions about microbial life on the planet—particularly life that may have existed on the warmer, wetter Mars of billions of years ago—were just beginning. In August 1996 this subject was thrust into the spotlight when we and a number of our colleagues at the National Aeronautics and Space Administration Johnson Space Center and at Stanford University announced that unusual characteristics in a meteorite known to have come from Mars could most reasonably be interpreted as the vestiges of ancient Martian bacterial life. The 1.9-kilogram, potato-sized meteorite, designated ALH84001, had been found in Antarctica in 1984.

Our theory was by no means universally embraced. Some researchers insisted that there were nonbiological explanations for the meteorite's peculiarities and that these rationales were more plausible than our biological explanation. We remain convinced that the facts and analyses that we outline in this essay point to the existence of a primitive form of life. Moreover, such life-forms may still exist on Mars if, as some researchers have theorized, pore spaces and cracks in rocks below the surface of the planet contain liquid water.

INHOSPITABLE PLANET

Conditions on Mars today are not hospitable to life as we know it. The planet's atmosphere consists of 95 percent carbon dioxide, 2.7 percent nitrogen, 1.6 percent argon and only trace amounts of oxygen and water vapor. Surface pressure is less than 1 percent of Earth's, and daily temperatures rarely exceed zero degrees Celsius, even in the planet's warmest regions in the summer. Most important, one of life's most fundamental necessities, liquid water, seems not to exist on the planet's surface.

Given these realities, it is perhaps not surprising that the two Viking space probes that settled on the planet's surface, in July and September of 1976, failed to find any evidence of life. The results cast doubt on—but did not completely rule out—the possibility that there is life on Mars. The landers, which were equipped to detect organic compounds at a sensitivity level of one part per billion, found none, either at the surface or in the soil several centimeters down. Similarly, three other experiments found no evidence of microbial organisms. Ultimately, researchers concluded that the possibility of life on Mars was quite low and that a more definite statement on the issue would have to await the analysis of more samples by future landers.

In addition, various meteorites found on Earth and known to be of Martian origin—including ALH84001 itself—offer tangible proof of Mars's watery past because they show unambiguous signs of having been altered by water. Specifically, some of these meteorites have been found to contain carbonates, sulfates, hydrates and clays, which can be formed, so far as planetary scientists know, only when water comes into contact with other minerals in the rock.

Of course, the entire argument hinges on ALH84001's having come from the red planet. Of this, at least, we can be certain. It is one of several meteorites found since the mid-1970s in meteorite-rich regions in Antarctica. In the early 1980s Donald D. Bogard and Pratt Johnson of the NASA Johnson Space Center began studying a group of meteorites found to contain minute bubbles of gas trapped within glass inside the rock. The glass is believed to have formed during impacts with meteoroids or comets while the rock was on the surface of Mars. Some of these glass-producing impacts apparently imparted enough energy to eject fragments out into space; from there, some of these rocks were captured by Earth's gravitational field. This impact scenario is the only one that planetary scientists believe can account for the existence on our world of bits of Mars.

Bogard and Johnson found that the tiny samples of gas trapped in the glass of some of the meteorites had the exact chemical and isotopic compositions as gases in the atmosphere of Mars, which had been measured by the Viking landers in 1976. The one-to-one correlation between the two gas samples—over a range of nine orders of magnitude—strongly suggests that these meteorites are from Mars. In all, five meteorites have been shown to contain samples of trapped Martian atmosphere. ALH84001 was not among the five so analyzed; however, its distribution of oxygen isotopes, minerology and other characteristics place it in the same group with the other five Martian rocks.

The distribution of oxygen isotopes within a group of meteorites has been the most convincing piece of evidence establishing that the rocks—including ALH84001—come from Mars. In the early 1970s Robert N. Clayton and his coworkers at the University of Chicago showed that the isotopes oxygen 16, oxygen 17 and oxygen 18 in the silicate materials within various types of meteorites have unique relative abundances. The finding was significant because it demonstrates that the bodies of our solar system formed from distinct regions of the solar nebula and thus have unique oxygen isotopic compositions. Using this isotopic "fingerprint," Clayton helped to show that a group of 12 meteorites, including ALH84001, are indeed closely related. The combination of trapped Martian atmospheric gases and the specific distribution of oxygen isotopes has led researchers to conclude that the meteorites must have come from Mars.

INVADER FROM MARS

Other analyses, mainly of radioisotopes, have enabled researchers to outline ALH84001's history from its origins on the red planet to the present day. The three key time periods of interest are the age of the rock (the length of time since it crystallized on Mars), how long the meteorite traveled in space and how long it has been on Earth. Analysis of three different sets of radioactive isotopes in the meteorite have established each of these time periods.

The length of time since the rock solidified from molten materials—the so-called crystallization age of the material—has been determined through the use of three different dating techniques. One uses isotopes of rubidium and strontium, another, neodymium and samarium, and the third, argon. All three methods indicated that the rock is 4.5 billion years old. By geologic standards the rock is extremely old; the 4.5-billion-year figure means that it crystallized within the first 1 percent of Mars's history. In comparison, the other 11 Martian meteorites that have been analyzed are all between 1.3 billion years old and 165 million years old. It is remarkable that a rock so old, and so little altered on Mars or during its residence in the Antarctic ice, became available for scientists to study.

The duration of the meteorite's space odyssey was determined through the analysis of still other isotopes, namely helium 3, neon 21 and argon 38. While a meteorite is in space, it is bombarded by cosmic rays and other high-energy particles. The particles interact with the nuclei of certain atoms in the meteorite, producing the three isotopes listed above. By studying the abundances and production rates of these cosmogenically produced isotopes, scientists can determine how long the meteorite was exposed to the high-energy flux and, therefore, how long the specimen was in space. Using this approach, researchers concluded that after being torn free from the planet, ALH84001 spent 16 million years in space before falling in the Antarctic.

To determine how long the meteorite lay in the Antarctic ice, A. J. Timothy Jull of the University of Arizona used carbon 14 dating. When silicates are exposed to cosmic rays in space, carbon 14 is produced. In time, the rates of production and decay of carbon 14 balance, and the meteorite becomes saturated with the isotope. The balance is upset when the meteorite falls from space and production of carbon 14 ceases. The decay goes on, however, reducing the amount in the rock by one half every 5,700 years. By determining the difference between the saturation level and the amount measured in the silicates, researchers can determine how long the meteorite has been on Earth. Jull's finding was that ALH84001 fell from space 13,000 years ago.

From the very moment it was discovered, the meteorite now known as ALH84001 proved unusual and intriguing. In 1984 U.S. geologist Roberta Score found the meteorite in the Far Western Icefield of the Allan Hills Region. Score recognized that the rock was unique because of its pale greenish-gray color.

CARBONATES ARE KEY

The most interesting aspect of ALH84001 are the carbonates, which exist as tiny discoids, like flattened spheres, 20 to 250 microns in diameter. They cover the walls of cracks in the meteorite and are oriented in such a way that they are flattened against the inside walls of the fractures. The globules were apparently deposited from a fluid saturated with carbon dioxide that percolated through the fracture after the silicates were formed. None of the other 11 meteorites known to have come from Mars have such globules.

It was within the carbonate globules that our research team found the assortment of unique features that led us to hypothesize that microbial organisms came into contact with the rock in the distant past. Basically, the case for ancient microbial life on Mars is built almost entirely around the globules.

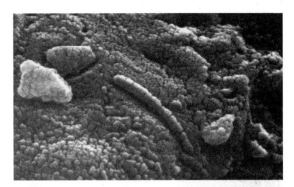

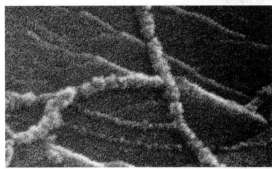

Top: This segmented object is 380 nanometers in length and was found in a carbonate globule in meteorite ALH84001. The minute structure resembles fossilized bacteria or microfossils, found on Earth.

Bottom: The vertically oriented object to the right is believed to be a microfossil. The object, which is also 380 nanometers long, was found 400 meters below Earth's surface (in Washington State) in a type of geologic formation known as Columbia River Basalt.

Individually, none of the features we found are strongly indicative of life. Collectively, however—and especially within the confines of the tiny discoids—the globules can be plausibly explained as the ancient vestiges of microbial life. The features fall into several categories of evidence. One category centers on the presence of tiny iron oxides and iron sulfide grains, which resemble those formed by terrestrial bacteria. The second group revolves around the presence of organic carbon molecules in and on the globules. Finally, unusual structures found within the globules bear a striking resemblance to bacteria fossils found on Earth. Another relevant piece of evidence suggests the globules formed from a water-rich fluid below 100 degrees C.

NASA Johnson Space Center researchers, along with Monica Grady of the British Museum of Natural History and workers at the Open University in the U.K., performed the first isotopic analysis of carbon and oxygen in the carbonate globules. The carbon analysis indicates that the globules have more carbon 13 than any carbonates found on Earth but just the right amount to have come from Mars.

Most carbon on Earth is made up of 98.9 percent carbon 12 and 1.1 percent carbon 13. Various reactions, however, can alter this ratio. For example, in general, a sample of carbon that has been a part of an organic chemical system—say, in plant matter—is somewhat more enriched in carbon 12, whereas carbon in limestone is relatively enriched in carbon 13. The carbon in the globules of ALH84001 is more enriched in carbon 13 than any natural materials on Earth. Moreover, the enrichment is different from that of the other 11 Martian meteorites. This fact suggests that the carbon in the globules—unlike the trace amounts of carbon seen in the other Martian meteorites—may have been derived from Mars's atmosphere.

Analysis of the distribution of oxygen isotopes in the carbonates can provide information about the temperature at which those minerals formed. The subject bears directly on the question of whether the carbonates were formed at temperatures that could support microbial life, because terrestrial organisms do not survive at temperatures above about 115 degrees C. The NASA–U.K. team analyzed the oxygen isotopes in the carbonate globules. Those findings strongly suggest that the globules formed at temperatures no higher than 100 degrees C.

We are extremely interested in the age of the carbonates, because it would allow us to estimate when microbial life left its mark on the rock that became ALH84001. Yet all we can say for sure is that the carbonates crystallized in the fractures in the meteorite some time after the rock itself crystallized. Various research groups have come up with ages ranging from 1.3 to 3.6 billion years; the data gathered so far, however, are insufficient to date the carbonate globules conclusively.

BIOMINERAL CLUES

The first category of evidence involves certain minerals found inside the carbonate globules; the type and arrangement of the minerals are similar, if not identical, to certain biominerals found on Earth. Inside, the globules are rich in magnesite ($MgCO_3$) and siderite ($FeCO_3$) small amounts of calcium and manganese carbonates. Fine-grained particles of magnetite (Fe_3O_4) and sulfides ranging in size from 10 to 100 nanometers on a side are present within the carbonate host. The magnetite crystals are cuboid, teardrop or irregular in shape. Individual crystals have well-preserved structures with little evidence of defects or trace impurities.

An analysis of the samples conducted with high-resolution transmission electron microscopy coupled with energy-dispersive spectroscopy indicates that the size, purity, morphology and crystal structures of all these magnetites are typical of magnetites produced by bacteria on Earth.

Terrestrial magnetite particles associated with fossilized bacteria are known as magnetofossils. These particles are found in a variety of sediments and soils and are classified according to size, as superparamagnetic (less than 20 nanometers on an edge) or single-domain (20 to 100 nanometers). The magnetites within ALH84001 are typically 40 to 60 nanometers on an edge.

Single-domain magnetite has been reported in ancient terrestrial limestones and is generally regarded as having been produced by bacteria. Most intriguing, some of the magnetites in ALH84001 are arranged in chains, not unlike pearls in a necklace. Terrestrial bacteria often produce magnetite in precisely this pattern, because as they biologically process iron and oxygen from the water, they produce crystals that naturally align themselves with the Earth's magnetic field.

ORGANIC CARBON MOLECULES

The presence of organic carbon molecules in ALH84001 constitutes the second group of clues. In recent years, researchers have found organic molecules not only in Martian meteorites but also in ones known to have come from the asteroid belt in interplanetary space, which could hardly support life. Nevertheless, the type and relative abundance of the specific organic molecules identified in ALH84001 are suggestive of life processes. The presence of indigenous organic molecules within ALH84001 is the first proof that such molecules have existed on Mars.

On Earth, when living organisms die and decay, they create hydrocarbons associated with coal, peat and petroleum. Many of these hydrocarbons belong to a class of organic molecules known as polycyclic aromatic hydrocarbons (PAHs). There are thousands of different PAHs. Their presence in a sample does not in itself demonstrate that biological processes occurred. It is the location and association of the PAHs in the carbonate globules that make their discovery so interesting.

In ALH84001 the PAHs are always found in carbonate-rich regions, including the globules. In our view, the relatively simple PAHs are the decay products of living organisms that were carried by a fluid and trapped when the globules were formed. In 1996 a team at the Open University showed that the carbon in the globules in ALH84001 has an isotopic composition suggestive of microbes that used methane as a food source. If confirmed, this finding will be one of the strongest pieces of evidence to date that the rock bears the imprint of biological activity.

In our 1996 announcement, Richard N. Zare and Simon J. Clemett of Stanford used an extremely sensitive analytical technique to show that ALH84001 contains a relatively small number of different PAHs, all of which have been identified in the decay products of microbes. Most important, the PAHs were found to be located inside the meteorite, where contamination is very unlikely to have occurred. This crucial finding supports the idea that the carbonates are Martian and contain the vestiges of ancient living organisms.

Ultrasensitive analysis of the distribution of the PAHs in ALH84001 indicated that the PAHs could not have come from Earth or from an extraterrestrial source—other than Mars.

Perhaps the most visually compelling piece of evidence that at least vestiges of microbes came into contact with the rock are objects that appear to be the fossilized remains of microbes themselves. Detailed examination of the ALH84001 carbonates using high-resolution scanning electron microscopy (SEM) revealed unusual features that are similar to those seen in terrestrial samples associated with biogenic activity. Close-up SEM views show that the carbonate globules contain ovoid and tube-shaped bodies. The objects are around 380 nanometers long, which means they could very well be the fossilized remains of bacteria. To pack in all the components that are normally required for a typical terrestrial bacterium to function, sizes larger than 250 nanometers seem to be required. Additional tubelike curved structures found in the globules are 500 to 700 nanometers in length.

NANOBACTERIA OR APPENDAGES?

Other objects found within ALH84001 are close to the lower size limit for bacteria. These ovoids are only 40 to 80 nanometers long; other, tube-shaped bodies range from 30 to 170 nanometers in length and 20 to 40 nanometers in diameter. These sizes are about a factor of 10 smaller than the terrestrial microbes that are commonly recognized as bacteria. Still, typical cells often have appendages that are generally quite small—in fact about the same size as these features observed within ALH84001. It may be possible that some of the features are fragments or parts of larger units within the sample.

ALH84001's numerous ovoid and elongated features are essentially identical in size and morphology to those of so-called nanobacteria on Earth. Fossilized bacteria found within subsurface basalt samples from the Columbia River basin in Washington State (see "Microbes Deep Inside the Earth," by James K. Fredrickson and Tullis C. Onstott; *Scientific American*, October 1996) have features that are essentially identical to some of those observed in the ovoids in ALH84001.

ALH84001 was present on Mars 4.5 billion years ago, when the planet was wetter, warmer and had a denser atmosphere. Therefore, we might expect to see evidence that the rock had been altered by contact with water. Yet the rock bears few traces of so-called aqueous alteration evidence. One such piece of evidence would be clay minerals, which are often produced by aqueous reactions. The meteorite does indeed contain phyllosilicate clay mineral, but only in trace amounts. It is not clear, moreover, whether the clay mineral formed on Mars or in the Antarctic.

Mars had liquid water on its surface early in its history and may still have an active groundwater system below the permafrost or cryosphere. If surface microorganisms evolved during a period when liquid water covered parts of Mars, the microbes might have spread to subsurface environments when conditions turned harsh on the surface. The surface of Mars contains abundant basalts that were undoubtedly fractured during the period of early bombardment in the first 600 million years of the planet's history. These fractures could serve as pathways for liquid water and could have harbored any biota that were adapting to the changing conditions on the planet.

Organisms may also have developed at hot springs or in underground hydrothermal systems on Mars where chemical disequilibriums can be maintained in environments somewhat analogous to those of the mineral-rich "hot smokers" on the seafloor of Earth.

Thus, it is entirely possible that if organisms existed on Mars in the distant past, they may still be there. Availability of water within the pore spaces of a subsurface reservoir would facilitate their survival. If the carbonates within ALH84001 were formed as early as 3.6 billion years ago and have biological origins, they may be the remnants of the earliest Martian life.

The analyses so far of ALH84001 are consistent with the meteorite's carbonate globules containing the vestiges of ancient microbial life. Studies of the meteorite are far from over, however. Whether or not these investigations confirm or modify our hypothesis, they will be invaluable learning experiences for researchers, who may get the opportunity to put the experience to use in coming years. We hope that in 2005 a "sample-return" mission will be launched to collect Martian rocks and soil robotically and return them to Earth.

Through projects such as the sample-return, we will finally begin to collect the kind of data that will enable us to determine conclusively whether life came into being on Mars. This kind of insight, in turn, may ultimately provide perspective on one of the greatest scientific mysteries: the prevalence of life in our universe.

Searching for Life on Other Planets

J. Roger, P. Angel and Neville J. Woolf

The possibility that we are not alone in the universe has fascinated people for centuries. In the 1600s Galileo Galilei peered into the night sky with his newly invented telescope, recognized mountains on the Moon and noted that other planets were spheres like Earth. About 60 years later other stargazers observed polar ice caps on Mars, as well as color variations on the planet's surface, which they believed to be vegetation changing with the seasons (the colors are now known to be the result of dust storms). But samples of Martian soil obtained in the 1970s by the Viking lander spacecraft lacked material evidence of any life. Indeed, the present conditions in the rest of our solar system seem to be generally incompatible with life like that found on Earth.

But our search for extraterrestrial life has recently been extended—we can now turn our attention to planets outside our own solar system. After years of looking, astronomers have turned up evidence of planets orbiting distant stars similar to our Sun. Planets around these and other stars may have evolved living organisms. Finding extraterrestrial life may seem a Herculean task, but within the next decade, we could build the equipment needed to locate planets with life-forms like the primitive ones on Earth.

The largest and most powerful telescope now in space, the Hubble Space Telescope, can just make out mountains on Mars. Pictures sharp enough to display geologic features of planets around other stars would require an array of space telescopes the size of the United States. Furthermore, as the late Carl Sagan of Cornell University has pointed out, pictures of Earth do not reveal the presence of life unless they are taken at very high resolution.

Taking photographs, however, is not the best way to start studying distant planets. Astronomers instead rely on the technique of spectroscopy to obtain most of their information. In spectroscopy, light originating from an object in space can be analyzed for unique markers that help researchers piece together characteristics such as the celestial body's temperature, atmospheric pressure and chemical composition.

The vital signs easiest to spot with spectroscopy are radio signals designed by extraterrestrials for interstellar communication. Such transmissions would be totally unlike natural phenomena; such unexpected features are examples of the kind of beacons that we must look for to locate intelligent life elsewhere. Yet sensitive scans of faraway star systems have not come across any signals, indicating only that extraterrestrials bent on interstellar radio communication are uncommon.

But planets may be home to noncommunicating life-forms, so we need to be able to find evidence of even the simplest organisms. To expand our capacity to locate distant planets and determine whether these worlds are inhabited, we have proposed a powerful and novel successor to Galileo's telescope that will, we believe, enable us to detect life on other planets.

The simplest forms of life on our planet altered the conditions on Earth in ways that a distant observer could perceive. Earth's humble blue-green algae do not operate radio transmitters, but they are chemical engineers par excellence. As algae became more widespread, they began adding large quantities of oxygen to the atmosphere. The production of oxygen is fundamental to carbon-based life: the simplest organisms take in water, nitrogen and carbon dioxide as nutrients and then release oxygen into the atmosphere as waste. Oxygen is a chemically reactive gas; without continued replenishment by algae and, later in Earth's evolution, by plants, its concentration would fall. Thus, the presence of large amounts of oxygen in a planet's atmosphere is the first indicator that some form of carbon-based life may exist there.

Oxygen leaves an unmistakable mark on the radiation emitted by a planet. For example, some of the sunlight that reaches Earth's surface is reflected through the atmosphere back toward space. Oxygen in the atmosphere absorbs some of this radiation, and thus an observer of Earth using spectroscopy to study the reflected sunlight could pick out the distinctive signature associated with oxygen.

In 1980 Toby C. Owen, then at the State University of New York at Stony Brook, suggested looking for oxygen's signal in the visible red light reflected by planets, as a sign of life there. Closer to home, Sagan reported in 1993 that the Galileo space probe recorded the distinctive spectrum of oxygen in the red region of visible light coming from Earth. Indeed, this indication of life's existence has been radiating a recognizable signal into space for at least the past 500 million years.

SEARCHING FOR ANOTHER EARTH

Our water-rich planet is obviously favorable to life. Water provides a solvent for the biochemical reactions of life to take place and serves as a source of needed hydrogen for living matter. Planets similar to Earth in size and distance from their Sun represent the most plausible homes for carbon-based life in other solar systems, primarily because liquid water could exist on these worlds. A planet's distance from its star determines its temperature— whether it will be too hot or too cold for liquid water.

We can easily estimate the "Goldilocks orbit"—the distance at which conditions are

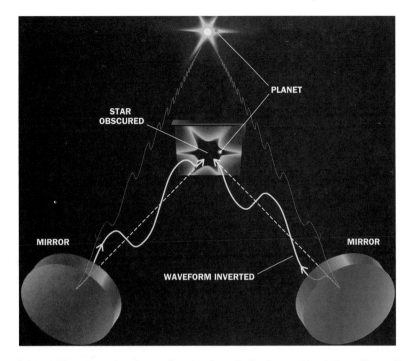

Canceling starlight enables astronomers to see dim planets typically obscured by stellar radiance. Two telescopes focused on the same star (top) can cancel out much of its light: one telescope inverts the light—making peaks into troughs and vice versa (right). When the inverted light is combined with the non-inverted starlight from the second telescope (left), the light waves interfere with one another, and the image of the star vanishes (center).

"just right" to generate and sustain life as it exists on Earth. For a large, hot star, 25 times as bright as our Sun, a hypothetical Earthlike planet would lie at about the distance that Jupiter circles the Sun. For a small, cool star, one tenth as bright as the Sun, the planet's orbit would resemble Mercury's course.

The best methods for detecting such bodies actually involve looking not at the planets themselves but at their stars. Astronomers watch for slight variations in a star's orbit or light emission that can be explained only by the presence of planets. Unfortunately, indirect observation of planets tells us little about their characteristics. Indeed, all indirect techniques reveal only a body's mass and position; ascertaining whether it carries inhabitants remains impossible.

SEEING INFRARED

Clearly, we need a different technique to reveal characteristics as specific as what chemicals can be found on a planet. Previously we mentioned that the visible radiation coming from a planet can confirm the presence of certain molecules, in particular oxygen, that we know support life. But distinguishing faint oxygen signals in light reflected by

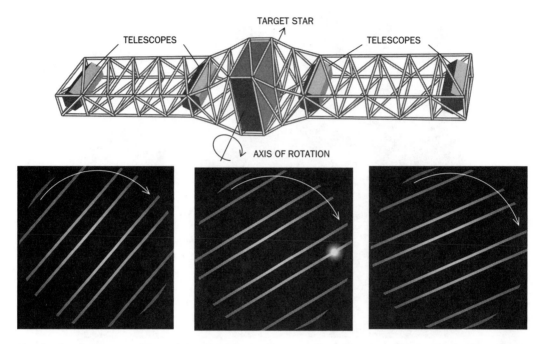

Rotating interferometer could reveal the existence of a planet around a distant star. The four telescopes arranged as shown above in the authors' proposed instrument would produce a composite view of the sky partially darkened by numerous bands; the star to be obscured would be hidden by one strip. As the instrument rotates about the line connecting the center of the device with the star, the dark bands will also rotate. A nearby planet would pass in and out of the bands (panels a–c). Scientists could then analyze the pattern of blinking to determine how far the planet is from its star.

a small planet around even a star in our own Sun's neighborhood would be extraordinarily difficult.

For example, the glow from a distant planet's Sun would outshine the planet by a factor of 10 billion. So hunting for planets can be as challenging as trying to pick out a glowworm sitting next to a searchlight, both of which are thousands of kilometers away. Even if we could pick out the light reflected by a planet, any oxygen features in its visible spectrum would be weak and remarkably hard to spot.

Faced with this quandary, in 1986 we proposed, along with Andrew Y. S. Cheng, that monitoring the mid-infrared wavelengths (longer than visible red wavelengths) emitted by a planet would be a better method for finding planets and looking for extraterrestrial life. This type of radiation—really the planet's radiated heat—has a wavelength 10 to 20 times longer than that of visible light. At these wavelengths, a planet emits about 40 times as many photons—particles of light—as it does at shorter wavelengths, and the nearby star would outshine the planet "only" 10 million times, a ratio 1,000 times more favorable than that which red light offers.

Moreover, three compounds that should appear together on inhabited planets—ozone (a form of oxygen usually located high in the atmosphere), carbon dioxide and water—are easily recognizable by examining the infrared spectrum. Once again, our solar system provides promising support for this technique: a survey of the infrared emissions of local planets reveals that only Earth displays the infrared signature of life. Although Earth, Mars and Venus all have atmospheres with carbon dioxide, only Earth shows the signature of plentiful water and ozone.

What kind of telescope do we need to locate Earthlike planets and pick up their infrared emissions? Some of today's ground-based telescopes can detect the strong infrared radiation emanating from stars. But the heat emitted by our atmosphere and by the telescope itself would completely swamp any sign of a planet. Even Antarctica is not nearly cold enough to enable us to pick out such a faint image: the telescope must be cooled to at least minus 225 degrees Celsius (about 50 kelvins). More troublesome, radiation passing through Earth's atmosphere is imprinted with exactly the features of ozone, carbon dioxide and water we hope to find on another planet. Obviously, we reasoned, we must move the telescope into space.

We can predict the performance of telescopes and thus know in advance what kind of image quality we can expect. For example, to monitor the infrared spectrum of an Earthlike planet circling, say, a star 30 light-years away, we would need an enormous space telescope, close to 60 meters in diameter. With current technology, the cost of such an instrument would rival the national debt. And even telescope enthusiasts such as ourselves regard the size of this device as daunting.

RETHINKING THE TELESCOPE

To develop a more reasonably sized telescope that would allow us to locate small, perhaps habitable, planets, we knew we would have to play some tricks with our instruments. One useful stratagem had been suggested by Ronald N. Bracewell of Stanford University. He showed how two small telescopes could be adapted to search for large, cool planets similar to Jupiter. The instrument he proposed consisted of two one-meter telescopes separated by 20 meters. Each telescope alone would have yielded blurred pictures that would never have enabled Bracewell to resolve the faint images of planets. But together the two devices could be arranged to observe distant worlds.

If he focused both telescopes on the same star, Bracewell envisaged that he would be able to invert the light waves from one telescope, flipping peaks into troughs and vice versa. Then he would combine the inverted light with light from the second telescope. Because the first image would be the reverse of the second, when Bracewell combined the two so that they overlapped precisely, the light from the star—both the core and the surrounding halo—would be canceled out. (The light would not disappear, of course; energy must be conserved. Instead the light from the star would be diverted to a separate part of the telescope.) Scientists refer to this type of device as an interferometer because it reveals details about the source of light by employing the interference of light waves.

The interferometer designed by Bracewell can obscure a star only if the star is perpendicular to the line joining the centers of each telescope. With such an arrangement, both telescopes receive exactly the same pattern of light waves from the star. If we sweep the instrument through the sky, stars will appear to blink as they move in and out of alignment.

A planet separated from its star by even a fairly small distance, however, will not be aligned with the device when its star is brought into alignment. The two telescopes will register the planet's signal at slightly different times, so the light waves from the planet will not cancel one another out. If light shines through the interferometer after we have canceled out the star's image, we know that some additional source of infrared radiation—perhaps a planet—exists near the star. We can analyze this signal by rotating the interferometer about the line joining the instrument and the star. The image will change intensity as the device rotates; planets should display a recognizable pattern of variation.

After working out the design for this interferometer, Bracewell realized that the main obstacle to locating a Jupiterlike planet would not be the overpowering light from a nearby star; it would instead be the heat radiated by dust particles in our solar system, referred to as zodiacal glow. The faint signal from a distant planet would be almost imperceptible against the background glare. Any hope of discovering a planet would require averaging data for at least one month to see through this glowing background.

To tackle these restrictions, a number of researchers, including the two of us, have been working on alternative strategies. In 1990 one of us (Angel) suggested that arranging four mirrors in a diamond pattern allows better cancellation of starlight. The instrument has one significant drawback, however. Because it is so effective at canceling out a star's light, the device can sometimes conceal a nearby planet as well.

Here the matter rested until 1995, when the National Aeronautics and Space Administration solicited from researchers a road map for the exploration of other solar systems. NASA selected three teams to investigate various methods for discovering planets around other stars. We assembled a team that included Bracewell, Léger and his colleague Jean-Marie Mariotti of the Paris Observatory, as well as some 20 other scientists and engineers. In particular, the two of us at the University of Arizona have been studying the potential of a new approach. We have designed an interferometer with two pairs of mirrors all arranged in a straight line. Each pair of mirrors will darken the star's main image, but significantly, each pair will also cancel the starlight leak of the other pair.

It turns out that because this interferometer cancels starlight very effectively, it can be made rather long, roughly 50 to 75 meters in length. The size of the instrument offers an important advantage: with this arrangement, the signals from planets are complex and unique. With the proper analysis, we can use the data from the interferometer to reconstruct an image of a distant solar system. As we envision the orbiting interferometer, it would point to a different star each day but could return to interesting systems for more extensive observations.

If pointed at our own solar system from a nearby star, the interferometer could pick out Venus, Earth, Mars, Jupiter and Saturn. And the data could be analyzed to determine the chemical composition of each planet's atmosphere. From our solar system, the device could easily study the newly discovered planet around 47 Ursae Majoris. More important, this interferometer could identify Earthlike planets elsewhere that would otherwise elude us, and the device can check all these planets for the presence of carbon dioxide, water and ozone.

Remarkably, the technology to assist in this discovery is at our fingertips. Soon we should be able to answer the centuries-old question, "Is life on Earth alone in the universe?"

The Search for Extraterrestrial Intelligence

Carl Sagan and Frank Drake

Is mankind alone in the universe? Or are there somewhere other intelligent beings looking up into their night sky from very different worlds and asking the same kind of question? Are there civilizations more advanced than ours, civilizations that have achieved interstellar communication and have established a network of linked societies throughout our galaxy? Such questions, bearing on the deepest problems of the nature and destiny of mankind, were long the exclusive province of theology and speculative fiction. Today, for the first time in human history, they have entered into the realm of experimental science.

From the movements of a number of nearby stars, we have now detected unseen companion bodies in orbit around them that are about as massive as large planets. From our knowledge of the processes by which life arose here on the Earth, we know that similar processes must be fairly common throughout the universe. Since intelligence and technology have a high survival value, it seems likely that primitive life-forms on the planets of other stars, evolving over many billions of years, would occasionally develop intelligence, civilization and a high technology. Moreover, we on the Earth now possess all the technology necessary for communicating with other civilizations in the depths of space. Indeed, we may now be standing on a threshold about to take the momentous step a planetary society takes but once: first contact with another civilization.

In our present ignorance of how common extraterrestrial life may actually be, any attempt to estimate the number of technical civilizations in our galaxy is necessarily unreliable. We do, however, have some relevant facts. There is reason to believe that solar systems are formed fairly easily and that they are abundant in the vicinity of the Sun. In our own solar system, for example, there are three miniature "solar systems": the satellite systems of the planets Jupiter, Saturn and Uranus. It is plain that however such systems are made, four of them formed in our immediate neighborhood.

The only technique we have at present for detecting the planetary systems of nearby stars is the study of the gravitational perturbations such planets induce in the motion of their parent star. Imagine a nearby star that over a period of decades moves measurably with respect to the background of more distant stars. Suppose it has a nonluminous companion that circles it. Both the star and the companion revolve around a common center of mass. The center of mass will trace a straight line against the stellar background, and thus the luminous star will trace a sinusoidal path. From the existence of the oscillation, we can deduce the existence of the companion. Furthermore, from the period and amplitude of the oscillation, we can calculate both the period and the mass of the companion. The technique is only sensitive enough, however, to detect the perturbations of a massive planet around the nearest stars.

We know that the master molecules of living organisms on the Earth are the proteins and the nucleic acids. The proteins are built up of amino acids, and the nucleic acids are built up of nucleotides. The Earth's primordial atmosphere was, like the rest of the universe, rich in hydrogen and in hydrogen compounds. From laboratory experiments we can determine the amount of amino acids produced per photon of ultraviolet radiation, and from our knowledge of stellar evolution, we can calculate the amount of ultraviolet radiation emitted by the Sun over the first billion years of the existence of the Earth. Those two rates enable us to compute the total amount of amino acids that were formed on the primitive Earth. Amino acids also break down spontaneously at a rate that is dependent on the ambient temperature. Hence, it is possible to calculate their steady-state abundance at the time of the origin of life. If amino acids in that abundance were mixed into the oceans of today, the result would be a 1 percent solution of amino acids. That is approximately the concentration of amino acids in the better brands of canned chicken bouillon, a solution that is alleged to be capable of sustaining life.

The origin of life is not the same as the origin of its constituent building blocks, but laboratory studies on the linking of amino acids into molecules resembling proteins and on the linking of nucleotides into molecules resembling nucleic acids are progressing well.

The laboratory experiments also yield a large amount of a brownish polymer that seems to consist mainly of long hydrocarbon chains. The spectroscopic properties of the polymer are similar to those of the reddish clouds on Jupiter, Saturn and Titan, the largest satellite of Saturn. Because the atmospheres of these objects are rich in hydrogen and are similar to the atmosphere of the primitive Earth, the coincidence is not surprising. It is nonetheless remarkable. Jupiter, Saturn and Titan may be enormous planetary laboratories engaged in prebiological organic chemistry.

Other evidence on the origin of life comes from the geologic record of the Earth. Thin sections of sedimentary rocks between 2.7 and 3.5 billion years old reveal the presence of small inclusions a hundredth of a millimeter in diameter. These inclusions have been identified by Elso S. Barghoorn of Harvard University and J. William Schopf of the University of California at Los Angeles as bacteria and blue-green algae. Bacteria and blue-green algae are evolved organisms and must themselves be the beneficiaries of a long evolutionary history. There are no rocks on the Earth or on the Moon, however, that are more than four billion years old; before that time the surface of both bodies is believed to have melted in the final stages of their accretion. Thus, the time available for the origin of life seems to have been short: a few hundred million years at the most. Since life originated on the Earth in a span much shorter than the present age of the Earth, we have additional evidence that the origin of life has a high probability, at least on planets with an abundant supply of hydrogen-rich gases, liquid water and sources of energy. Since those conditions are common throughout the universe, life may also be common.

Until we have discovered at least one example of extraterrestrial life, however, that conclusion cannot be considered secure. Such an investigation was one of the objectives of the Viking mission that landed on the surface of Mars in 1976 to conduct the first rigorous search for life on another planet. The Viking lander carried three separate experiments on the metabolism of hypothetical Martian microorganisms, one experiment on the organic chemistry of the Martian surface material and a camera system that might just conceivably detect macroscopic organisms if they exist. The results were inconclusive.

Intelligence and technology have developed on the Earth about halfway through the stable period in the lifetime of the Sun. There are obvious selective advantages to intelligence and technology, at least up to the present evolutionary stage. Barring such disasters, the physical environment of the Earth will remain stable for many more billions of years. It is possible that the number of individual steps required for the evolution of intelligence and technology is so large and improbable that not all inhabited planets evolve technical civilizations. It is also possible—some would say likely—that civilizations tend to destroy themselves at about our level of technological development. On the other hand, if there are 100 billion suitable planets in our galaxy, if the origin of life is highly probable, if there are billions of years of evolution available on each such planet and if even a small fraction of technical civilizations pass safely through the early stages of technological adolescence, the number of technological civilizations in the galaxy today might be very large.

It is obviously a highly uncertain exercise to attempt to estimate the number of such civilizations. The opinions of those who have considered the problem differ significantly. Our best guess is that there are a million civilizations in our galaxy at or beyond the Earth's present level of technological development. If they are distributed randomly through space, the distance between us and the nearest civilization should be about 300 light-years. Hence, any information conveyed between the nearest civilization and our own will take a minimum of 300 years for a one-way trip and 600 years for a question and a response.

Electromagnetic radiation is the fastest and also by far the cheapest method of establishing such contact. Interstellar space vehicles cannot be excluded a priori, but in all cases they would be a slower, more expensive and more difficult means of communication.

```
0 0 0 0 0 0 1 0 1 0 1 0 1 0 0 0 0 0 0 0 0 0 0 0 0 0 0 1 0 1 0 0 0 0 0 1 0 1 0
0 0 0 0 0 0 1 0 0 1 0 0 0 1 0 0 0 1 0 0 1 0 0 1 0 1 1 0 0 1 0 1 0 1 0 1
0 1 0 1 0 1 0 1 0 1 0 0 1 0 0 1 0 0 0 0 0 0 0 0 0 0 0 0 0 0 0 0 0 0 0 0
0 0 0 0 0 0 0 0 0 0 0 0 0 0 1 1 0 0 0 0 0 0 0 0 0 0 0 0 0 0 0 0 0 0 0 0
1 1 0 1 0 0 0 0 0 0 0 0 0 0 0 0 0 0 0 0 1 1 0 1 0 0 0 0 0 0 0 0 0 0
0 0 0 0 0 0 0 0 1 0 1 0 1 0 0 0 0 0 0 0 0 0 0 0 0 0 0 0 0 1 1 1 1 1 0
0 0 0 0 0 0 0 0 0 0 0 0 0 0 0 0 0 0 0 0 0 0 0 0 0 0 0 0 1 1 0 0 0 0
1 1 1 0 0 0 1 1 0 0 0 0 1 1 0 0 0 1 0 0 0 0 0 0 0 0 0 0 0 1 1 0 0 1 0
0 0 0 1 1 0 1 0 0 0 1 1 0 0 0 1 1 0 0 0 0 1 1 0 1 0 1 1 1 1 1 0 1 1 1 1 1
0 1 1 1 1 1 0 1 1 1 1 1 0 0 0 0 0 0 0 0 0 0 0 0 0 0 0 0 0 0 0 0 0 0 0
0 1 0 0 0 0 0 0 0 0 0 0 0 0 0 0 0 1 0 0 0 0 0 0 0 0 0 0 0 0 0 0 0 0
0 0 0 0 0 0 0 0 0 0 1 0 0 0 0 0 0 0 0 0 0 0 0 0 0 0 0 1 1 1 1 1 1 0 0
0 0 0 0 0 0 0 0 0 0 1 1 1 1 0 0 0 0 0 0 0 0 0 0 0 0 0 0 0 0 0 0 0 0 0
0 0 1 1 0 0 0 0 1 1 0 0 0 0 1 1 1 0 0 0 1 1 0 0 0 1 0 0 0 0 0 0 1 0 0 0
0 0 0 0 0 0 1 0 0 0 0 1 1 0 1 0 0 0 0 1 1 0 0 0 1 1 1 0 0 1 1 0 1 0 1 1 1
1 1 0 1 1 1 1 1 0 1 1 1 1 1 0 1 1 1 1 0 0 0 0 0 0 0 0 0 0 0 0 0 0 0
0 0 0 0 0 0 0 0 0 1 0 0 0 0 0 0 1 1 0 0 0 0 0 0 0 0 1 0 0 0 0 0 0 0 0
0 0 1 1 0 0 0 0 0 0 0 0 0 0 0 0 0 0 0 1 0 0 0 0 0 1 1 0 0 0 0 0 0 0 0 0
1 1 1 1 1 1 0 0 0 0 0 1 1 0 0 0 0 0 0 1 1 1 1 0 0 0 0 0 0 0 0 0 0 1 1 0
0 0 0 0 0 0 0 0 0 0 0 0 0 1 0 0 0 0 0 0 0 1 0 0 0 0 0 0 0 0 1 0 0 0 0 0 1
0 0 0 0 0 0 1 1 0 0 0 0 0 0 0 1 0 0 0 0 0 0 0 1 1 0 0 0 0 1 1 0 0 0 0 0 0
1 0 0 0 0 0 0 0 0 0 0 1 1 0 0 0 1 0 0 0 0 1 1 0 0 0 0 0 0 0 0 0 0 0 0
0 1 1 0 0 1 1 0 0 0 0 0 0 0 0 0 0 0 0 1 1 0 0 0 1 0 0 0 0 1 1 0 0 0 0 0
0 0 0 0 1 1 0 0 0 0 1 1 0 0 0 0 0 0 1 0 0 0 0 0 0 1 0 0 0 0 0 0 1 0 0 0
0 0 0 0 0 1 0 0 0 0 1 0 0 0 0 0 0 0 1 1 0 0 0 0 0 0 0 1 0 0 0 1 0 0 0
0 0 0 0 0 1 1 0 0 0 0 0 0 0 1 0 0 0 1 0 0 0 0 0 0 0 0 1 0 0 0 0 0 0
1 0 0 0 0 0 1 0 0 0 0 0 0 0 1 0 0 0 0 0 0 1 0 0 0 0 0 0 1 0 0 0 0 0 0
0 0 0 0 0 0 1 1 0 0 0 0 0 0 0 0 1 1 0 0 0 0 0 0 0 0 1 1 0 0 0 0 0 0 0
0 1 0 0 0 1 1 1 0 1 0 1 1 0 0 0 0 0 0 0 0 0 1 0 0 0 0 0 0 0 1 0 0 0 0
0 0 0 0 0 0 0 0 0 1 0 0 0 0 0 0 1 1 1 1 0 0 0 0 0 0 0 0 0 0 0 1 0 0 0
0 1 0 1 1 1 0 1 0 0 1 0 1 1 0 1 1 0 0 0 0 0 0 1 0 0 1 1 1 0 0 1 0 0 0 1 1 1
1 1 1 1 0 1 1 1 0 0 0 0 1 1 1 0 0 0 0 0 1 1 0 1 1 1 1 0 0 0 0 0 0 0 0 1 0
1 0 0 0 0 0 1 1 1 0 1 1 0 0 1 0 0 0 0 0 1 0 1 0 0 0 0 0 1 1 1 1 1 1 0 0
1 0 0 0 0 0 0 1 0 1 0 0 0 0 0 0 1 1 0 0 0 0 0 1 0 0 0 0 0 1 1 0 1 1 0 0 0
0 0 0 0 0 0 0 0 0 0 0 0 0 0 0 0 0 0 0 0 0 0 0 0 0 0 0 0 0 0 0 1 1 1 0 0
0 0 0 1 0 0 0 0 0 0 0 0 0 0 0 0 0 1 1 1 0 1 0 1 0 0 0 1 0 1 0 1 0 1 0 1
0 1 0 0 1 1 1 0 0 0 0 0 0 0 0 0 1 0 1 0 1 0 1 0 0 0 0 0 0 0 0 0 0 0 0
0 0 1 0 1 0 0 0 0 0 0 0 0 0 0 0 0 1 1 1 1 0 0 0 0 0 0 0 0 0 0 0 0 0 0
0 0 0 1 1 1 1 1 1 1 1 0 0 0 0 0 0 0 0 0 0 0 1 1 1 0 0 0 0 0 0 0 1 1 1
0 0 0 0 0 0 0 0 0 1 1 0 0 0 0 0 0 0 0 0 0 1 1 0 0 0 0 0 0 0 1 1 0 1 0 0
0 0 0 0 0 0 0 1 0 1 1 0 0 0 0 0 1 1 0 0 1 1 0 0 0 0 0 0 0 1 1 0 0 1 1 0 0
0 0 1 0 0 0 1 0 1 0 0 0 0 0 1 0 1 0 0 0 1 0 0 0 1 0 0 0 1 0 0 1 0 0 0 1
0 0 1 0 0 0 1 0 0 0 0 0 0 0 1 0 0 0 1 0 1 0 0 0 1 0 0 0 0 0 0 0 0 0 0
0 1 0 0 0 0 1 0 0 0 0 1 0 0 0 0 0 0 0 0 0 0 0 1 0 0 0 0 0 0 0 0 0 1 0 0
0 0 0 0 0 0 0 0 0 0 0 0 1 0 0 1 0 1 0 0 0 0 0 0 0 0 0 0 1 1 1 1 0 0 1 1
1 1 1 0 1 0 0 1 1 1 1 0 0 0
```

Arecibo Message in binary code was transmitted in 1974 toward the Great Cluster in Hercules from the 1,000-foot antenna at Arecibo. The message is decoded by breaking up the characters into 73 consecutive groups of 23 characters each and arranging the groups in sequence one under the other, reading from right to left and then from top to bottom. The result is a visual message that can be more easily interpreted by making each 0 of the numerical binary code represent a white square and each 1 of the code a black square.

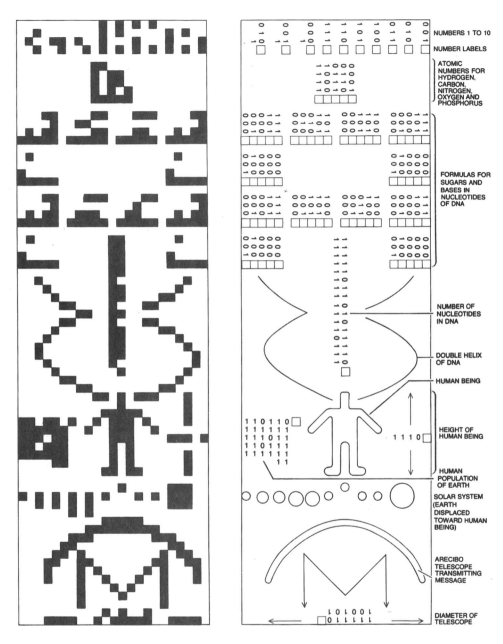

Arecibo Message in a binary version is decoded. Each number that is used is marked with a label that indicates its start. When all the digits of a number cannot be fitted into one line, the digits for which there is no room are written under the least significant digit. (The message must be oriented in three different ways for all the numbers shown to be read.) The chemical formulas are those for the components of the DNA molecule: the phosphate group, the deoxyribose sugar and the organic bases thymine, adenine, guanine and cytosine. Both the height of the human being and the diameter of the Aricebo telescope are given in units of the wavelength that is used to transmit the message, namely, 12.6 centimeters.

Since we have achieved the capability for interstellar radio communication only in the past few decades, there is virtually no chance that any civilization we come in contact with will be as backward as we are. There also seems to be no possibility of dialogue except between very long-lived and patient civilizations. In view of these circumstances, which should be common to and deducible by all the civilizations in our galaxy, it seems to us quite possible that one-way radio messages are being beamed at the Earth at this moment by radio transmitters on planets in orbit around other stars.

To intercept such signals, we must guess or deduce the frequency at which the signal is being sent, the width of the frequency band, the type of modulation and the star transmitting the message. Although the correct guesses are not easy to make, they are not as hard as they might seem.

The frequency of the spin-flip transition of hydrogen at 1,420 megahertz was first suggested as a channel for interstellar communication in 1959 by Philip Morrison and Giuseppe Cocconi. Such a channel may be too noisy for communication precisely because hydrogen, the most abundant interstellar gas, absorbs and emits radiation at that frequency. The number of other plausible and available communication channels is not large, so that determining the right one should not be too difficult.

We cannot use a similar logic to guess the bandwidth that might be used in interstellar communication. The narrower the bandwidth is, the farther a signal can be transmitted before it becomes too weak for detection. On the other hand, the narrower the bandwidth is, the less information the signal can carry. A compromise is therefore required between the desire to send a signal the maximum distance and the desire to communicate the maximum amount of information. Perhaps simple signals with narrow bandwidths are sent to enhance the probability of the signals' being received. Perhaps information-rich signals with broad bandwidths are sent in order to achieve rapid and extensive communication. The broad-bandwidth signals would be intended for those enlightened civilizations that have invested major resources in large receiving systems.

We do not, of course, know now which star we should listen to. The most conservative approach is to turn our receivers to stars that are rather similar to the Sun, beginning with the nearest. Two nearby stars, Epsilon Eridani and Tau Ceti, both about 12 light-years away, were the candidates for Project Ozma, the first search with a radio telescope for extraterrestrial intelligence, conducted by one of us (Drake) in 1960. Project Ozma, named after the ruler of Oz in L. Frank Baum's children's stories, was "on the air" for four weeks at 1,420 megahertz. The results were negative. Since then there have been a number of other studies. In spite of some false alarms to the contrary, none has been successful. The lack of success is not unexpected. If there are a million technical civilizations in a galaxy of some 200 billion stars, we must turn our receivers to 200,000 stars before we have a fair statistical chance of detecting a single extraterrestrial message. So far we have listened to only a few more than 200 stars. In other words, we have mounted only .1 percent of the required effort.

Our present technology is entirely adequate for both transmitting and receiving messages across immense interstellar distances. For example, if the 1,000-foot radio telescope at the Arecibo Observatory in Puerto Rico were to transmit information at the rate of one bit (binary digit) per second with a bandwidth of one hertz, the signal could be received by an identical radio telescope anywhere in the galaxy. By the same token, the Arecibo telescope could detect a similar signal transmitted from a distance hundreds of times greater than our estimate of 300 light-years to the nearest extraterrestrial civilization.

A search of hundreds of thousands of stars in the hope of detecting one message would require remarkable dedication and would probably take several decades. It seems unlikely that any existing major radio telescope would be given over to such an intensive program to the exclusion of its usual work.

So far we have been discussing the reception of messages that a civilization would intentionally transmit to the Earth. An alternative possibility is that we might try to "eavesdrop" on the radio traffic an extraterrestrial civilization employs for its own purposes. Such radio traffic could be readily apparent. On the Earth, for example, a new radar system employed with the telescope at the Arecibo Observatory for planetary studies emits a narrow-bandwidth signal that, if it were detected from another star, would be between one million and 10 billion times brighter than the Sun at the same frequency. In addition, because of radio and television transmission, the Earth is extremely bright at wavelengths of about a meter. If the planets of other civilizations have a radio brightness comparable to the Earth's from television transmission alone, they should be detectable. Because of the complexity of the signals and the fact that they are not beamed specifically at the Earth, however, the receiver we would need in order to eavesdrop would have to be much more elaborate and sensitive than any radio-telescope system we now possess.

One such system has been devised in a preliminary way by Bernard M. Oliver of the Hewlett-Packard Company, who directed a study sponsored by the Ames Research Center of the National Aeronautics and Space Administration. The system, known as Cyclops, would consist of an enormous radio telescope connected to a complex computer system. The computer system would be designed particularly to search through the data from the telescope for signals bearing the mark of intelligence, to combine numerous adjacent channels in order to construct signals of various effective bandwidths and to present the results of the automatic analyses for all conceivable forms of interstellar radio communication in a way that would be intelligible to the scientists involved in the project.

To construct a radio telescope of enormous aperture as a single antenna would be prohibitively expensive. The Cyclops system would instead capitalize on our ability to connect many individual antennas to act in unison. This concept is already the basis of the Very Large Array in New Mexico. This array consists of 27 antennas, each 82 feet in diameter, arranged in a Y-shaped pattern whose three arms are each 10 miles long. The Cyclops system would be much larger. Its current design calls for 1,500 antennas each 100 meters in diameter, all electronically connected to one another and to the computer system. The Cyclops array would be as compact as possible but nonetheless would cover perhaps 25 square miles.

The effective signal-collecting area of the system would be hundreds of times the area of any existing radio telescope, and it would be capable of detecting even relatively weak signals such as television transmissions from civilizations several hundred light-years away. Moreover, it would be the instrument par excellence for receiving signals specifically directed at the Earth.

The estimated cost of the Cyclops system, ranging up to $10 billion, may make it prohibitively expensive for the time being. Moreover, the argument in favor of eavesdropping is not completely persuasive. Half a century ago, before radio transmissions were commonplace, the Earth was quiet at radio wavelengths. Half a century from now, because of the development of cable television and communications satellites that relay signals in a narrow beam, the Earth may again be quiet. Thus, perhaps for only a century out of billions of years do planets such as the Earth appear remarkably bright at radio wavelengths. The odds of our discovering a civilization during that short period in its history may not be good enough to justify the construction of a system such as Cyclops.

How could we be sure that a particular radio signal was deliberately sent by an intelligent being? It is easy to design a message that is unambiguously artificial. The first 30 prime numbers, for example, would be difficult to ascribe to some natural astrophysical phenomenon. A simple message of this kind might be a beacon or announcement signal. A subsequent informative message could have many forms and could consist of an enormous number of bits. One method of transmitting information, beginning simply and progressing to more elaborate concepts, is pictures.

One final approach in the search for extraterrestrial intelligence deserves mention. If there are indeed civilizations thousands or millions of years more advanced than ours, it is entirely possible that they could beam radio communications over immense distances, perhaps even over the distances of intergalactic space. We do not know how many advanced civilizations there might be compared with the number of more primitive Earthlike civilizations, but many of these older civilizations are bound to be in galaxies older than our own. For this reason the most readily detectable radio signals from another civilization may come from outside our galaxy. The relatively small number of such extragalactic transmitters might be more than compensated for by the greater strength of their signals. At the appropriate frequency they could even be the brightest radio signals in the sky. Therefore, an alternative to examining the nearest stars of the same spectral type as the Sun is to examine the nearest galaxies. Spiral galaxies such as the Great Nebula in Andromeda are obvious candidates, but the elliptical galaxies are much older and more highly evolved and could conceivably harbor a large number of extremely advanced civilizations.

Should we be sending messages ourselves? It is obvious that we do not yet know where we might best direct them. One message has already been transmitted to the Great Cluster in Hercules by the Arecibo radio telescope, but only as a kind of symbol of the capabilities of our existing radio technology. Any radio signal we send would be detectable over interstellar distances if it is more than about 1 percent as bright as the Sun at the same frequency. Actually something close to 1,000 such signals from our everyday internal communications have left the Earth every second for the past two decades. This electromagnetic frontier of mankind is now some 20 light-years away, and it is moving outward at the speed of light. Its spherical wave front, expanding like a ripple from a disturbance in a pool of water and inadvertently carrying the news that human beings have achieved the capacity for interstellar discourse, envelops about 20 new stars each year.

How much do we care? Enough to devote an appreciable effort with existing telescopes to search for life elsewhere in the universe? Enough to take a major step such as Project Cyclops that offers a greater chance of carrying us across the threshold, to communicate at last with a variety of extraterrestrial beings who, if they exist, would inevitably enrich mankind beyond imagination? The real question is not how, because we know how; the question is when. If enough of the beings of the Earth cared, the threshold might be crossed within the lifetime of most of those alive today.

Giant Planets Orbiting Faraway Stars

Geoffrey W. Marcy and R. Paul Butler

No doubt humans have struggled with the question of whether we are alone in the universe since the beginning of consciousness. Today, armed with evidence that planets do indeed orbit other stars, astronomers wonder more specifically: What are those planets like? Of the 100 billion stars in our Milky Way Galaxy, how many harbor planets? Among those planets, how many constitute arid deserts or frigid hydrogen balls? Do some contain lush forests or oceans fertile with life? For the first time in history, astronomers can now address these questions concretely. During the past two and a half years, researchers have detected eight planets orbiting sunlike stars. In October 1995 Michel Mayor and Didier Queloz of Geneva Observatory in Switzerland reported finding the first planet. Observing the star 51 Pegasi in the constellation Pegasus, they noticed a telltale wobble, a cyclical shifting of its light toward the blue and red ends of the spectrum. The timing of this Doppler shift suggests that the star wobbles because of a closely orbiting planet, which revolves around the star fully every 4.2 days—at a whopping speed of 482,000 kilometers (299,000 miles) an hour, more than four times faster than the Earth orbits the Sun.

Another survey of 107 sunlike stars, performed by our team at San Francisco State University and the University of California at Berkeley, has turned up six more planets. Of those, one planet circling the star 16 Cygni B was independently discovered by astronomers William D. Cochran and Artie P. Hatzes of the University of Texas McDonald Observatory on Mount Locke in western Texas.

Detection of an eighth planet was reported in April 1997, when a nine-member team led by Robert W. Noyes of Harvard University detected a planet orbiting the star Rho Coronae Borealis. A ninth large object, which orbits the star known by its catalogue number HD114762, has also been observed—an object first detected in 1989 by astronomer David W. Latham of the Harvard-Smithsonian Center for Astrophysics and his collaborators. But this bulky companion has a mass more than 10 times that of Jupiter—large, though not unlike another large object discovered around the star 70 Virginis, a similar object with a mass 6.8 times that of Jupiter. The objects orbiting both HD114762 and 70 Virginis are so large that most astronomers are not sure whether to consider them big planets or small brown dwarfs, entities whose masses lie between those of a planet and a star.

DETECTING EXTRASOLAR PLANETS

Finding extrasolar planets has taken a long time because detecting them from Earth, even using current technology, is extremely difficult. Unlike stars, which are fueled by nuclear reactions, planets faintly reflect light and emit thermal infrared radiation. In our solar system, for example, the Sun outshines its planets about one billion times in visible light and one million times in the infrared. Because of the distant planets' faintness, astronomers have had to devise special methods to locate them. The current leading approach is the Doppler planet-detection technique, which involves analyzing wobbles in a star's motion.

Here's how it works. An orbiting planet exerts a gravitational force on its host star, a force that yanks the star around in a circular or oval path—which mirrors in miniature the planet's orbit. Like two twirling dancers tugging each other in circles, the star's wobble reveals the presence of orbiting planets, even though we cannot see them directly.

The trouble is that this stellar motion appears very small from a great distance. Someone gazing at our Sun from 30 light-years away would see it wobbling in a circle whose radius measures only one seventh of one millionth of one degree. In other words, the Sun's tiny, circular wobble appears only as big as a quarter viewed from 10,000 kilometers away.

Yet the wobble of the star is also revealed by the Doppler effect of the starlight. As a star sways to and fro relative to Earth, its light waves become cyclically stretched, then compressed—shifting alternately toward the red and blue ends of the spectrum. From that cyclical Doppler shifting, astronomers can retrace the path of the star's wobble and, from Newton's law of motion, compute their masses, orbits and distances from their host stars. The cyclical Doppler shift itself remains extremely tiny: stellar light waves shrink and expand by only about one part in 10 million because of the pull of a large, Jupiterlike planet. The Sun, for example, wobbles with a speed of only about 12.5 meters per second, pivoting around a point just outside its surface. To detect planets around other stars, measurements must be highly accurate, with errors in stellar velocities below 10 meters per second.

Using the Doppler technique, our group can now measure stellar motions with an accuracy of plus or minus three meters per second—a leisurely bicycling speed. To do this, we use an iodine absorption cell—a bottle of iodine vapor—placed near a telescope's focus. Starlight passing through the iodine is stripped of specific wavelengths, revealing tiny shifts in its remaining wavelengths. So sensitive is this technique that we can measure wavelength changes as small as one part in 100 million.

As recorded by spectrometers and analyzed by computers, a star's light reveals the tell-tale wobble produced by its orbiting companions. For example, Jupiter, the largest

ORBIT OF STAR AND PLANET
AS VIEWED FROM SIDE

STAR

PLANET

ORBIT OF STAR AND PLANET
AS VIEWED FROM TOP

A planet orbiting its host star caus-es the star to wobble. Although Earth-based astronomers have not yet been able to see an orbiting planet they can deduce its size, mass and distance from its host by analyzing the to-and-fro oscillation of that star's light.

planet in our solar system, is one thousandth the mass of the Sun. Therefore, every 11.8 years (the span of Jupiter's orbital period) the Sun oscillates in a circle that is one thou-sandth the size of Jupiter's orbit. The other eight planets also cause the Sun to wobble, albeit by smaller amounts. Take Earth, having a mass 1/318 that of Jupiter and an orbit five times closer: it causes the Sun to move a mere nine centimeters a second.

Yet some uncertainty about each extrasolar planet's mass remains. Orbital planes that astronomers view edge-on will give the true mass of the planet. But tilted orbital planes reduce the Doppler shift because of a smaller to-and-fro motion, as witnessed from Earth. This effect can make the mass appear smaller than it is. Without knowing a plan-et's orbital inclination, astronomers can compute only the least possible mass for the planet; the actual mass could be larger.

Thus, using the Doppler technique to analyze light from about 300 stars similar to the Sun—all within 50 light-years of Earth—astronomers have turned up eight planets sim-ilar in size and mass to Jupiter and Saturn. Specifically, their masses range from about a half to seven times that of Jupiter, their orbital periods span 3.3 days to three years and their distances from their host stars extend from less than one twentieth of Earth's distance to the Sun to more than twice that distance.

To our surprise, the eight newly found planets exhibit two unexpected characteristics.

First, unlike planets in our solar system, which display circular orbits, two of the new planets move in eccentric, oval orbits around their hosts. Second, five of the new planets orbit very near their stars—closer, in fact, than Mercury orbits the Sun. Exactly why these huge planets orbit so closely—some skim just over their star's blazing coronal gases—remains unclear. These findings are mysterious, given that the radius of Jupiter's orbit is five times larger than that of Earth. These observations, in turn, provoke questions about our own solar system's origin, prompting some astronomers to revise the standard explanation of planet formation.

RECONSIDERING HOW PLANETS FORM

What we have learned about the nine planets in our own solar system has constituted the basis for the conventional theory of planet formation. The theory holds that planets form from a flat, spinning disk of gas and dust that bulges out of a star's equatorial plane, much as pizza dough flattens when it is tossed and spun. This model shows the disk's material orbiting circularly in the same direction and plane as our nine planets do today. Based on this theory, planets cannot form too close to the star, because there is too little disk material, which is also too hot to coalesce. Nor do planets clump extremely far from the star, because the material is too cold and sparse.

Considering what we now know, such expectations about planets in the rest of the universe seem narrow-minded. The planet orbiting the star 47 Ursae Majoris in the Big Dipper constellation stands as the only one resembling what we expected, with a minimum bulk of 2.4 Jupiter-masses and a circular orbit with a radius of 2.1 astronomical units (AU)—1 AU representing the 150-million-kilometer distance from Earth to the Sun. Only a bit more massive than Jupiter, this planet orbits in a circle farther from its star than Mars does from the Sun. If placed in our solar system, this new planet might appear as Jupiter's big brother.

But the remaining planetary companions around other stars baffle us. The two planets with oval orbits have eccentricities of 0.68 and 0.40. (An eccentricity of zero is a perfect circle, whereas an eccentricity of 1.0 is a long, slender oval.) In contrast, in our solar system the greatest eccentricities appear in the orbits of Mercury and Pluto, both about 0.2; all other planets show nearly circular orbits (eccentricities less than 0.1).

These eccentric orbits have prodded astronomers to scratch their heads and revise their theories. Within two months of the first planet sighting, theorists hatched new ideas and adjusted the standard planet formation theory.

For instance, astronomers Pawel Artymowicz of the University of Stockholm and Patrick M. Cassen of the National Aeronautics and Space Administration Ames Research Center recalculated the gravitational forces at work when planets emerge from disks of gas and dust seen swirling around young, sunlike stars. Their calculations

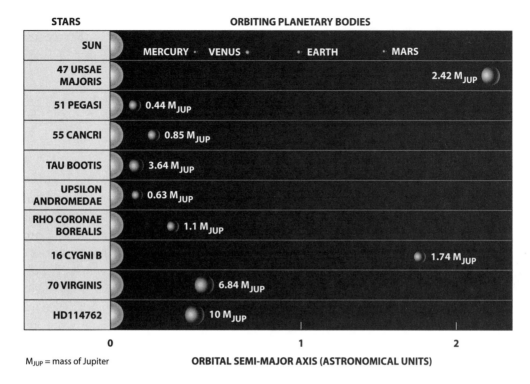

STARS	ORBITING PLANETARY BODIES
SUN	MERCURY · VENUS · · EARTH · MARS
47 URSAE MAJORIS	2.42 M_JUP
51 PEGASI	0.44 M_JUP
55 CANCRI	0.85 M_JUP
TAU BOOTIS	3.64 M_JUP
UPSILON ANDROMEDAE	0.63 M_JUP
RHO CORONAE BOREALIS	1.1 M_JUP
16 CYGNI B	1.74 M_JUP
70 VIRGINIS	6.84 M_JUP
HD114762	10 M_JUP

M_{JUP} = mass of Jupiter

0   1   2

ORBITAL SEMI-MAJOR AXIS (ASTRONOMICAL UNITS)

Planetary objects orbiting distant stars include eight planets, plus HD114762, which—with its large mass—may be a planet or a brown dwarf. These planets show a wide range of orbital distances and eccentricities, which has prompted theorists to revise standard planet-formation theories.

show that gravitational forces exerted by protoplanets—planets in the process of forming—on the gaseous, dusty disks create alternating spiral "density waves." Resembling the "arms" of spiral galaxies, these waves exert forces back on the forming planets, driving them from circular motion. Over millions of years, planets can easily wander from circular orbits into eccentric, oval ones.

A second theory also accounts for large orbital eccentricities. Suppose, for instance, that Saturn had grown much larger than it actually is. Conceivably, all four giant planets in our solar system—Jupiter, Saturn, Uranus and Neptune—could have swelled into bigger balls if our original protoplanetary disk had contained more mass or had existed longer. In this case, the solar system would contain four superplanets, exerting gravitational forces on one another, perturbing one another's orbits and causing them to intersect.

Eventually, some of the superplanets might be gravitationally thrust inward, others outward, an unlucky few even ejected from the planetary system. Like balls ricocheting on a billiards table, the scattered giant planets might adopt extremely eccentric orbits, as we now observe for three of the new planets. Interestingly, this billiards model for eccentric planets shows that we should be able to detect the massive planets causing eccentric orbits—planets perhaps orbiting farther out than the planets we have detected thus far. A variation on this theme suggests that a companion star, rather than other planets, might gravitationally scatter planet orbits.

The most bizarre of the new planets are the four so-called 51 Peg planets, which show orbital periods shorter than 15 days. The four members of this class are 51 Peg itself, Tau Bootis, 55 Cancri and Upsilon Andromedae, which have orbital periods of just 4.2, 3.3, 14.7 and 4.6 days, respectively.

These orbits are all small, with radii less than one tenth the distance between Earth and the Sun—indeed, less than one third of Mercury's distance from the Sun. Yet these planets are as big as, or bigger than, the largest planet in our solar system. They range in mass from 0.44 of Jupiter's mass for 51 Peg to 3.64 of Jupiter's mass for Tau Bootis. Their Doppler shifts suggest that these planets orbit in circles.

MYSTERIOUS 51 PEGASI–TYPE PLANETS

The 51 Peg planets defy conventional planet formation theory, which predicts that giant planets such as Jupiter, Saturn, Uranus or Neptune would form in the cooler outskirts of a protoplanetary disk, at least five times the distance from Earth to the Sun.

To account for these planetary oddities, a revised planet formation theory is making the rounds in theorists' circles. Astronomers Douglas N. C. Lin and Peter Bodenheimer, both of the University of California at Santa Cruz, and Derek C. Richardson of the University of Washington extend the standard model by arguing that a young protoplanet precipitating out of a massive protoplanetary disk will carve a groove in the disk, separating it into inner and outer sections. According to their theory, the inner disk dissipates energy because of dynamical friction, causing the disk material and the protoplanet to spiral inward and eventually plunge into the host star.

A planet's salvation stems from the young star's rapid rotation, spinning every five to 10 days. Approaching its star, a planet would cause tides on the star to rise, just as the Moon raises tides on Earth. With the young star rotating faster than the protoplanet orbiting the star, the star would tend to sprout a bulge whose gravity would tug the planet forward. This effect would tend to whip the protoplanet into a larger orbit, halting its deathly inward spiral.

In this model, the protoplanet hangs poised in a stable orbit, delicately balanced between the disk's drag and the rotating star's forward tug. Even before the discovery of the 51 Peg planets, Lin predicted that Jupiter should have spiraled into the Sun during its formation. If this were so, then why did Jupiter survive? Perhaps our solar system contained previous "Jupiters" that did indeed spiral into the Sun, leaving our Jupiter as the sole survivor.

Why, we wonder, does no large 51 Peg–like planet orbit close to our Sun? Perhaps Jupiter formed near the end of our protoplanetary disk's lifetime. Or the protoplanetary disk may have lacked enough gas and dust to exert sufficient tidal drag. Perhaps protoplanetary disks come in a wide range of masses, from a few Jupiter-masses to hundreds of Jupiter-masses. In that case, the diversity of new planets may correspond to different disk masses or disk lifetimes, perhaps even to different environments, including the presence or absence of nearby radiation-emitting stars.

On the other hand, astronomer David F. Gray of the University of Western Ontario in Canada has challenged the existence of the 51 Peg planets altogether. Gray argues that the alleged planet-bearing stars are themselves oscillating—almost like wobbling water balloons. In his view, the cyclical Doppler shifts in these stars stem from inherent stellar wobbles, not planets tugging at stars.

Armed with new data, astronomers now largely dismiss the existence of the oscillations. The strongest argument against the oscillations stems from the single period and frequency seen in the Doppler variations from the star. Most oscillating systems, such as tuning forks, display a set of harmonics, or several different oscillations occurring at different frequencies, rather than just one frequency. But the 51 Peg stars show only one period each, quite unlike harmonic oscillations.

Moreover, ordinary physical models predict that the strongest wobbles would occur at higher frequencies than those of the observed oscillations of these stars. In addition, the 51 Peg stars show no variations in brightness, suggesting that their sizes and shapes are not changing.

PLANETARY COMPARISONS

Although we are tempted to compare the eight new planets with our own nine, the comparison is, unfortunately, quite challenging. No one can draw firm conclusions from only eight new planets. So far our ability to spot other types of planets remains limited. At present, our instruments cannot even detect Earth-size companions. Although the extrasolar planets found to date have orbital periods no longer than three years, this finding does not necessarily represent planetary systems in general. Rather it arises from the fact that astronomers have searched for other planets with better techniques for only about a decade. With more time and improved Doppler precision, more planets with longer orbital periods may be found.

Curiously, finding these new planets proves that our own history could easily have played out quite differently. Suppose that gravitational scattering of planets occurs commonly in planetary systems. We see in our own solar system evidence that during its first billion years, planetesimals—fragmentary bodies of rock and ice—hurtled through space. Our cratered moon and Uranus's highly tilted axis—nearly perpendicular to the axes of all its neighbors—show that collisions were common, some involving planet-size objects. The neatly carved orbits of our now stable solar system emerged from the collision-happy orbits of its youth.

We should consider ourselves lucky that Jupiter ended up in a nearly circular orbit. If it had careened into an oval orbit, Jupiter might have scattered Earth, thwacking it out of the solar system. Without stable orbits for Earth and Jupiter, life might never have emerged.

THE FUTURE OF PLANET HUNTING

In July 1996 we began a second Doppler survey of 400 stars, using the 10-meter Keck telescope at Mauna Kea Observatory in Hawaii. Mayor and Queloz of Geneva Observatory recently tripled the size of their Northern Hemisphere Doppler survey to about 400 stars, and soon they will begin a Southern Hemisphere survey of 500 more stars. Within the next year, Doppler surveys of several hundred additional stars will begin at the nine-meter Hobby-Eberly Telescope located at McDonald Observatory.

By the year 2000 two Keck telescopes on Mauna Kea and a binocular telescope at the University of Arizona will become optical interferometers, precise enough to image extrasolar planets. NASA plans to launch at least three spaceborne telescopes to detect planets in infrared light.

One proposed NASA space-based interferometer, a second-generation telescope known as the Terrestrial Planet Finder, should obtain pictures of candidate habitable planets orbiting distant stars. Arguably the greatest telescope ever conceived, Planet Finder could spot other Earths, starting in about 2010. Using a spectrometer, it could analyze light from far-off planets to determine the chemical makeup of their atmospheres—data to determine if biological activity is proceeding. This monumental, spaceborne telescope would span a football field and sport four huge mirrors. Drawing from the data on planets found so far, we believe other planets orbit similar stars, many the size of Jupiter, some the size of Earth. It may be that as many as 10 percent of all stars in our galaxy host planetary companions. Based on this estimate, 10 billion planets would exist in our Milky Way galaxy alone.

Seeking the ideal Earthlike planet on which life could flourish, astronomers will search for planets that are neither too cold nor too hot, temperate enough to sustain liquid water to serve as the mixer and solvent for organic chemistry and biochemistry. Planets with the perfect blend of molecular constituents orbiting at just the right distance from the Sun enjoy what astronomers call a "Goldilocks" orbit.

Seeing such a planet would spawn an endless stream of questions: Does its atmosphere contain oxygen, nitrogen and carbon dioxide, like Earth's, or sulfuric acid and CO_2, the deadly combination on Venus? Is there a protective ozone layer, or is the surface scorched by harmful ultraviolet rays? Even if a planet has oceans, does the water have a pH neutral enough to permit cells to grow?

There may even exist some other biology that thrives on sulfuric acid—even starves without it. Indeed, if primitive life does arise on another Earth, does it always evolve toward intelligence, or is our human technology some fluke of Darwinian luck? Are we humans a rare quirk of nature, destined to appear on Earthlike planets only once in a universe that otherwise teems with primitive life?

Amazing as it seems, answers to some of these questions may arise during our lifetimes, using tools such as telescopes already in existence or on the drawing board. We can only barely imagine what the next generation will see in our reconnaissance of the galactic neighborhood. Human destiny lies in exploring the galaxy and finding our roots, biologically and chemically, out among the stars.

End of the Proterozoic Eon

Andrew H. Knoll

Living organisms have inhabited the surface of our planet for nearly four billion years. Yet the plants and animals that define our everyday existence have far more recent origins. The ancestors of modern trees and terrestrial animals first colonized land only about 450 million years ago. In the oceans, animals have a longer record, but macroscopic invertebrates did not appear even there until about 580 million years ago—roughly 85 percent of the way through life's history. The earliest animals, which are collectively referred to as the Ediacaran fauna (after the Ediacara Hills in southern Australia), have intrigued paleontologists since their discovery more than 50 years ago.

The surprisingly young age of the fossils presents a most interesting puzzle. If life is so ancient, why did animals appear so late in the evolutionary day? Why—once the basic blueprint of life was drawn—did animals not emerge for more than three billion years? Alternatively, is the fossil record misleading? Is it possible that animals are far older than the record suggests?

Our time has been well spent. We now know that the Ediacaran radiation was indeed abrupt and that the geologic floor to the animal fossil record is both real and sharp. More important, we have reason to believe that the emergence of animals was closely linked to unprecedented changes in the Earth's physical environment, including a significant increase in atmospheric oxygen that may have made the evolution of large animals possible.

Before I present the evidence on which we have based such conclusions, our findings must be placed within the framework of geologic time. Earth history is conventionally divided into three eons. The oldest is the Archean, which encompasses Earth history from its origin until 2.5 billion years ago; the most recent eon is the Phanerozoic, which began with the expansion of skeleton-forming organisms 540 million years ago and continues to the present day. Separating them is the Proterozoic eon, which lasted for 2.1 billion years. It is near the end of the Proterozoic when the events described in this essay took place.

Many paleontologists have been enticed by the mystery of early animal evolution. But while many of my colleagues have concentrated on identifying and classifying the first animal fossils, my goal has been to place the fossils in the context of a wider pattern of late Proterozoic biological and environmental change. Doing so, however, required that

I find exceptionally well preserved sedimentary deposits, ones laid down just before the Ediacaran radiation.

Fortunately, a few such records exist. One of the best can be found in the glaciated mountains of Spitsbergen, a small island halfway between the northern tip of Norway and the North Pole. Here glaciers have exposed about 7,000 meters (22,000 feet) of gently folded but essentially unmetamorphosed sedimentary rocks that reflect shallow ocean conditions from about 600 to 850 million years ago. Throughout the past decade, Keene Swett of the University of Iowa and I have analyzed these rocks for signs of biological and environmental change.

As it turns out, Spitsbergen rocks provide an unmatched portrait of the Earth and its biota as they existed just before the Ediacaran radiation. Indeed, the richness of the Spitsbergen fossil record has enabled us to make a number of significant discoveries. To begin with, we found the Spitsbergen fossils not only represent a variety of habitats but belong to morphologically and taxonomically diverse taxa. In addition, both prokaryotic and eukaryotic cell types are present in the sediments. Prokaryotes are generally simple organisms whose cells lack nuclei and other organelles; they are represented by bacteria, including the cyanobacteria, or blue-green "algae." According to Julian W. Green, a former student in my laboratory, who is now at the University of South Carolina at Spartanburg, many of the prokaryotes from Spitsbergen and related areas exhibit characteristics of morphology, development and behavior (as inferred from their orientations in the sediments) that render them virtually indistinguishable from cyanobacteria and other bacteria that live in comparable habitats today.

Eukaryotes, which include single-celled protozoa and algae as well as multicellular plants, animals and fungi, differ from prokaryotes in having nuclei bounded by a membrane; most also have energy-yielding processes localized in organelles such as mitochondria and chloroplasts. Some Spitsbergen eukaryotes resemble modern prasinophyte (green) algae, whereas others bear closer resemblance to the so-called chromophyte algae such as the dinoflagellates that are ubiquitous in modern oceans.

We also discovered that not all Spitsbergen fossils are unicellular. Nicholas Butterfield, a graduate student in my laboratory, has found beautifully preserved multicellular algae (seaweeds) in strata that are approximately 800 million years old. Several major groups of algae are represented, including species that once formed extensive carpets on the quiet, subtidal seafloor.

Despite the presence of multicellular algae and diverse single-celled eukaryotes, there are no indications of animal life in the Spitsbergen sediments. Tracks, trails or burrows normally associated with animal activity simply have not been found in these rocks or in other beds of comparable age. This finding, combined with the fact that many of the

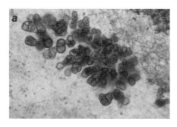

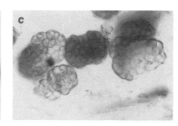

Diverse organisms, including photosynthetic cyanobacteria, protists and seaweeds, can be found in Upper Troterozoic sediments, indicating that many of life's major evolutionary events had occurred by then. Represented here are Synodophycus euthemos, probably cyanobacterium (a); Polybesurus bipartitus, a cyanobacterium that formed crusts in tidal environments (b); Hyella dichotoma, a cyanobacterium that bored through carbonate sediments (c); Trachyhystrichosphaera vidalii, a large, unicellular alga (d); and a multicellular green alga or seaweed (as yet unnamed) (e). Macroscopic animals, such as Dickinsonia costata (f), did not appear until near the end of the Proterozoic, 580 million years ago.

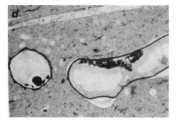
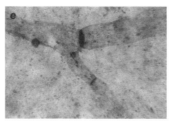

sediments are finely laminated, offers compelling evidence that animals large enough to disrupt fine-scale bedding did not exist at that time.

What factors might have deterred the emergence of large animals until so late in the evolutionary day? The Proterozoic radiation of single-celled algae and protozoa might suggest that evolution of the eukaryotic cell was a limiting factor, but such a notion is put to rest by the discovery of eukaryotic fossils in sediments much older than those in Spitsbergen. Simple spheroidal vesicles that are reasonably interpreted as resting cysts produced by algae have been found in rocks that are roughly twice as old as the Spitsbergen strata. In addition, steranes, which are the geologic form of sterols (molecules thought to be synthesized exclusively by nucleated cells), have been identified in petroleum deposits at least 1.7 billion years old by Roger E. Summons and his colleagues at the Australian Bureau of Mineral Resources, Geology and Geophysics. Clearly, eukaryotic cells arose more than a billion years before the Ediacaran radiation and may have arisen much earlier. Carl Woese of the University of Illinois at Urbana-Champaign and others have inferred from molecular comparisons of living species that eukaryotes arose fully as early as prokaryotes.

In addition, Spitsbergen and older rocks show that the evolution of multicellularity did not directly trigger the Ediacaran radiation. Multicellular eukaryotes (seaweeds) have been found in 1.4-billion-year-old strata by Du Rulin of the Hebei Institute of Geology.

What physical events might have facilitated the evolution of macroscopic animals? Specifically, what kind of environmental barrier might have separated a world inhabited by unicellular organisms and seaweeds from one harboring large animals? A credible answer to this question was offered more than 30 years ago by J. Ralph Nursall of the University of Alberta, who proposed that throughout most of Earth history (until the time of the Ediacaran radiation), levels of atmospheric oxygen were too low to permit metabolic activity by macroscopic invertebrates. Since then, the idea has been championed by many paleontologists and biologists, most notably by Preston Cloud of the University of California at Santa Barbara, who has long argued that important clues to biological evolution reside in the geochemical record of sediments.

The most widely cited "oxygen control" hypothesis was proposed more than 25 years ago by Lloyd V. Berkner and Lauriston C. Marshall of the Graduate Research Center of the Southwest in Dallas. They suggested that oxygen did not rise above 1 percent of present-day atmospheric levels until the end of the Proterozoic eon. Only then would aerobic metabolism have been possible and would sufficient ozone have accumulated in the atmosphere to absorb the Sun's lethal ultraviolet radiation. Most researchers still consider 1 percent of present-day levels to be a critical threshold for biological activity, but it is now clear that this threshold was crossed at least 1.3 billion years before the Ediacaran event, ruling out a direct relation to the appearance of large animals. Nevertheless, atmospheric oxygen cannot be disregarded as a factor in metazoan evolution.

To begin with, although 1 percent of present-day atmospheric levels represents a critical threshold for oxygen-dependent cells and minute, architecturally simple animals, macroscopic animals demand much higher oxygen concentrations. Physiological functions that require significantly higher oxygen levels include collagen synthesis, exercise metabolism and oxygenation of the body's tissues.

Taking such functions into account, Bruce Runnegar of the University of California at Los Angeles has estimated that the relatively simple invertebrates found in Ediacaran assemblages would have needed oxygen at concentrations equal to or greater than 6 to 10 percent of present-day atmospheric levels. His estimate assumes that Ediacaran animals had well-developed circulatory systems capable of transporting oxygen efficiently to their tissues. But it is likely that the first macroscopic animals lacked such sophisticated circulation and instead oxygenated their cells by means of simple diffusion. In this case, substantially higher oxygen levels (perhaps comparable to those of today) would have been necessary to sustain macroscopic animals.

Recognition that the first large animals had high oxygen requirements allows us to reformulate the oxygen control hypothesis, substituting a higher threshold value than the one envisioned by Berkner and Marshall. Nevertheless, the hypothesis is based entirely on the physiological needs of living organisms. Although such data are consistent with a late Proterozoic increase in atmospheric oxygen, they do not prove that such an increase actually occurred.

Desiring such proof, my colleagues and I set out to examine the late Proterozoic sedimentary record for concrete evidence of changing oxygen levels. Although the oxygen content of Proterozoic air cannot be measured directly, our data do show that the Earth underwent a number of profound physical changes near the end of the Proterozoic, quite likely including a significant rise in atmospheric oxygen.

Before our findings can be interpreted, the source of free oxygen must be determined. Most scientists agree that the amount produced by nonbiological processes is negligible; virtually all free oxygen comes from photosynthesis, the process by which green plants, algae and cyanobacteria use the Sun's energy to convert carbon dioxide and water to sugars (which the cells then sequester) and oxygen (which is released as a by-product). Most of the time, atmospheric oxygen appears to be in a steady state. That is, the amount of oxygen generated by photosynthesis is balanced by the amount of oxygen consumed by biological and geologic activities. Such activities include respiration (the process by which organisms use oxygen to derive energy from organic molecules), weathering (the oxidation of reduced sulfur, iron and other materials in exposed rocks) and oxidation of reduced gases given off by organisms and volcanoes. Only when the amount of oxygen released during photosynthesis exceeds the amount consumed by oxidation will oxygen levels increase.

It seems reasonable that a build-up of atmospheric oxygen will ensue if more and more photosynthetic organic matter is produced. But this is not necessarily so, because the oxygen produced by photosynthesis is usually consumed by higher rates of respiration or weathering. Oxygen levels are most likely to rise not when more photosynthetic matter is produced but when more is buried in sediments. Recall that during photosynthesis organic matter is synthesized and oxygen released. Because removal of organic matter by burial decreases the amount of organic material available for respiration, the net result is a buildup of oxygen. (Of course, oxygen will accumulate in the atmosphere only if it is not consumed by weathering and other oxidation reactions.)

To our good fortune, the relation between oxygen production and organic carbon burial provided a way to assess environmental change during the late Proterozoic. If oxygen levels increased during that period, we could expect to find the increase reflected in higher rates of organic carbon burial. Although such rates are difficult to measure directly, the isotopic composition of carbon in ancient carbonates and organic matter provides a useful estimate.

Let me explain why. The element carbon has two stable isotopes: ^{12}C, which contains six protons and six neutrons and makes up about 99 percent of all carbon atoms, and ^{13}C, which has an extra neutron and is thus heavier. (A very small fraction of carbon atoms has eight neutrons, forming the radioactive isotope ^{14}C; because this isotope decays to nitrogen with a half-life of only a few thousand years, it does not figure in discussions of Proterozoic carbon.)

Carbon occurs principally in carbonate minerals such as calcite and aragonite, as well as in dolomite (which also contains magnesium) and organic matter. Once formed, the ratio of ^{13}C to ^{12}C in these materials changes only slightly over time. The ratio depends mainly on isotopic fractionation associated with the preferential uptake of ^{12}C (the lighter isotope) during photosynthesis, but to some degree it also depends on the relative fluxes of carbonate and organic matter in and out of sediments.

Another clue exists in the form of strontium. Like carbon, strontium occurs in several isotopic forms, of which two—^{87}Sr and ^{86}Sr—are of concern here. In contrast to carbon, however, the ratio of these two isotopes in seawater (and in carbonates precipitated from seawater) depends on continental erosion (which generally supplies strontium in a high ^{87}Sr-to-^{86}Sr ratio) and on hydrothermal input associated with the spreading of ocean ridges (which typically provides a low ^{87}Sr-to-^{86}Sr ratio).

Ján Veizer of the University of Ottawa has measured the isotopic composition of strontium in carbonates that are 600 to 850 million years old and found the ratio of ^{87}Sr to 86Sr in them to be unusually low. He concluded the late Proterozoic must have been a time of anomalously heightened hydrothermal activity. Equally important, Veizer's data show a distinct shift toward higher ^{87}Sr-to^{86}Sr ratios not long before the Ediacaran radiation.

My Harvard colleagues Stein B. Jacobsen, Louis Derry and Yemane Asmerom and I have found similar strontium ratios in carbonates from Spitsbergen and elsewhere, strengthening support for the idea that unusually strong hydrothermal activity swept the Earth during the late Proterozoic.

It is now clear that the end of the Proterozoic was beset by change. Carbon isotope ratios indicate that high (albeit fluctuating) rates of organic carbon burial prevailed during much of the eon's last 300 million years. Strontium isotope ratios suggest there was strong hydrothermal activity in the oceans, which appears related to continental breakup and mountain building. Finally, the coincidence of increased iron formation, glacial activity and fluctuating carbon isotope ratios suggests the oceans underwent episodic stagnation (accompanied by oxygen depletion in deep waters) at the same time the planet was experiencing considerable climatic change.

Complete understanding of how these phenomena relate to one another remains elusive. Recently, however, James C. G. Walker of the University of Michigan and I devised a computer model to test how such changes may have affected one another. But our model raised one concern: reduced volcanic gases produced in association with intense hydrothermal activity could have consumed all the oxygen released by carbon burial. Indeed, the high rates of organic carbon burial suggested by late Proterozoic carbon isotope ratios may actually have been associated with a decrease in oxygen levels. (This would explain why animals—if they existed prior to 600 million years ago—would have been tiny.) Fortunately, strontium isotope data indicate that intense hydrothermal activity ended roughly 600 million years ago, and, according to our model, oxygen levels also increased rapidly at this time, which coincides with the diversification of large animals.

Available evidence thus links the Ediacaran radiation to a late Proterozoic increase in atmospheric oxygen. Moreover, these events now seem embedded within an even larger framework of tectonic, climatic and biogeochemical change. My guess is that the fundamental driver of late Proterozoic change was tectonic. Once the physiological barrier created by a limited oxygen supply was removed, the first microscopic metazoans were free to evolve into the macroscopic forms that quickly came to dominate the animal world.

At present, this hypothesis, like others before it, must be regarded as heuristic rather than gospel. Yet it articulates an explicit set of relations between the Earth and its biota and so makes predictions that can be tested by further study. Whether or not my particular view of the late Proterozoic will withstand such testing remains unclear. What is important is that we have begun to ask new questions about evolution on a dynamic Earth. In this case, we are starting with a vivid sense that the modern world arose when biogeochemical cycles linked the physical and biological Earth in profound change at the end of the long Proterozoic eon.

Chapter Eight:
The Microverse

Introduction

When I was a teenager, I thought I knew enough about astronomy to teach other youngsters. In 1966 at the Adirondack Science Camp, and the following summer at Camp Skycrest in Pennsylvania, I introduced the stars and planets to children. One of those children was a precocious boy named Alex Scheeline. I next bumped into Alex at a public lecture I was giving in 1994. He has a family of his own and is professor of chemistry at the University of Illinois at Urbana-Champaign, and he could teach me about aspects of the universe I had not even considered all those years ago.

When we start out in life, young scientists tend to see the world as divided into discrete areas. In 1967 I never thought that a course in chemistry would relate in any way to my love of the night sky. Alex did, though, and the result is his essay on how the basic structure of the universe is set up.

This chapter deals not with the structure of the universe, or even how a planetary atmosphere is formed and maintained. But if it were not for our understanding of atoms and molecules, we would not have a clue as to how these vast structures work. Through this chapter on the "microverse," we bring our journey from the universe to the quark to an end.

The Discovery of the Top Quark

Tony M. Liss and Paul L. Tipton

In March 1995 scientists gathered at Fermilab—the Fermi Laboratory in Batavia, Illinois, near Chicago—to witness a historic event. In back-to-back seminars, physicists from rival experiments within the lab announced the discovery of a new particle, the top quark. A decades-long search of particle physics had come to an end.

The top quark is the sixth, and quite possibly the last, quark. Along with leptons—the electron and its relatives—quarks are the building blocks of matter. The lightest quarks, designated "up" and "down," make up the familiar protons and neutrons. Along with the electrons, these make up the entire periodic table. Heavier quarks (such as the charm, strange, top and bottom quarks) and leptons, though abundant in the early moments after the big bang, are now commonly produced only in accelerators. The standard model describes the interactions among these building blocks. It requires that leptons and quarks each come in pairs, often called generations.

Physicists had known that the top must exist since 1977, when its partner, the bottom, was discovered. But the top proved exasperatingly hard to find. Although a fundamental particle with no discernible structure, the top quark turns out to have a mass of 175 billion electron volts (GeV)—as much as an atom of gold and far greater than most theorists had anticipated. The proton, made of two ups and one down, has a mass of just under 1 GeV. (The electron volt is a unit of energy, related to mass via $E = mc^2$.)

Creating a top quark thus required concentrating immense amounts of energy into a minute region of space. Physicists do this by accelerating two particles and having them smash into each other. Out of a few trillion collisions at least a handful, experimenters hoped, would cause a top quark to be created out of energy from the impact. What we did not know was how much energy it would take.

Although particles can be created from nothing but energy, certain features, such as electrical charge, cannot—these are "conserved." A top quark cannot be born all by itself. The easiest way to make a top is along with an antitop—identical in mass but with opposite signs for other properties, so that conserved quantities cancel out.

In 1985, when the Fermilab collider was first activated, the search for the top had already been going on for eight years. Over the years the hunt moved on to different accelerators as they came into operation with ever more energetic particle beams. In the early 1980s at CERN, the European laboratory for particle physics near Geneva, beams of protons and antiprotons hitting one another at energies up to 315 GeV generated two new particles, the W and the Z.

Finding it would still be a difficult feat. When protons and antiprotons hit one another at high energies, the actual collision is between their internal quarks and gluons. Each quark or gluon carries just a modest fraction of the total energy of its host proton or antiproton, yet the collision must be energetic enough to generate top quarks. Such collisions are rare, and the higher the required energy—that is, the higher the top mass—the rarer they are.

By 1988 the collider at Fermilab was just coming into its own with our young CDF (Collider Detector at Fermilab). A brief flurry of the intense competition brought the decade to a close without a top but with the knowledge that its mass could be no lower than 77 GeV.

By this time the competition was between CDF and a new experiment across the accelerator ring at Fermilab, called D0 (pronounced "dee zero," after its location on the ring). Both CDF and D0 are international collaborations of more than 400 physicists. There are also numerous engineers, technicians and support personnel. The rival teams are independent of each other and never collaborate on their analyses.

The critical part of a high-energy experiment is the detector, which records the debris from a collision. Based on the best theoretical calculations, we expected that about one out of every 10 billion collisions would produce a top quark. The rest, though interesting for a host of other projects, would be a complicated backdrop from which the top would have to be extracted.

Over the course of a decade, both the CDF and D0 collaborations constructed enormous, complicated instruments, with hundreds of thousands of channels of electronics, in order to isolate the top's "signature"—the trace it would leave in the detectors. Whereas the CDF detector emphasizes the ability to track accurately the paths of individual particles in a magnetic field (in order to measure their momenta), the D0 device relies on an extremely precise segmented calorimeter, which measures the energy from each collision.

The top and antitop, once produced, decay almost instantly. Unlike the up and down quarks, which are stable, the top quark has a lifetime of only about 10^{-24} second. The standard model predicts that if heavy enough, the top quark will decay nearly all the time into a W and a bottom quark. So a top and antitop, if created, should generate two Ws, a bottom and an antibottom.

Unfortunately, neither the Ws nor the bottom quarks can be directly observed. The W's

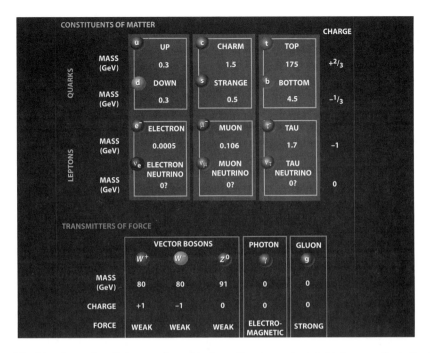

CONSTITUENTS OF MATTER						CHARGE
QUARKS	MASS (GeV)	u UP 0.3	c CHARM 1.5	t TOP 175		+2/3
	MASS (GeV)	d DOWN 0.3	s STRANGE 0.5	b BOTTOM 4.5		−1/3
LEPTONS	MASS (GeV)	e ELECTRON 0.0005	μ MUON 0.106	τ TAU 1.7		−1
	MASS (GeV)	νe ELECTRON NEUTRINO 0?	νμ MUON NEUTRINO 0?	ντ TAU NEUTRINO 0?		0

TRANSMITTERS OF FORCE	VECTOR BOSONS			PHOTON	GLUON
	W^+	W^-	Z^0	γ	g
MASS (GeV)	80	80	91	0	0
CHARGE	+1	−1	0	0	0
FORCE	WEAK	WEAK	WEAK	ELECTRO-MAGNETIC	STRONG

Matter consists of two types of particles: quarks and leptons. These are associated into generations. Up and down quarks, for instance, occur along with electrons inside atoms; they are members of the first generation. Much heavier quarks such as the top and bottom are created only in accelerators. For each quark or lepton, there is an antiquark or antilepton with opposite charge (not shown).

Force is transmitted by different set of particles; the W, Z, photon and gluons. The W and Z "bosons" transmit the weak nuclear force, involved in radioactive decays. For instance, an up quark may change into a down quark by emitting a W particle, which then decays into a quark or lepton pair. The photon transmits the electromagnetic force, which at high energies is unified with the weak force. The gluons transmit the strong force that binds up and down quarks into protons and neutrons. An extra particle that is believed to exist, the Higgs, has not yet been found.

lifetime is about the same as the top's. The bottom, too, is unstable, though much longer lived than the top. Moreover, individual—or "bare"—quarks are never seen. The strong force, which binds the quarks together, ensures that quarks always appear stuck together with other quarks and antiquarks—in pairs called mesons or in triplets called baryons. (Protons and neutrons are examples of baryons.) When a quark emerges from a collision, it gets "dressed up" by a cloud of other quarks and antiquarks. What is observed is a jet, a directed beam of particles that have roughly the same direction of motion as the original quark.

A BARRAGE OF JETS

The W can decay into a quark and an antiquark from the same generation, such as an up and an antidown. In this case, the quark and antiquark show up in a particle detector as two jets.

By the time we started taking data in August 1992, we had pushed the top mass limit up to 91 GeV. This represented a milestone. The W mediates interactions between quarks in the same generation—and so between the top and the bottom. If the top were light enough—below about 75 GeV—a W might have produced a top by decaying to it, along with an antibottom. But now we knew that the only way we could find a top was by creating a top-antitop pair.

Among the most striking features of a top "event" are the jets produced by bottom quarks. The bottom quark travels in a jet as part of a meson or baryon, then decays roughly half a millimeter from where it was generated. In 1992 we started to track the particles in jets very precisely using a special instrument placed right on top of the region where the beams collide [see "The Silicon Microstrip Detector," by Alan M. Litke and Andreas S. Schwarz; *Scientific American*, May 1995]. This silicon vertex detector could locate the path of a particle to within 15 microns. By finding most of the tracks in a jet and extrapolating them backward, we hoped to find the point where the bottom quark decayed—and thereby identify it as a bottom jet.

The silicon technology was new, and we were concerned about the effects of trillions of particles passing through it. We knew that the entire detector could be fried in a fraction of a second if an accelerator glitch spilled the beams into it. Even as we were learning how to use the new vertex detector, the D0 collaboration was commissioning its own new detector on the opposite side of the accelerator ring.

In October 1992, just three months later, we saw our first hint of the top—an event characterized by a highly energetic muon and electron, lots of missing momentum and at least two jets. We analyzed that one event in excruciating detail, finally concluding that it was probably the real thing. But a single event was not enough; we needed to observe the top in several different ways to make sure we were not being fooled by "background," events randomly mimicking the top signature. We began to analyze the data even more avidly than before, but when nothing particularly spectacular showed up, we knew we were in for a long haul.

Nearly a year into the run, the mass limit was pushed to 108 GeV by CDF and later to 131 GeV by D0, and we were still searching. Then, in July 1993, the entire CDF collaboration presented the results of their ongoing analyses.

Of the trillion or so collisions created within CDF, we had isolated 12 events that seemed to involve the creation of a top-antitop pair. Other physical processes can imitate the signature of such an event, and we had to estimate their likelihood. After months of effort, we estimated that roughly 5.7 of these background events were to be expected. The probability that background alone was responsible for these 12 events was about one in 400, leaving a small chance that no tops had been observed.

We subjected the 12 events to exhaustive analysis. One crucial study involved an attempt to "reconstruct" the top mass. By adding up the energies in the jets and leptons emitted by a (presumed) top-antitop pair, we could arrive at a value for the mass of the top. If the events were indeed from such a pair, the derived masses should fall close to some one value—the true top mass. In contrast, background events should give a much broader distribution. The mass indeed clustered in a narrow range, implying a top mass of about 175 GeV. To many of us, this was convincing evidence that we were not being fooled by background.

CDF held a seminar and press conference at Fermilab to announce the findings. The D0 collaboration presented its results as well. Although consistent with CDF's, the D0 data showed little compelling evidence for top quarks except for the one exceptional event recorded early in their run. The group had, however, assumed a low value for the top mass and as a consequence had not designed its search optimally.

Within weeks D0 had finished its reanalysis (for a heavier top) and were observing some signs of it as well. Meanwhile both teams set about collecting more data. To confirm the finding, we would need at least twice as many top events. We wrote a new algorithm for using the vertex detector to detect top candidates, putting to good use our previous experience. Once we had enough data, we processed them with the completed algorithm. It was almost immediately obvious that we indeed had the top.

The final presentations showed overwhelming evidence for the top quark from both CDF and D0. Both teams reported a probability of less than one in 500,000 that their top quark candidates could be explained by background alone.

The extremely large mass of the top—the current value is 175.6 GeV—suggests that it may be fundamentally different from the other quarks, and may lead us past the standard model. Although successful, this model leaves many questions unanswered.

Within the standard model the weak interaction, mediated by the W and Z particles, and the electromagnetic interaction, transmitted by photons, are unified into a single "electroweak" interaction at very high energies. Such energies existed in the very early universe. In the low-energy world in which we live, the electromagnetic and weak interactions behave very differently. The mechanism behind the "breaking" of their initial symmetry is not known, but in the simplest model it is caused by a new particle called the Higgs.

At high energies, when the symmetry exists, the W, Z, photon, leptons and quarks are all massless. At lower energies, when the symmetry breaks, the W and the Z interact with the Higgs and become massive. The quarks and leptons also acquire masses in the process. But whereas the W and Z masses can be calculated from the standard model, the quark and lepton masses have to be inserted by means of adjustable parameters that describe how strongly each type of quark or lepton interacts, or "couples," with the Higgs.

For an electron, which is very light, the interaction strength is 3×10^{-6}. For a top quark, it is almost exactly unity. This relatively strong coupling with the Higgs, and to some extent the mystique associated with a value of unity, suggests that the top quark may have a special role. Certainly the top's great mass makes it the most influential quark, in terms of its interactions with other particles. A very precise measurement of the top's mass, for example, along with that of a W, would lead to a prediction for the Higgs's mass.

OVER THE TOP

The sheer enormousness of the top's mass makes its decays fertile ground for new particle searches. Some theorists have speculated that a few of the events collected by CDF may contain supersymmetric particles (see "Is Nature Supersymmetric?" by Howard E. Haber and Gordon L. Kane; *Scientific American*, June 1986). Supersymmetry is a postulated symmetry that assigns as yet undiscovered partners to every particle in the standard model. If such partners exist and are lighter than the top, they might show up in top events.

Supersymmetry predicts not just one Higgs but a family of four or more. If they exist and are lighter than the top, some of these particles could be found in top decays. CDF and D0 have both mounted searches for these hypothetical particles, so far with null results.

Another critical question is whether quarks, especially the massive top, are really fundamental particles with no substructure. Recently the CDF collaboration measured the rate at which high-energy jets are produced at Fermilab's collider, finding that it is higher than expected. Very energetic scattering at wide angles (reminiscent of Rutherford scattering, which revealed that the atom has a nucleus) offers insights into the structure of the colliding objects. One possible interpretation of our results is that the excess jets are caused by collisions of even smaller objects within quarks—something not observed by any other experiment.

So radical a conclusion, which would completely change the theory of quarks, can be reached only if we can rule out all other possibilities. An "excessive" production of jets could be coming from subtle inaccuracies in the predictions. For now we must conclude that the top quark, though massive, is indeed fundamental; it has no parts.

The Fermilab accelerator is being revamped, and both CDF and D0 collaborations are dramatically improving their detectors. The accelerator upgrades will allow top quarks to be produced at 20 times the previous rate, and the detector upgrades will improve the efficiency of identifying top quarks. The net result is that both groups will find tops

30 times faster than before, allowing a more detailed look at the top's characteristics.

In a few years, physicists will start using the top to try to answer the many questions that still remain about matter and the forces that govern the physical world. What new tenets of physics may arise beyond what we now know is a matter of active speculation that will end only when measurements start to unravel the workings of nature.

Quantum Philosophy

John Horgan

In ancient Greece, Plato tried to think and talk his way to the truth in extended dialogues with his disciples. Today physicists such as Leonard Mandel of the University of Rochester operate in a somewhat different fashion. He and his students, who are more likely to wear T-shirts and laserproof goggles than robes and sandals, spend countless hours bent over a large metal table trying to align a laser with a complex network of mirrors, lenses, beam splitters and light detectors.

Yet the questions they address in their equipment-jammed laboratory are no less profound than those contemplated by Plato in his grassy glade. What are the limits of human knowledge? Is the physical world shaped in some sense by our perception of it? Is there an element of randomness in the universe, or are all events predetermined?

Mandel, being inclined toward understatement, offers a more modest description of his mission. "We are trying to understand the implications of quantum mechanics," he says. "The subject is very old, but we are still learning."

Indeed, it has been nearly a century since Max Planck proposed that electromagnetic radiation comes in tidy bundles of energy called quanta. Building on this seemingly tenuous supposition, scientists erected what is by far the most successful theory in the history of science. In addition to yielding theories for all the fundamental forces of nature except gravity, quantum mechanics has accounted for such disparate phenomena as the shining of stars and the order of the periodic table. From it have sprung technologies ranging from nuclear reactors to lasers.

Still, quantum theory has deeply disturbing implications. For one, it shattered traditional notions of causality. The elegant equation devised by Erwin Schrödinger in 1926 to describe the unfolding of quantum events offered not certainties, as Newtonian mechanics did, but only an undulating wave of possibilities. Werner Heisenberg's uncertainty principle then showed that our knowledge of nature is fundamentally limited—as soon as we grasp one part, another part slips through our fingers.

The founders of quantum physics wrestled with these issues. Albert Einstein, who in 1905 showed how Planck's electromagnetic quanta, now called photons, could explain the photoelectric effect (in which light striking metal induces an electric current), insisted later that a more detailed, wholly deterministic theory must underlie the vagaries of quantum mechanics. Arguing that "God does not play dice," he designed imaginary "thought" experiments to demonstrate the theory's "unreasonableness." Defenders of the theory such as Niels Bohr, armed with thought experiments of their own, asserted that Einstein's objections reflected an obsolete view of reality. It is not the job of scientists, Bohr chided his friend, "to prescribe to God how He should run the world."

The goal of the quantum truth-seekers is not to build faster computers or communications devices—although that could be an outcome of the research. And few expect to "disprove" a theory that has been confirmed in countless experiments. Instead their goal is to lay bare the curious reality of the quantum realm. "For me, the main purpose of doing experiments is to show people how strange quantum physics is," says Anton Zeilinger of the University of Innsbruck, who is both a theorist and experimentalist. "Most physicists are very naive; most still believe in real waves or particles."

So far the experiments are confirming Einstein's worst fears. Photons, neutrons and even whole atoms act sometimes like waves, sometimes like particles, but they actually have no definite form until they are measured. Measurements, once made, can also be erased, altering the outcome of an experiment that has already occurred. A measurement of one quantum entity can instantaneously influence another far away. This odd behavior can occur not only in the microscopic realm but even in objects large enough to be seen with the naked eye.

These findings have spurred a revival of interest in "interpretations" of quantum mechanics, which attempt to place it in a sensible framework. But the current interpretations seem anything but sensible. Some conjure up multitudes of universes. Others require belief in a logic that allows two contradictory statements to be true. "Einstein said that if quantum mechanics is right, then the world is crazy," says Daniel Greenberger, a theorist at the City College of New York. "Well, Einstein was right. The world is crazy."

The root cause of this pathology is the schizophrenic personality of quantum phenomena, which act like waves one moment and particles the next. The mystery of wave-particle duality is an old one, at least in the case of light. No less an authority than Newton proposed that light consisted of "corpuscles," but a classic experiment by Thomas Young in the early 1800s convinced most scientists that light was essentially wavelike.

Young aimed a beam of light through a plate containing two narrow slits, illuminating a screen on the other side. If the light consisted of particles, just two bright lines should have appeared on the screen. Instead a series of lines formed. The lines could be explained only by assuming that the light was propagating as waves, which were split into pairs of wavelets by the two-slit apparatus. The pattern on the screen was formed by the overlapping, or interference, of the wavelet pairs. The screen was bright where crests coincided and dark where crests met troughs, canceling each other out.

But more recent two-slit experiments suggest that Newton was also right. Modern photodetectors (which exploit the photoelectric effect explained by Einstein) can show individual photons plinking against the screen behind the slits in a particular spot at a particular time—just like particles. But as the photons continue striking the screen, the interference pattern gradually emerges, a sure sign that each individual photon went through both slits, like a wave.

Moreover, if the researcher either leaves just one slit at a time open or moves the detectors close enough to the two slits to determine which path a photon took, the photons go through one slit or the other, and the interference pattern disappears. Photons, it seems, act like waves as long as they are permitted to act like waves, spread out through space with no definite position. But the moment someone asks where the photons are—by determining which slit they went through or making them hit a screen—they abruptly become particles.

Actually, wave-particle duality is even more baffling than this explanation suggests, as John A. Wheeler of Princeton University demonstrated with a thought experiment he devised in 1980. "Bohr used to say that if you aren't confused by quantum physics, then you haven't really understood it," remarks Wheeler, who studied under Bohr in the 1930s and went on to become one of the most adventurous explorers of the quantum world.

In the two-slit experiments, the physicist's choice of apparatus forces the photon to choose between going through both slits, like a wave, or just one slit, like a particle. But what would happen, Wheeler asked, if the researcher could somehow wait until after the light had passed the two slits before deciding how to observe it?

PSYCHIC PHOTONS

The fallacy giving rise to such speculations, Wheeler explains, is the assumption that a photon had some physical form before the astronomer observed it. Either it was a wave or a particle; either it went both ways around the quasar or only one way. Actually, Wheeler says, quantum phenomena are neither waves nor particles but are intrinsically undefined until the moment they are measured. In a sense, the British philosopher Bishop Berkeley was right when he asserted two centuries ago that "to be is to be perceived."

Reflecting on quantum mechanics some 60 years ago, the British physicist Sir Arthur Eddington complained that the theory made as much sense as Lewis Carroll's poem "Jabberwocky," in which "slithy toves did gyre and gimble in the wabe." Unfortunately, the jargon of quantum mechanics is rather less lively. An unobserved quantum entity is said to exist in a "coherent superposition" of all the possible "states" permitted by its "wave function." But as soon as an observer makes a measurement capable of distinguishing between these states, the wave function "collapses," and the entity is forced into a single state.

Yet even this deliberately abstract language contains some misleading implications. One is that measurement requires direct physical intervention. Physicists often explain the uncertainty principle in this way: in measuring the position of a quantum entity, one inevitably knocks it off its course, losing information about its direction and about its phase, the relative position of its crests and troughs.

Most experiments do in fact involve intrusive measurements. For example, blocking one path or the other or moving detectors close to the slits obviously disturbs the photons' passage in the two-slit experiments, as does placing a detector along one route of the delayed-choice experiment. But an experiment done in 1991 by Mandel's team at the University of Rochester shows that a photon can be forced to switch from wavelike to particlelike behavior by something much more subtle than direct intervention.

The experiment relies on a parametric down-converter, an unusual lens that splits a photon of a given energy into two photons whose energy is half as great. Although the device was developed in the 1960s, the Rochester group pioneered its use in tests of quantum mechanics. In the experiment, a laser fires light at a beam splitter. Reflected photons are directed to one down-converter, and transmitted photons go to another down-converter. Each down-converter splits any photon impinging on it into two lower-frequency photons, one called the signal and the other called the idler. The two down-converters are arranged so that the two idler beams merge into a single beam. Mirrors steer the overlapping idlers to one detector and the two signal beams to a separate detector.

This design does not permit an observer to tell which way any single photon went after encountering the beam splitter. Each photon therefore goes both right and left at the beam splitter, like a wave, and passes through both down-converters, producing two signal wavelets and two idler wavelets. The signal wavelets generate an interference pattern at their detector. The pattern is revealed by gradually lengthening the distance that signals from one down-converter must go to reach the detector. The rate of detection then rises and falls as the crests and troughs of the interfering wavelets shift in relation to each other, going in and out of phase.

Now comes the odd part. The signal photons and the idler photons, once emitted by the down-converters, never again cross paths; they proceed to their respective detectors independently of each other. Nevertheless, simply by blocking the path of one set of idler photons, the researchers destroy the interference pattern of the signal photons. What has changed?

342

In 1948, John A. Wheeler (center) and Robert Oppenheimer (right) are greeted in Paris by a French colleague.

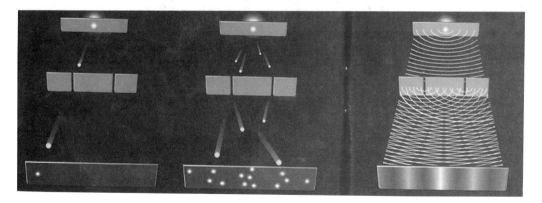

Two-slit experiments reveal that photons, the quantum entities giving rise to light and other forms of electro-magnetic radiation, act both like particles and like waves. A single photon will strike the screen in a particular place, like a particle (left). But as more photons strike the screen, they begin to create an interference pattern (center). Such a pattern could occur only if each photon had actually gone through both slits, like a wave (right).

The answer is that the observer's potential knowledge has changed. He can now determine which route the signal photons took to their detector by comparing their arrival times with those of the remaining, unblocked idlers. The original photon can no longer go both ways at the beam splitter, like a wave, but must either bounce off or pass through, like a particle.

The comparison of arrival times need not actually be performed to destroy the interference pattern. The mere "threat" of obtaining information about which way the photon traveled, Mandel explains, forces it to travel only one route. "The quantum state reflects not only what we know about the system but what is in principle knowable," Mandel says.

Can the threat of obtaining incriminating information, once made, be retracted? In other words, are measurements reversible? Many theorists, including Bohr, thought not, and the phrase "collapse of the wave function" reflects that belief. But since 1983 Marian O. Scully, a theorist at the University of New Mexico, has argued that it should be possible to gain information about the state of a quantum phenomenon, thereby destroying its wavelike properties, and then restore those properties by "erasing" the information.

Several groups working with optical interferometry, including Mandel's, claim to have demonstrated what Scully has dubbed a "quantum eraser." The group that has come closest, according to Scully, is one led by Raymond Y. Chiao of the University of California at Berkeley.

Chiao's group passed a beam of light through a down-conversion crystal, generating two identical photons. After being directed by mirrors along separate paths, the two photons crossed paths again at a half-silvered mirror and then entered two detectors. Because it was impossible to know which photon ended up in which detector, each photon seemed to go both ways. As in Mandel's experiment, the interference pattern was revealed by lengthening one arm of the interferometer; a device called a coincidence counter showed the simultaneous firings of the two photon detectors rising and falling as the two wavelets entering each detector went in and out of phase.

Then the workers added a device to the interferometer that shifted the polarization of one set of photons by 90 degrees. If one thinks of a ray of light as an arrow, polarization is the orientation of the plane of the arrowhead. One of the peculiarities of polarization is that it is a strictly binary property; photons are always polarized either vertically or horizontally. The altered polarization served as a tag; by putting polarization detectors in front of the simple light detectors at the end of the routes, one could determine which route each photon had taken. The two paths were no longer indistinguishable, and so the interference pattern disappeared.

Finally, Chiao's group inserted two devices that admitted only light polarized in one direction just in front of the detectors. The paths were indistinguishable again, and the interference pattern reappeared. Unlike Humpty-Dumpty, a collapsed wave function can be put back together again.

SPOOKY ACTION

Such possibilities provoke consternation in some quarters. Edwin T. Jaynes of Washington University, a prominent theorist whose work helped to inspire Scully to conceive the quantum eraser, has nonetheless dubbed it "medieval necromancy." Scully was so pleased by Jaynes's remark that he included it in a recent article on the quantum eraser.

Necromancy cannot hold a candle to nonlocality. Einstein, Boris Podolsky and Nathan Rosen first drew attention to this bizarre quantum property (which is now often called the EPR effect in their honor) in 1935 with a thought experiment designed to prove that quantum mechanics was hopelessly flawed. What would happen, Einstein and his colleagues asked, if a particle consisting of two protons decayed, sending the protons in opposite directions? According to quantum mechanics, as long as both protons remain unobserved their properties remain indefinite, in a superposition of all possible states; that means each one travels in all possible directions.

But because of their common origin, the properties of the protons are tightly correlated, or "entangled." For example, through simple conservation of momentum, one knows that if one proton heads north, the other must have headed south. Consequently, measuring the momentum of one proton instantaneously determines the momentum of the other proton—even if it has traveled to the opposite end of the universe. Einstein said that this "spooky action at a distance" was incompatible with any "realistic" model of reality; all the properties of each proton must be fixed from the moment they first fly apart.

Until the early 1960s, most physicists considered the issue entirely academic, since no one could imagine how to resolve it experimentally. Then, in 1964, John S. Bell of CERN, the European laboratory for particle physics, showed that quantum mechanics predicted stronger statistical correlations between entangled particles than the so-called local realistic theory that Einstein preferred. Bell's papers triggered a flurry of laboratory work, culminating in a classic (but not classical) experiment performed a decade ago by Alain Aspect of the University of Paris.

Instead of the momentum of protons, Aspect analyzed the polarization of pairs of photons emitted by a single source toward separate detectors. Measured independently, the polarization of each set of photons fluctuated in a seemingly random way. But when the two sets of measurements were compared, they displayed an agreement stronger than could be accounted for by any local realistic theory—just as Bell had predicted. Einstein's spooky action at a distance was real.

Still more ambitious EPR experiments have been proposed but not yet carried out. Greenberger, Zeilinger and Michael Horne of Stonehill College have shown that three or more particles sprung from a single source will exhibit much stronger nonlocal correlations than those between just two particles. Bernard Yurke and David Stoler of AT&T Bell Laboratories have even suggested a way in which three particles emitted from separate locations can exhibit the EPR effect.

A die-hard realist might dismiss the experiments described above, since they all involve that quintessence of ineffability, light. But electrons, neutrons, protons and even whole atoms—the stuff our own bodies are made of—also display pathological behavior. Researchers observed wavelike behavior in electrons through indirect means as early as the 1920s, and they began carrying out two-slit experiments with electrons several decades ago.

SUPERPOSED PHILOSOPHERS

A new round of electron experiments may be carried out soon if Yakir Aharonov of Tel Aviv University has his way. Noting that superposition is generally inferred from observations of large numbers of particles, Aharonov contends that a single electron bound to a hydrogen atom could be detected smeared out in a relatively large cavity—say, 10 centimeters across—by very delicately scattering photons off it.

Aharonov has not yet published his idea—"I am a very fast thinker but a very slow writer," he says—and some physicists he has discussed it with are skeptical. On the other hand, many were skeptical in 1958, when Aharonov and David Bohm of the University of London suggested a way in which a magnetic field could influence an electron that, strictly speaking, lay completely beyond the field's range. The so-called Aharonov-Bohm effect has now been confirmed in laboratories.

Since the mid-1970s various workers have done interference experiments with neurons, which are almost 2,000 times heavier than electrons. Some 15 years ago [in 1977] for example, Samuel A. Werner of the University of Missouri at Columbia and others found that the interference pattern formed by neurons diffracted along two paths by a sculpted silicon crystal could be altered simply by changing the interferometer's orientation relative to the Earth's gravitational field. It was the first demonstration that the Schrödinger equation holds true under the sway of gravity.

Investigators have begun doing interferometry with whole atoms only in the past few years. Such experiments are extraordinarily difficult. Atoms cannot pass through lenses or crystals, as photons, electrons and even neutrons can. Moreover, since the wavelength of an object is inversely proportional to its mass and velocity, the particle must move slowly for its wavelength to be detectable. Yet workers such as David E. Pritchard of the Massachusetts Institute of Technology have created the equivalent of beam splitters, mirrors and lenses for atoms out of metal plates with precisely machined grooves and even standing waves of light, formed when a wave of light reflects back on itself in such a way that its crests and troughs match precisely.

Pritchard says physicists may one day be able to pass biologically significant molecules such as proteins or nucleic acids through an interferometer. In principle, one could even observe wavelike behavior in a whole organism, such as an amoeba. There are some obstacles, though: the amoeba would have to travel very slowly, so slowly, in fact, that it would take some three years to get through the interferometer, according to Pritchard. The experiment would also have to be conducted in an environment completely free of gravitational or other influences—that is, in outer space.

Getting a slightly larger and more intelligent organism, for instance, a philosopher, to take two paths through a two-slit apparatus would be even trickier. "It would take longer than the age of the universe," Pritchard says.

While physicists may never nudge a philosopher into a superposition of states, they are hard at work trying to induce wavelike behavior in objects literally large enough to see. The research has rekindled interest in a famous thought experiment posed by Schrödinger in 1935. In a version altered by John Bell, the EPR theorist, to be more palatable to animal lovers, a cat is placed in a box containing a lump of radioactive matter, which has a 50 percent chance of emitting a particle in a one-hour period. When the particle decays, it triggers a Geiger counter, which in turn causes a flask of milk to pour into a bowl, feeding the cat. (In Schrödinger's version, a hammer smashes a flask of poison gas, killing the cat.)

Common sense dictates that a cat cannot have a stomach both empty and full. But quantum mechanics dictates that after one hour, if no one has looked in the box, the radioactive lump and so the cat exist in a superposition of indistinguishable states; the former is both decayed and undecayed, and the latter is both hungry and full.

Various resolutions to the paradox have been suggested. Wojciech H. Zurek, a theorist at Los Alamos National Laboratory, contends that as a quantum phenomenon propagates, its interaction with the environment inevitably causes its superposed states to become distinguishable and thus to collapse into a single state. Mandel of the University of Rochester thinks this view is supported by his experiment, in which the mere potential for knowledge of a photon's path destroyed its interference pattern. After all, one can easily learn whether the cat has been fed—say, by making the box transparent—without actually disturbing it.

All the recent experiments, completed and proposed, have hardly led to a consensus on what exactly quantum mechanics means. If only by default, the "orthodox" view of quantum mechanics is still the one set forth in the 1920s by Bohr. Called the Copenhagen interpretation, its basic assertion is that what we observe is all we can know; any speculation about what a photon, an atom or even a Squid "really is" or what it is doing when we're not looking is just that—speculation.

To be sure, the Copenhagen interpretation has come under attack from theorists in recent years, most notably from John Bell, author of the brilliant proof of the divergence between "realistic" and quantum predictions for EPR experiments.

Bell's exhortations helped to revive interest in a realistic theory originally proposed in the 1950s by Bohm. In Bohm's view, a quantum entity such as an electron does in fact exist in a particular place at a particular time, but its behavior is governed by an unusual field, or pilot wave, whose properties are defined by the Schrödinger wave function. The hypothesis does allow one quantum quirk, nonlocality, but it eliminates another, the indefiniteness of position of a particle. Its predictions are identical to those of standard quantum mechanics.

Bell also boosted the standing of a theory developed by Giancarlo Ghirardi and Tullio Weber of the University of Trieste and Alberto Rimini of the University of Pavia and refined more recently by Philip Pearle of Hamilton College. By adding a nonlinear term to the Schrödinger equation, the theory causes superposed states of a system to converge into a single state as the system approaches macroscopic dimensions, thereby eliminating the Schrödinger's cat paradox, among other embarrassments.

Yet another view currently enjoying some attention, although not as a result of Bell's efforts, is the many-worlds interpretation, which was invented in the 1950s by Hugh Everett III of Princeton. The theory sought to answer the question of why, when we observe a quantum phenomenon, we see only one outcome of the many allowed by its wave function. Everett proposed that whenever a measurement forces a particle to make a choice, for instance, between going left or right in a two-slit apparatus, the entire universe splits into two separate universes; the particle goes left in one universe and right in the other.

Although the theory was long dismissed as more science fiction than science, it has been revived in a modified form by Murray Gell-Mann of the California Institute of Technology and James B. Hartle of the University of California at Santa Barbara. They call their version the many-histories interpretation and emphasize that the histories are "potentialities" rather than physical actualities.

THE IT FROM BIT

Other philosophers call for a sea change in our very modes of thought. After Einstein introduced his theory of relativity, notes Jeffrey Bub, a philosopher at the University of Maryland, "we threw out the old Euclidean notion of space and time, and now we have a more generalized notion." Quantum theory may demand a similar revamping of our concepts of rationality and logic, Bub says. Boolean logic, which is based on either-or propositions, suffices for a world in which an atom goes either through one slit or the other, but not both slits. "Quantum mechanical logic is non-Boolean," he comments. "Once you understand that, it may make sense." Bub concedes, however, that none of the so-called quantum logic systems devised so far has proved very convincing.

A different kind of paradigm shift is envisioned by Wheeler. The most profound lesson of quantum mechanics, he remarks, is that physical phenomena are somehow defined by the questions we ask of them. "This is in some sense a participatory universe," he says. The basis of reality may not be the quantum, which despite its elusiveness is still a physical phenomenon, but the bit, the answer to a yes-or-no question, which is the fundamental currency of computing and communications. Wheeler calls his idea "the it from bit."

Following Wheeler's lead, various theorists are trying to recast quantum physics in terms of information theory, which was developed to maximize the amount of information transmitted over communications channels. Already these investigators have found that Heisenberg's uncertainty principle, wave-particle duality and nonlocality can be formulated more powerfully in the context of information theory, according to William K. Woorters of Williams College, a former Wheeler student who is pursuing the it-from-bit concept.

If that doesn't work, there is always Aharonov's time machine. The machine, which is based not only on quantum theory but also on general relativity, is a massive sphere that can rapidly expand or contract. Einstein's theory predicts that time will speed up for an occupant of the sphere as it expands and gravity becomes proportionately weaker, and time will slow down as the sphere contracts. If the machine and its occupant can be induced into a superposition of states corresponding to different sizes and so different rates of time, Aharonov says, they may "tunnel" into the future. The occupant can then disembark, ask physicists of the future to explain the mysteries of quantum mechanics and then bring the answers—assuming there are any—back to the present. Until then, like Plato's benighted cave dwellers, we can only stare at the shadows of quanta flickering on the walls of our cave and wonder what they mean.

Atoms, Molecules and the Early Universe

Alexander Scheeline

People take the material world for granted. Since long before there were sentient beings, matter largely had the form we currently see: dense bodies such as stars and planets, molecules, the largely empty vacuum of space and energy released mostly from nuclear fusion in stars. But where did the molecules, atoms and subatomic particles come from? Why does matter stick together at all? Why, when supernovae are such rare events, do we have significant amounts of elements sufficiently heavy that they could only be formed under conditions far more energetic than those occurring in most stars? The early universe was an incredibly hot place, so that in the time immediately following the big bang, conditions were quite different from those encountered in everyday life. Some of those conditions can be recreated (on a small scale) in a laboratory setting, so that the condensation of ordinary matter from the primordial universe can be extrapolated from quite modest experiments. Observation of the cosmos provides data consistent with the main aspects of these experiments. We can "hear" the echo of the big bang in the microwave radiation background, we can see the shadows of distant molecules through their absorption of light emitted by more distant galaxies and quasars, and we can sense atoms in the outer layers of stars by seeing the light they emit and absorb. But to make sense of any of this, we have to start by describing the most plentiful type of matter in the universe. This is neither solid, liquid nor gas. It is plasma.

In a plasma, particles of opposite charge, which one would normally think would attract each other and stick together, in fact do not form neutral species. Think of particles with one charge (say, positive) as a bucket and particles of the opposite charge (negative) as baseballs. If you place the baseball in the bucket, it will stay there indefinitely. If you throw the ball into the bucket, it may bounce right out if you threw the ball hard enough. The same principle applies to charged particles: if they are moving fast enough, they will bounce off each other rather than sticking. Aside from the plasma in stars, the most common plasma most people see is lightning, a nitrogen-electron plasma. While fluorescent lights also contain plasma, what one actually sees is not the plasma, but rather the fluorescence of a coating inside the glass tube that is excited by the plasma.

In the primordial universe, everything was so hot (which is to say, all the particles were moving so fast) that nothing stuck to anything. Everything was plasma. It was so hot that not even quarks and gluons had condensed into muons, let alone into protons or neutrons. There were no atoms, thus no molecules, and the universe expanded, at least in part due to the huge pressure of the densely energetic plasma. In some sense, it was like shooting a bullet from a gun. The pressure of the hot gas moves matter outward and, as it expands, it cools off.

INITIAL ATOM FORMATION

The strongest interactions known to physics are those holding the nuclei of atoms together. This "strong nuclear force" is much stronger than either gravity, magnetic forces, or electrostatic forces, the forces we see in everyday life. However, they act only over very short ranges, distances of the order of the diameter of atomic nuclei. As the universe cooled, the strong nuclear force caused primordial protons and neutrons to form from quarks and gluons, and then to condense. Then progressively weaker interactions caused additional condensation as the cooling proceeded. So, the universe at this stage contained an immense amount of plasma strikingly similar to the interior of stars, with lots of bare protons, some protons and neutrons bound into nuclei and just the right number of electrons so that the universe as a whole was neutral. (For over a century, astronomers have searched for evidence that there is any charge imbalance on a large scale in the universe. They have found none.)

Perhaps this is a good place to note that "hot" and "cold" are terms whose intuitive meanings can be refined by describing how they are reflected in the motion of particles. Louis Boltzmann and James Clerk Maxwell focused our understanding of these terms. Something that is hot is energetic and so moves rapidly. Something that is cold has little energy and so moves slowly. For condensed matter (solids and liquids), heat corresponds to vibrational or rotational motion of the particles involved. The hotter a liquid is, the more the molecules in that liquid vibrate and tumble. But all the matter does not move with the same speed. For a hot system, there is a broader distribution of speeds (or vibrations and rotations) than for cold matter. Thus, an ice cube traveling at thousands of miles an hour is still cold—all the water in the cube is moving together, but the molecules move slowly with respect to each other. It is only when the ice cube collides with something else that the bulk kinetic energy gets turned into heat —the rapid random motion of the molecules that made up the cube prior to the collision. When Comet Shoemaker-Levy 9 hit Jupiter, the pieces of the comet went from subfreezing temperatures to thousands of degrees in the moment of impact.

So the primordial universe cooled by having the plasma expand, emit electromagnetic radiation and convert relative motion into radiation whose wavelengths ranged from light-years to femtometers. This continuum is now seen as the microwave background of radio astronomy. When the cooling had proceeded far enough, electrons could start sticking to nuclei, and atoms were formed. But this did not happen quickly, for a reason that Isaac Newton understood 300 years ago. In any collision, both momentum and energy must be conserved. If there are three bodies colliding, this is fairly easy to arrange. But if two bodies collide, it is much harder. This is familiar to anyone who plays billiards. Ignoring friction with the table, it is very tricky to get two balls to collide in such a way that they end up moving together—most of the time they bounce off each other. The same is true when an electron approaches a bare nucleus. The electron can go into orbit around that nucleus from only a limited range of angles and relative velocities. In fact, it is almost always the case that when an electron is captured by a nucleus, light is emitted (and light carries away the excess energy and momentum). This process is called "radiative recombination." The reader of this chapter might think, "Shouldn't that be radiative combination? Where did the 're' come from?" Such electron capture processes were seen and understood in laboratories before they were observed astronomically. In the laboratory, neutral species are ionized by heating them up, and they reform as they cool down. So the term comes from the earthbound experiments, not the early universe.

Until the background radiation had cooled to temperatures less than several tens of thousands of degrees, recombination acted as a reversible process. Light emitted during radiative recombination of one atom might be absorbed by a second atom, causing it to ionize. The universe was thus "optically dense;" light emitted at one point was reabsorbed somewhere else, so that no light escaped ahead of the expanding plasma. But continued expansion and cooling finally brought the mean temperature of the universe into the range of a few thousand degrees. Light emitted by radiative recombination was then sufficiently diffuse that, on average, it was not reabsorbed by other atoms. The universe became transparent, as we see it today. And, on average, atoms stayed together rather than perpetually reionizing and recombining.

The vast majority of nuclear particles in the primordial universe were single protons, so that capturing one electron produced a hydrogen atom. Where nuclei involved more than one proton, more than one electron had to be captured to make a neutral atom. Multiple electron capturing was slow for many reasons. First, radiative recombination is nearly as inefficient for multielectron atoms as for hydrogen, so that many collisions between forming atoms and electrons were needed to make completed atoms. Second, the hydrogen had already scavenged most of the universe's electrons, so collisions with electrons were rarer for heavy atoms needing more than one electron than for the original dense hydrogen. Last, the universe continued to expand, so the plasma became more and more diffuse and the atoms more dilute. It is the same as trying to head for the exits in a room with 100 people in it. If there are 100 exits and the room is the size of a stadium, you can head straight for an exit without bumping into anyone. But if there is standing room only and just one door, it will be a slow and highly collisional process!

MOLECULES APPEAR

Even as atoms were still forming from nuclei and electrons, atoms and ions (that is, atoms with fewer electrons than protons, or in a few cases with excess electrons) collided with each other. Just as some electron-nucleus collisions resulted in atom formation, so atom-atom and atom-ion collisions could result in diatomic molecules or molecular ions. Laboratory apparatus known as molecular beams can be used to carry out experiments on these types of processes, and the 1986 Nobel Prize in Chemistry went to Dudley R. Hirschbach, Yuan T. Lee and John C. Polanyi for the invention of such technology. Because atoms are bigger than nuclei, the requisite collisions are more frequent, even when the concentrations of atoms are lower than were the concentrations in the primordial plasma. It is a bit like the contrast between trying to collide a bullet with a cannonball (electron with nucleus) versus trying to collide two cars at an intersection (atom with atom, or molecule with atom). The cars are bigger, they are moving more slowly and thus they hit each other more readily.

Once diatomic molecules exist, forming even bigger molecules becomes at once much easier and, paradoxically, much less likely. If a collision between a diatomic molecule and another atom or ion occurs, chemical reaction is quite likely because now there is a way to get rid of excess momentum or energy without having to immediately emit a photon. The rotation of the molecule can absorb excess momentum, and vibration can absorb excess energy. In extreme cases, the electrons can jump into a highly energetic state, absorbing both momentum and energy. If one has a cold, dense plasma, such molecule formation is common. But, by the time small molecules were forming in the early universe, the expansion had proceeded so far that collisions became unlikely, and thus it remains to this day. The interstellar medium contains a variety of diatomic and polyatomic molecules and ions, but the molecules are quite dilute. Gravity can concentrate these molecules into the dust clouds we see near forming stars, and the aftermath of supernovae re-creates on a small scale the cooling plasma medium that initiated molecule formation eons ago.

Because the early distribution of elements emphasized hydrogen and helium, with minor amounts of lithium, beryllium and boron, only very simple compounds could form: H_2, LiH, BeH_2, B_2H_6 and larger boranes. Molecular ions also formed (most prominently H_2+). The molecules most commonly seen in interstellar space today, which prominently include carbon, nitrogen, and oxygen, could not form until stars had formed, lived and died in the form of supernovae, spreading the heavier elements through galaxies.

The distribution of atoms, ions and molecules was not entirely homogeneous in space. Density inhomogeneities behave in a manner called "self-organizing." That is, if one starts with perfect homogeneity, and any fluctuation occurs, that fluctuation tends to self-amplify. Regions of low density become less dense, while those of high density become more dense. The higher density regions formed into galaxies, and the highest density regions within these protogalaxies became stars. The low density regions became spaces between galaxies and, on a smaller scale, between stars. Cosmologists worry about the origins of large scale inhomogeneities during the big bang, but heterogeneity on a microscopic scale is inherent in any plasma or molecular gas. If matter were a continuum instead of atomic and molecular, the sky would be black instead of blue. The fact that the sky has color is a demonstration of microheterogeneity. Thus, even if the early universe was not lumpy, at some point the inhomogeneities in the cooling primordial plasma would have led to the lumpy universe we now see, with concentrations of matter widely separated by apparent emptiness.

PROBING INTERSTELLAR CHEMISTRY

This picture of gradual cooling, collision, and condensation has been presented as if such behavior was inevitable. In fact, science does not deal in inevitability; it looks at observations and tries to make sense of them. What observations can be made that provide data that would lead to the above description? Spectroscopy, the science of the interaction of light with matter, is the primary source for both astronomical observation and laboratory modeling of what occurred. Every sort of light, from low frequency, low energy, long wavelength radio waves through high frequency, high energy x rays can reveal something of the behavior of the universe. We will limit discussion here to how we see evidence of atoms and molecules far from Earth. The key idea is that neither molecules nor atoms typically emit or absorb light of just any energy. The energy must be of precisely the right magnitude. Atoms emit and absorb across narrow spectral "lines"; molecules emit and absorb over closely grouped lines known as "bands." Atomic emission and absorption is due to the electrons changing energy, while molecular emission and absorption is due either to electrons changing energy, the molecule as a whole changing the amount of vibration occurring, or the molecule as a whole changing its rotation or tumbling motion.

While line and band emission was observed in the 19th century and some arithmetic relationships among the energies of the lines and bands computed, there was no real understanding of why these relationships occurred. The first insight came from Max Planck's attempt to merge the Maxwell-Boltzmann understanding of heat and equilibrium with the behavior of a closed, heated cavity known as a "black body." On December 14, 1900, he submitted a paper to *Annalen der Physik* in which he explained the distribution of energy in such a closed cavity as a function of temperature. The form of the distribution is exactly that seen in the microwave background throughout the sky (the microwave background is a rather chilly -3 degrees above absolute zero!). But in order to make everything work, he had to postulate that light had at least some characteristics of a particle, and that the energy of that particle was related to its color or wavelength. The redder, longer wavelength light had less energy per particle (dubbed a "photon" by Einstein five years later) than bluer, shorter wavelengths. By 1930 quantum mechanics, the branch of physics that studies small objects that share particulate and wavelike properties, had developed to the point that the electronic, rotational, and vibrational energy of molecules could be explained in a consistent manner. For large systems, the actual computations could not be done until large digital computers were built decades later, but atomic and molecular structure were completely understood, at least qualitatively. Since the molecules could only exist stably for discrete energies, transitions between stable molecular states could also only occur at specific energies. Thus, a hydrogen atom can emit green light with the distance between crests of the wave of which each photon is constituted of 486.1 billionths of a meter (486.1 nanometers), but not 486.0 or 486.2. Such green light is readily seen in optical telescopes, and filters specifically designed to isolate this "H_b" light are readily available. Putting vibrational energy into a hydrogen molecule (H_2) requires energy equivalent to that in light with a wavelength of 2.27 micrometers (about the width of a transistor in an early microcomputer, or the width of 10 transistors in a Pentium), and setting that molecule to tumbling takes light with a wavelength of 82 micrometers (or integer fractions of 82 micrometers such as 41 micrometers, 27 micrometers, and so on).

Vibrations typically have energies that fall in the infrared region of the spectrum. Astronomical infrared data can only be obtained from telescopes high above the troposphere's water vapor or from telescopes in space. The 10-meter Keck telescope on Mauna Kea and a number of satellite-mounted telescopes are capable of seeing vibrating molecules many light-years away. Rotation is a very low-energy process, with wavelengths corresponding to radio waves. Thus, the massive radio telescopes at Arecibo in Puerto Rico, Soccoro, New Mexico, and elsewhere can sniff out radiating tumbling molecules in addition to seeing radio wave emissions from plasmas, quasars and neutron stars.

While much of what is learned from radioastronomy deals with molecules, one particularly interesting signal comes not from molecules but from hydrogen atoms. Both protons and electrons have magnetic moments (they act like small magnets) and can have their magnetic poles either coaligned or antialigned. The energy difference between these two magnetic states corresponds to a radio wave with a wavelength of 21 cm in the absence of an external magnetic field. By scanning the heavens with a radio receiver tuned to this frequency, we pick up the highly improbable but nevertheless detectable signal of hydrogen atoms flipping their spins. So, even in the coldest parts of the heavens, we can sense the presence of the simplest of atoms.

The real fun is figuring out how the lines and bands can be interpreted to definitively indicate what atoms and molecules are "out there." For materials that are common on Earth, spectra (measurements of emission or absorption as a function of photon energy, wavelength or frequency) were readily obtained. The orange color in fireworks is atomic emission from sodium or calcium. Green comes from copper and copper oxide. Red emission comes from lithium or strontium, and blue comes from burning potassium. (You may never be able to watch fireworks without thinking about interstellar space again!) So, for the common elements, simply comparing spectra obtained in flames or laboratory plasmas to what is seen through telescopes provides a simple analysis of the composition of hot bodies. Alternatively, if there is a cloud of cold atoms between a hot star and our telescopes, we see dark absorption lines where the atom cloud absorbs the light from the star before it gets to us. This process cuts both ways—helium was identified in the Sun before it was observed on Earth. While many people worked on identifying the spectra of small molecules, the most commonly cited work is that of Gerhard Herzberg who tabulated the properties of many species. By 1950, anything that occurred in large quantities on Earth or could be simply made and that also occurred in the heavens was identified. Much progress has been made since on identifying a wide variety of molecules and molecular ions, both in the vicinity of stars and in the interstellar medium, whether or not they were present at primordial times.

Chapter Nine:
Speculation on Endings and Beginnings

Introduction

With this chapter, we return to the large-scale structure of the universe. This time, however, we are armed with different questions. How big is the universe? How did the universe get to be its present size? Did it undergo some kind of dramatic inflation soon after its birth? And when did it all begin?

These questions are very different from those that concern our understanding of other parts of astronomy, like the planets. They cannot be answered by a spacecraft visit to Neptune, although another spacecraft, the Hubble Space Telescope, is providing us with strong evidence.

The story of that evidence began in 1786, when John Goodricke, a young deaf-mute, discovered the variation of Delta Cephei. Stars like Delta Cephei vary like clockwork. As a Cepheid star expands in size, its fades in brightness, and as it contracts it brightens. At the start of this century, Henrietta Leavitt, an assistant at Harvard College Observatory, studied the cycles of about 25 Cepheids in the Small Magellanic Cloud, one of the closest galaxies to us. In 1912 she found that the brighter their average magnitudes were, the longer were their periods of variation. Then astronomer Harlow Shapley made a great intuitive leap about all the Cepheids at all distances: for any two Cepheids with the same period of variation, the one with the brighter average magnitude is closer to us. Shapley had turned this period-luminosity relationship, as he called it, into an elegant yardstick for measuring distances in space.

With the Hubble Space Telescope, astronomers are now able to study Cepheids in galaxies farther and farther away. It is possible that one of the most basic questions in all science—how old is the universe?—might be answered thanks to these distant stars. The mighty Hubble Space Telescope now has the cosmic tools to answer this basic question. That's a long way to come in two centuries, when a young boy who could neither speak nor hear looked up in the sky, found a star and set us on the road to understanding.

The Inflationary Universe

Alan H. Guth and Paul J. Steinhardt

In the past few years [prior to 1984] certain flaws in the standard big bang theory of cosmology have led to the development of a new model of the very early history of the universe. The model, known as the inflationary universe, agrees precisely with the generally accepted description of the observed universe for all times after the first 10^{-30} second. For this first fraction of a second, however, the story is dramatically different. According to the inflationary model, the universe had a brief period of extraordinarily rapid inflation, or expansion, during which its diameter increased by a factor perhaps 10^{50} times larger than had been thought. In the course of this stupendous growth spurt, all the matter and energy in the universe could have been created from virtually nothing. If the new model is correct, the observed universe is only a very small fraction of the entire universe.

The inflationary model has many features in common with the standard big bang model. In both models the universe began between 10 and 15 billion years ago as a primeval fireball of extreme density and temperature, and it has been expanding and cooling ever since. This picture has been successful in explaining many aspects of the observed universe, including the redshifting of the light from distant galaxies, the cosmic microwave background radiation and the primordial abundances of the lightest elements. All these predictions have to do only with events that presumably took place after the first second, when the two models coincide.

There were few serious attempts to describe the universe during its first second. The temperature in this period is thought to have been higher than 10 billion kelvins, and little was known about the properties of matter under such conditions. Cosmologists are now attempting to understand the history of the universe back to 10^{-45} second after its beginning. (At even earlier times the energy density would have been so great that Einstein's general theory of relativity would have to be replaced by a quantum theory of gravity, which so far does not exist.) When the standard big bang model is extended to these earlier times, various problems arise. First, it becomes clear that the model requires a number of stringent, unexplained assumptions about the initial conditions of the universe. In addition most of the new theories of elementary particles imply that the standard model would lead to a tremendous overproduction of the exotic particles called magnetic monopoles. (Each such monopole corresponds to an isolated north or south magnetic pole.)

The inflationary universe was invented to overcome these problems. The equations that describe the period of inflation have a very attractive feature: from almost any initial conditions the universe evolves to precisely the state that had to be assumed as the initial one in the standard model. Moreover, the predicted density of magnetic monopoles becomes small enough to be consistent with observations.

The standard big bang model is based on several assumptions. First, it is assumed that the fundamental laws of physics do not change with time and that the effects of gravitation are correctly described by Einstein's general theory of relativity. It is also assumed that the early universe was filled with an almost perfectly uniform, expanding, intensely hot gas of elementary particles in thermal equilibrium. The gas filled all of space, and the gas and space expanded together at the same rate. When they are averaged over large regions, the densities of matter and energy have remained nearly uniform from place to place as the universe has evolved. It is further assumed that any changes in the state of the matter and the radiation have been so smooth that they have had a negligible effect on the thermodynamic history of the universe. The violation of the last assumption is a key to the inflationary universe model.

Unlike the successes of the big bang model, all of which pertain to events a second or more after the big bang, the problems all concern times when the universe was much less than a second old. One set of problems has to do with the special conditions the model requires as the universe emerged from the big bang.

The first problem is the difficulty of explaining the large-scale uniformity of the observed universe. The large-scale uniformity is most evident in the microwave background radiation, which is known to be uniform in temperature to about one part in 10,000. In the standard model the universe evolves much too quickly to allow this uniformity to be achieved by the usual processes whereby a system approaches thermal equilibrium. The reason is that no information or physical process can propagate faster than a light signal. At any given time there is a maximum distance, known as the horizon distance, that a light signal could have traveled since the beginning of the universe. In the standard model the sources of the microwave background radiation observed from opposite directions in the sky were separated from each other by more than 90 times the horizon distance when the radiation was emitted. Since the regions could not have communicated, it is difficult to see how they could have evolved conditions so nearly identical.

The puzzle of explaining why the universe appears to be uniform over distances that are large compared with the horizon distance is known as the horizon problem. It is not a genuine inconsistency of the standard model; if the uniformity is assumed in the initial conditions, the universe will evolve uniformly. The problem is that one of the most

TYPE OF UNIVERSE	RATIO OF ENERGY DENSITY TO CRITICAL DENSITY (Ω)	SPATIAL GEOMETRY	VOLUME	TEMPORAL EVOLUTION
CLOSED	>1	POSITIVE CURVATURE (SPHERICAL)	FINITE	EXPANDS AND RECOLLAPSES
OPEN	<1	NEGATIVE CURVATURE (HYPERBOLIC)	INFINITE	EXPANDS FOREVER
FLAT	1	ZERO CURVATURE (EUCLIDEAN)	INFINITE	EXPANDS FOREVER. BUT EXPANSION RATE APPROACHES ZERO

Three types of universe, classified as closed, open and flat, can arise from the standard big bang model (under the usual assumption that the equations of general relativity are not modified by the addition of a cosmological term). The distinction between the different geometries depends on the quantity designated Ω, the ratio of the energy density of the universe to some critical density, whose value depends in turn on the rate of expansion of the universe. The value of Ω today is known to lie between 0.1 and 2, which implies that its value a second after the big bang was equal to 1 to within one part in 10^{15}. The failure of the standard big bang model to explain why Ω began so close to 1 is called the flatness problem.

salient features of the observed universe—its large-scale uniformity—cannot be explained by the standard model; it must be assumed as an initial condition.

Even with the assumption of large-scale uniformity, the standard big bang model requires yet another assumption to explain the nonuniformity observed on smaller scales. To account for the clumping of matter into galaxies, clusters of galaxies, super-clusters of clusters and so on, a spectrum of primordial inhomogeneities must be assumed as part of the initial conditions. The fact that the spectrum of inhomogeneities has no explanation is a drawback in itself, but the problem becomes even more pronounced when the model is extended back to 10^{-45} second after the big bang. The incipient clumps of matter develop rapidly with time as a result of their gravitational self-attraction, and so a model that begins at a very early time must begin with very small inhomogeneities. To begin at 10^{-45} second, the matter must start in a peculiar state of extraordinary but not quite perfect uniformity. A normal gas in thermal equilibrium would be far too inhomogeneous, because of the random motion of particles. This peculiarity of the initial state of matter required by the standard model is called the smoothness problem.

Another subtle problem of the standard model concerns the energy density of the universe. According to general relativity, the space of the universe can in principle be curved, and the nature of the curvature depends on the energy density. If the energy density exceeds a certain critical value, which depends on the expansion rate, the universe is said to be closed: space curves back on itself to form a finite volume with no boundary. (A familiar analogy is the surface of a sphere, which is finite in area and has no boundary.) If the energy density is less than the critical density, the universe is open: space curves but does not turn back on itself, and the volume is infinite. If the energy density is just equal to the critical density, the universe is flat: space is described by the familiar Euclidean geometry (again with infinite volume).

The ratio of the energy density of the universe to the critical density is a quantity cosmologists designate by the Greek letter Ω (omega). The value $\Omega = 1$ (corresponding to a flat universe) represents a state of unstable equilibrium. If W was ever exactly equal to 1, it would remain exactly equal to 1 forever. If Ω differed slightly from 1 an instant after the big bang, however, the deviation from 1 would grow rapidly with time. Given this instability, it is surprising that Ω is measured today as being between 0.1 and 2. (Cosmologists are still not sure whether the universe is open, closed or flat.) In order for Ω to be in this rather narrow range today, its value a second after the big bang had to equal 1 to within one part in 10^{15}. The standard model offers no explanation of why W began so close to 1 but merely assumes the fact as an initial condition. This shortcoming of the standard model, called the flatness problem, was first pointed out in 1979 by Robert H. Dicke and P. James E. Peebles of Princeton University.

Perhaps the most important development in the theory of elementary particles [since 1974] has been the notion of grand unified theories, the prototype of which was proposed in 1974 by Howard M. Georgi and Sheldon Lee Glashow of Harvard University. The theories are difficult to verify experimentally because their most distinctive predictions apply to energies far higher than the energies that can be reached with today's particle accelerators. Nevertheless, the theories have some experimental support, and they unify the understanding of elementary particle interactions so elegantly that many physicists find them extremely attractive.

The basic idea of a grand unified theory is that what were perceived to be three independent forces—the strong, the weak and the electromagnetic—are actually parts of a single unified force. In the theory a symmetry relates one force to another. Since experimentally the forces are very different in strength and character, the theory is constructed so that the symmetry is spontaneously broken in the present universe.

A spontaneously broken symmetry is one that is present in the underlying theory describing a system but is hidden in the equilibrium state of the system. For example, a liquid described by physical laws that are rotationally symmetric is itself rotationally symmetric: the distribution of molecules looks the same no matter how the liquid is turned. When the liquid freezes into a crystal, however, the atoms arrange themselves along crystallographic axes, and the rotational symmetry is broken. One would expect that if the temperature of a system in a broken-symmetry state were raised, it could

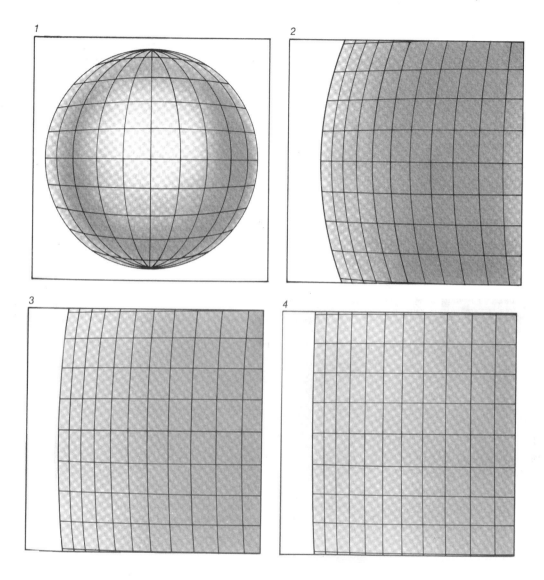

Solution of the flatness problem is illustrated by these drawings of an inflating sphere. The illustration shows how a flat spatial geometry (which corresponds to a value of Ω equal to 1) can be produced by the inflationary scenario in a simple and natural way. In each successive frame the sphere is inflated by a factor of three (and the number of grid lines on the surface is increased by three). The surface curvature quickly becomes undetectable at this scale.

undergo a kind of phase transition to a state in which the symmetry is restored, just as a crystal can melt into a liquid. Grand unified theories predict such a transition at a critical temperature of roughly 10^{27} kelvins.

One novel property of the grand unified theories has to do with the particles called baryons, a class whose most important members are the proton and the neutron. In all physical processes observed up to now, the number of baryons minus the number of antibaryons does not change; in the language of particle physics, the total baryon number of the system is said to be conserved. A consequence of such a conservation law is that the proton must be absolutely stable; because it is the lightest baryon, it cannot decay into another particle without changing the total baryon number. Experimentally the lifetime of the proton is known to exceed 10^{31} years.

Grand unified theories imply that baryon number is not exactly conserved. At low temperature, in the broken-symmetry phase, the conservation law is an excellent approximation, and the observed limit on the proton lifetime is consistent with at least many versions of grand unified theories. At high temperature, however, processes that change the baryon number of a system of particles are expected to be quite common.

One direct result of combining the big bang model with grand unified theories is the successful prediction of the asymmetry of matter and antimatter in the universe. It is thought that all the stars, galaxies and dust observed in the universe are in the form of matter rather than antimatter; their nuclear particles are baryons rather than antibaryons. It follows that the total baryon number of the observed universe is about 10^{78}. Before the advent of grand unified theories, when baryon number was thought to be conserved, this net baryon number had to be postulated as yet another initial condition of the universe. When grand unified theories and the big bang picture are combined, however, the observed excess of matter over antimatter can be produced naturally by elementary particle interactions at temperatures just below the critical temperature of the phase transition.

A serious problem that results from combining grand unified theories with the big bang picture is that a large number of defects are generally formed during the transition from the symmetric phase to the broken-symmetry phase. The defects are created when regions of symmetric phase undergo a transition to different broken-symmetry states. In an analogous situation, when a liquid crystallizes, different regions may begin to crystallize with different orientations of the crystallographic axes. The domains of different crystal orientation grow and coalesce, and it is energetically favorable for them to smooth the misalignment along their boundaries. The smoothing is often imperfect, however, and localized defects remain.

In the grand unified theories there are serious cosmological problems associated with pointlike defects, which correspond to magnetic monopoles, and surfacelike defects, called domain walls. Both are expected to be extremely stable and extremely massive. (The monopole can be shown to be about 10^{16} times as heavy as the proton.) A domain of correlated broken-symmetry phase cannot be much larger than the horizon distance at that time, and so the minimum number of defects created during the transition can be estimated. The result is that there would be so many defects after the transition that their mass would dominate the energy density of the universe and thereby speed up its subsequent evolution. The cosmic microwave background radiation would reach its present temperature of three kelvins only 30,000 years after the big bang instead of 10 billion years, and all the successful predictions of the big bang model would be lost. Thus, any successful union of grand unified theories and the big bang picture must incorporate some mechanism to drastically suppress the production of magnetic monopoles and domain walls.

All modern particle theories, including the grand unified theories, are examples of quantum field theories. The best-known field theory is the one that describes electromagnetism. According to the classical (nonquantum) theory of electromagnetism developed by James Clerk Maxwell in the 1860s, electric and magnetic fields have a well-defined value at every point in space, and their variation with time is described by a definite set of equations. Maxwell's theory was modified early in the 20th century in order to achieve consistency with the quantum theory. In the classical theory, it is possible to increase the energy of an electromagnetic field by any amount; in the quantum theory, however, the increases in energy can come only in discrete lumps, the quanta, which in this case are called photons. The photons have both wavelike and particlelike properties, but in the lexicon of modern physics they are usually called particles. In general the formulation of a quantum field theory begins with a classical theory of fields, and it becomes a theory of particles when the rules of the quantum theory are applied.

The detailed mechanism of spontaneous symmetry breaking in grand unified theories is simpler in many ways than the analogous mechanism in crystals. In a grand unified theory, spontaneous symmetry breaking is accomplished by including in the formulation of the theory a special set of fields known as Higgs fields (after Peter W. Higgs of the University of Edinburgh). The symmetry is unbroken when all the Higgs fields have a value of zero, but it is spontaneously broken whenever at least one of the Higgs fields acquires a nonzero value. Furthermore, it is possible to formulate the theory in such a way that a Higgs field has a nonzero value in the state of lowest energy density, which in this context is known as the true vacuum. At temperatures greater than about 10^{27} kelvins, thermal fluctuations drive the equilibrium value of the Higgs field to zero, resulting in a transition to the symmetric phase.

We have now assembled enough background information to describe the inflationary model of the universe, beginning with the form in which it was first proposed by one of us (Guth) in 1980. Any cosmological model must begin with some assumptions about the initial conditions, but for the inflationary model the initial conditions can be rather arbitrary. One must assume, however, that the early universe included at least some regions of gas that were hot compared with the critical temperature of the phase transition and that were also expanding. In such a hot region the Higgs field would have a value of zero. As the expansion caused the temperature to fall it would become thermodynamically favorable for the Higgs field to acquire a nonzero value, bringing the system to its broken-symmetry phase.

For some values of the unknown parameters of the grand unified theories, this phase transition would occur very slowly compared with the cooling rate. As a result, the system could cool to well below 10^{27} kelvins with the value of the Higgs field remaining at zero. This phenomenon, known as supercooling, is quite common in condensed-matter physics; water, for example, can be supercooled to more than 20 degrees below its freezing point, and glasses are formed by rapidly supercooling a liquid to a temperature well below its freezing point.

As the region of gas continued to supercool, it would approach a peculiar state of matter known as a false vacuum. The temperature, and hence the thermal component of the energy density, could rapidly decrease, and the energy density of the state would be concentrated entirely in the Higgs field. A zero value for the Higgs field implies a large energy density for the false vacuum. In the classical form of the theory, such a state would be absolutely stable, even though it would not be the state of lowest energy density. States with a lower energy density would be separated from the false vacuum by an intervening energy barrier, and there would be no energy available to take the Higgs field over the barrier.

In the quantum version of the model the false vacuum is not absolutely stable. Under the rules of the quantum theory, all the fields would be continually fluctuating. As was first described by Sidney R. Coleman of Harvard, a quantum fluctuation would occasionally cause the Higgs field in a small region of space to "tunnel" through the energy barrier, nucleating a "bubble" of the broken-symmetry phase. The bubble would then start to grow at a speed that would rapidly approach the speed of light, converting the false vacuum into the broken-symmetry phase.

The most peculiar property of the false vacuum is probably its pressure, which is both large and negative. To understand why this is so, consider again the process by which a bubble of true vacuum would grow into a region of false vacuum. The growth is favored energetically because the true vacuum has a lower energy density than the false vacuum. The growth also indicates, however, that the pressure of the true vacuum must be higher than the pressure of the false vacuum, forcing the bubble wall to grow outward. Because the pressure of the true vacuum is zero, the pressure of the false vacuum must be negative.

The negative pressure would not result in mechanical forces within the false vacuum, because mechanical forces arise only from differences in pressure. Nevertheless, there would be gravitational effects. Under ordinary circumstances, the expansion of the region of gas would be slowed by the mutual gravitational attraction of the matter within it. According to general relativity, the pressure also contributes to the attraction; to be specific, the gravitational force is proportional to the energy density plus three times the pressure. For the false vacuum, the contribution made by the pressure would overwhelm the energy density contribution and would have the opposite sign. Therefore, the bizarre notion of negative pressure leads to the even more bizarre effect of a gravitational force that is effectively repulsive. As a result, the expansion of the region would be accelerated and the region would grow exponentially, doubling in diameter during each interval of approximately 10^{-34} second.

This period of accelerated expansion is called the inflationary era, and it is the key element of the inflationary model of the universe. According to the model, the inflationary era continued for 10^{-32} second or longer, and during this period the diameter of the universe increased by a factor of 10^{50} or more. It is assumed that after this colossal expansion the transition to the broken-symmetry phase finally took place. The energy density of the false vacuum was then released, resulting in a tremendous amount of particle production. The region was reheated to a temperature of almost 10^{27} kelvins. (In the language of thermodynamics the energy released is called the latent heat; it is analogous to the energy released when water freezes.) From this point onward, the region would continue to expand and cool at the rate described by the standard big bang model. A volume the size of the observable universe would lie well within such a region.

The horizon problem is avoided in a straightforward way. In the inflationary model the observed universe evolves from a region that is much smaller in diameter (by a factor of 10^{50} or more) than the corresponding region in the standard model. Before inflation begins, the region is much smaller than the horizon distance, and it has time to homogenize and reach thermal equilibrium. This small homogeneous region is then inflated to become large enough to encompass the observed universe. Thus, the sources of the microwave background radiation arriving today from all directions in the sky were once in close contact: they had time to reach a common temperature before the inflationary era began.

The flatness problem is also evaded in a simple and natural way. The equations describing the evolution of the universe during the inflationary era are different from those for the standard model, and it turns out that the ratio Ω is driven rapidly toward 1, no matter what value it had before inflation. This behavior is most easily understood by recalling that a value of $\Omega = 1$ corresponds to a space that is geometrically flat. The rapid expansion causes the space to become flatter just as the surface of a balloon becomes flatter when it is inflated. The mechanism driving Ω toward 1 is so effective that one is led to an almost rigorous prediction: the value of Ω today should be very accurately equal to 1.

In the form in which the inflationary model was originally proposed, it had a crucial flaw: under the circumstances described, the bubbles would always remain in finite clusters disconnected from one another and each cluster would be dominated by a single largest bubble. Almost all the energy in the cluster would be initially concentrated in the surface of the largest bubble, and there is no apparent mechanism to redistribute energy uniformly. Such a configuration bears no resemblance to the observed universe.

For almost two years after the invention of the inflationary universe model, it remained a tantalizing but clearly imperfect solution to a number of important cosmological problems. Near the end of 1981, however, a new approach was developed by A. D. Linde of the P. N. Lebedev Physical Institute in Moscow and independently by Andreas Albrecht and one of us (Steinhardt) of the University of Pennsylvania. This approach, known as the new inflationary universe, avoids all the problems of the original model while maintaining all its successes.

The scenario begins just as it does in the original inflationary model. Again one must assume the early universe had regions that were hotter than about 10^{27} kelvins and were also expanding. In these regions, thermal fluctuations would drive the equilibrium value of the Higgs fields to zero, and the symmetry would be unbroken. As the temperature fell, it would become thermodynamically favorable for the system to undergo a phase transition in which at least one of the Higgs fields acquired a nonzero value, resulting in a broken-symmetry phase. As in the previous case, however, the rate of this phase transition would be extremely low compared with the rate of cooling. The system would supercool to a negligible temperature with the Higgs field remaining at zero, and the resulting state would again be considered a false vacuum.

The important difference in the new approach is the way in which the phase transition would take place. Quantum fluctuations or small residual thermal fluctuations would cause the Higgs field to deviate from zero. In the absence of an energy barrier the value of the Higgs field would begin to increase steadily; the rate of increase would be much like that of a ball rolling down a hill of the same shape as the curve of the energy density function, under the influence of a frictional drag force. Since the energy density curve is almost flat near the point where the Higgs field vanishes, the early stage of the evolution would be very slow. As long as the Higgs field remained close to zero, the energy density would be almost the same as it is in the false vacuum. As in the original scenario, the region would undergo accelerated expansion, doubling in diameter every

10^{-34} second or so. In this case, however, the expansion would cease to accelerate when the value of the Higgs field reached the steeper part of the curve. By computing the time required for the Higgs field to evolve, the amount of inflation can be determined. An expansion factor of 10^{50} or more is quite plausible, but the actual factor depends on the details of the particle theory one adopts.

So far the description of the phase transition has been slightly oversimplified. There are actually many different states of broken symmetry, just as there are many possible orientations for the axes of a crystal. There are a number of Higgs fields, and the various broken-symmetry states are distinguished by the combination of Higgs fields that acquire nonzero values. Since the fluctuations that drive the Higgs fields from zero are random, different regions of the primordial universe would be driven toward different broken-symmetry states, each region forming a domain with an initial radius of roughly the horizon distance. At the start of the phase transition the horizon distance would be about 10^{-24} centimeter. Once the domain formed, with the Higgs fields deviating slightly from zero in a definite combination, it would evolve toward one of the stable broken-symmetry states and would inflate by a factor of 10^{50} or more. The size of the domain after inflation would then be greater than 10^{26} centimeters. The entire observable universe, which at that time would be only about 10 centimeters across, would be able to fit deep inside a single domain.

Note that the crucial flaw of the original inflationary model is deftly avoided. Roughly speaking, the isolated bubbles that were discussed in the original model are replaced here by the domains. The domains of the slow-rollover transition would be surrounded by other domains rather than by false vacuum, and they would tend not to be spherical. The term "bubble" is therefore avoided. The key difference is that in the new inflationary model each domain inflates in the course of its formation, producing a vast, essentially homogeneous region within which the observable universe can fit.

Since the reheating temperature is near the critical temperature of the phase transition in the grand unified theory, the matter-antimatter asymmetry could be produced by particle interactions just after the phase transition. The production mechanism is the same as the one predicted by grand unified theories for the standard big bang model. In contrast to the standard model, however, the inflationary model does not allow the possibility of assuming the observed net baryon number of the universe as an initial condition; the subsequent inflation would dilute any initial baryon number density to an imperceptible level. Thus, the viability of the inflationary model depends crucially on the viability of particle theories, such as the grand unified theories, in which baryon number is not conserved.

One can now grasp the solutions to the cosmological problems discussed above. The horizon and flatness problems are resolved by the same mechanisms as in the original inflationary universe model. In the new inflationary scenario the problem of monopoles and domain walls can also be solved. Such defects would form along the boundaries separating domains, but the domains would have been inflated to such an enormous size that the defects would lie far beyond any observable distance. (A few defects might be generated by thermal effects after the transition, but they are expected to be negligible in number.)

Thus, with a few simple ideas the improved inflationary model of the universe leads to a successful resolution of several major problems that plague the standard big bang picture: the horizon, flatness, magnetic monopole and domain wall problems. Unfortunately the necessary slow-rollover transition requires the fine-tuning of parameters; calculations yield reasonable predictions only if the parameters are assigned values in a narrow range. Most theorists (including both of us) regard such fine-tuning as implausible. The consequences of the scenario are so successful, however, that we are encouraged to go on in the hope that we may discover realistic versions of grand unified theories in which such a slow-rollover transition occurs without fine-tuning.

The successes already discussed offer persuasive evidence in favor of the new inflationary model. Moreover, it was recently discovered that the model may also resolve an additional cosmological problem not even considered at the time the model was developed: the smoothness problem. The generation of density inhomogeneities in the new inflationary universe was addressed in the summer of 1982 at the Nuffield Workshop on the Very Early Universe by a number of theorists, including James M. Bardeen of the University of Washington, Stephen W. Hawking of the University of Cambridge, So-Young Pi of Boston University, Michael S. Turner of the University of Chicago, A. A. Starobinsky of the L. D. Landau Institute of Theoretical Physics in Moscow and the two of us. It was found that the new inflationary model, unlike any previous cosmological model, leads to a definite prediction for the spectrum of inhomogeneities. Basically the process of inflation first smoothes out any primordial inhomogeneities that might have been present in the initial conditions. Then, in the course of the phase transition, inhomogeneities are generated by the quantum fluctuations of the Higgs field in a way that is completely determined by the underlying physics. The inhomogeneities are created on an exceedingly small scale of length, where quantum phenomena are important, and they are then enlarged to an astronomical scale by the process of inflation.

The new inflationary model also predicts the magnitude of the density inhomogeneities, but the prediction is quite sensitive to the details of the underlying particle theory. Unfortunately, the magnitude that results from the simplest grand unified theory is far too large to be consistent with the observed uniformity of the cosmic microwave background. This inconsistency represents a problem, but it is not yet known whether the simplest grand unified theory is the correct one. In particular, the simplest grand unified theory predicts a lifetime for the proton that appears to be lower than present experimental limits. On the other hand, one can construct more complicated grand unified theories that result in density inhomogeneities of the desired magnitude. Many investigators imagine that with the development of the correct particle theory the new inflationary model will add the resolution of the smoothness problem to its list of successes.

In short, the inflationary model of the universe is an economical theory that accounts for many features of the observable universe lacking an explanation in the standard big bang model. The beauty of the inflationary model is that the evolution of the universe becomes almost independent of the details of the initial conditions, about which little if anything is known. It follows, however, that if the inflationary model is correct, it will be difficult for anyone to ever discover observable consequences of the conditions existing before the inflationary phase transition. Similarly, the vast distance scales created by inflation would make it essentially impossible to observe the structure of the universe as a whole. Nevertheless, one can still discuss these issues, and a number of remarkable scenarios seem possible.

The simplest possibility for the very early universe is that it actually began with a big bang, expanded rather uniformly until it cooled to the critical temperature of the phase transition and then proceeded according to the inflationary scenario. Extrapolating the big bang model back to zero time brings the universe to a cosmological singularity, a condition of infinite temperature and density in which the known laws of physics do not apply. The instant of creation remains unexplained. A second possibility is that the universe began (again without explanation) in a random, chaotic state. The matter and temperature distributions would be nonuniform, with some parts expanding and other parts contracting. In this scenario certain small regions that were hot and expanding would undergo inflation, evolving into huge regions easily capable of encompassing the observable universe. Outside these regions there would remain chaos, gradually creeping into the regions that had inflated.

Recently there has been some serious speculation that the actual creation of the universe is describable by physical laws. In this view the universe would originate as a quantum fluctuation, starting from absolutely nothing. The idea was first proposed by Edward P. Tryon of Hunter College of the City University of New York in 1973, and it was put forward again in the context of the inflationary model by Alexander Vilenkin of Tufts University in 1982. In this context "nothing" might refer to empty space, but Vilenkin uses it to describe a state devoid of space, time and matter. Quantum fluctuations of the structure of space-time can be discussed only in the context of quantum gravity, and so these ideas must be considered highly speculative until a working theory of quantum gravity is formulated. Nevertheless, it is fascinating to contemplate that physical laws may determine not only the evolution of a given state of the universe but also the initial conditions of the observable universe.

As for the structure of the universe as a whole, the inflationary model allows for several possibilities. (In all cases, the observable universe is a very small fraction of the universe as a whole; the edge of our domain is likely to lie 10^{35} or more light-years away.) The first possibility is that the domains meet one another and fill all space. The domains are then separated by domain walls, and in the interior of each wall is the symmetric phase of the grand unified theory. Protons or neutrons passing through such a wall would decay instantly. Domain walls would tend to straighten with time. After 10^{35} years or more, smaller domains (possibly even our own) would disappear, and larger domains would grow.

From a historical point of view, probably the most revolutionary aspect of the inflationary model is the notion that all the matter and energy in the observable universe may have emerged from almost nothing. This claim stands in marked contrast to centuries of scientific tradition in which it was believed that something cannot come from nothing. The tradition, dating back at least as far as the Greek philosopher Parmenides in the 5th century B.C., has manifested itself in modern times in the formulation of a number of conservation laws, which state that certain physical quantities cannot be changed by any physical process.

Since the observed universe apparently has a huge baryon number and a huge energy, the idea of creation from nothing has seemed totally untenable to all but a few theorists. With the advent of grand unified theories, however, it appears quite plausible that baryon number is not conserved. As a result, only the conservation of energy needs further consideration.

The total energy of any system can be divided into a gravitational part and a nongravitational part. The gravitational part (that is, the energy of the gravitational field itself) is negligible under laboratory conditions, but cosmologically it can be quite important. The nongravitational part is not by itself conserved; in the standard big bang model it decreases drastically as the early universe expands, and the rate of energy loss is proportional to the pressure of the hot gas. During the era of inflation on the other hand, the region of interest is filled with a false vacuum that has a large negative pressure. In this case, the nongravitational energy increases drastically. Essentially all the nongravitational energy of the universe is created as the false vacuum undergoes its accelerated expansion. This energy is released when the phase transition takes place, and it eventually evolves to become stars, planets, human beings and so forth. Accordingly, the inflationary model offers what is apparently the first plausible scientific explanation for the creation of essentially all the matter and energy in the observable universe.

Under these circumstances, the gravitational part of the energy is somewhat ill defined, but crudely speaking one can say that the gravitational energy is negative and that it precisely cancels the nongravitational energy. The total energy is then zero and is consistent with the evolution of the universe from nothing.

If grand unified theories are correct in their prediction that baryon number is not conserved, there is no known conservation law that prevents the observed universe from evolving out of nothing. The inflationary model of the universe provides a possible mechanism by which the observed universe could have evolved from an infinitesimal region. It then becomes tempting to go one step further and speculate that the entire universe evolved from literally nothing.

The Expansion Rate and Size of the Universe

Wendy L. Freedman

Our Milky Way and all other galaxies are moving away from one another as a result of the big bang, the fiery birth of the universe. When did the colossal expansion begin? Will the universe expand forever, or will gravity eventually halt its expansion and cause it to collapse back on itself?

For decades, cosmologists have attempted to answer such questions by measuring the universe's size scale and expansion rate. To accomplish this task, astronomers must determine both how fast galaxies are moving and how far away they are. Techniques for measuring the velocities of galaxies are well established, but estimating the distances to galaxies has proved far more difficult. Recently the superb resolution of the Hubble Space Telescope has extended and strengthened the calibration of the extragalactic distance scale, leading to new estimates of the expansion rate.

At present, several lines of evidence point toward a high expansion rate, implying that the universe is relatively young, perhaps only 10 billion years old. The evidence also suggests that the expansion of the universe may continue indefinitely. Still, many astronomers and cosmologists do not yet consider the evidence definitive. We actively debate the merits of our techniques.

An accurate measurement of the expansion rate is essential not only for determining the age of the universe and its fate but also for constraining theories of cosmology and models of galaxy formation. Furthermore, the expansion rate is important for estimating fundamental quantities, from the density of the lightest elements (such as hydrogen and helium) to the amount of nonluminous matter in galaxies, as well as clusters of galaxies. Because we need accurate distance measurements to calculate the luminosity, mass and size of astronomical objects, the issue of the cosmological distance scale, or the expansion rate, affects the entire field of extragalactic astronomy.

Astronomers began measuring the expansion rate of the universe some 70 years ago. In 1929 the eminent astronomer Edwin P. Hubble of the Carnegie Institution's observatories made the remarkable observation that the velocity of a galaxy's recession is proportional to its distance. His observations provided the first evidence that the entire universe is expanding.

THE HUBBLE CONSTANT

Hubble was the first to determine the expansion rate. Later this quantity became known as the Hubble constant: the recession velocity of the galaxy divided by its distance. A very rough estimate of the Hubble constant is 100 kilometers per second per megaparsec. (Astronomers commonly represent distances in terms of megaparsecs, where one megaparsec is the distance light travels in 3.26 million years). A galaxy at 500 megaparsecs therefore moves at about 50,000 kilometers per second, or more than 100 million miles per hour!

For seven decades, astronomers have hotly debated the precise value of the expansion rate. Hubble originally obtained a value of 500 kilometers per second per megaparsec (km/s/Mpc). After Hubble's death in 1953, his protége, Allan R. Sandage, also at Carnegie, continued to map the expansion of the universe. As Sandage and others made more accurate and extensive observations, they revised Hubble's original value downward into the range of 50 to 100 km/s/Mpc, thereby indicating a universe far older and larger than suggested by the earliest measurements.

During the past two decades, new estimates of the Hubble constant have continued to fall within this same range, but preferentially toward the two extremes. Notably, Sandage and his longtime collaborator Gustav A. Tammann of the University of Basel have argued for a value of 50 km/s/Mpc, whereas the late Gérard de Vaucouleurs of the University of Texas advocated a value of 100 km/s/Mpc. The controversy has created an unsatisfactory situation in which scientists have been free to choose any value of the Hubble constant between the two extremes.

In principle, determining the Hubble constant is simple, requiring only a measurement of velocity and distance. Measuring a galaxy's velocity is straightforward: Astronomers disperse light from a galaxy and record its spectrum. A galaxy's spectrum has discrete spectral lines, which occur at characteristic wavelengths caused by emission or absorption of elements in the gas and stars making up the galaxy. For a galaxy receding from Earth, these spectral lines shift to longer wavelengths by an amount proportional to the velocity—an effect known as redshift.

If the measurement of the Hubble constant is so simple in principle, then why has it remained one of the outstanding problems in cosmology for almost 70 years? In practice, measuring the Hubble constant is extraordinarily difficult, primarily for two reasons. First, although we can measure their velocities accurately, galaxies interact gravitationally with their neighbors. In so doing, their velocities become perturbed, inducing "peculiar" motions that are superimposed onto the general expansion of the universe. Second, establishing an accurate distance scale has turned out to be much more difficult than anticipated. Consequently, an accurate measure of the Hubble constant requires us not only to establish an accurate extragalactic distance scale but also to do this already difficult task at distances great enough that peculiar motions of galaxies are small compared with the overall expansion, or Hubble flow. To determine the distance to a galaxy, astronomers must choose from a variety of complicated methods. Each has its advantages, but none is perfect.

MEASURING DISTANCES TO GALAXIES

Astronomers can most accurately measure distances to nearby galaxies by monitoring a type of star commonly known as a Cepheid variable. Over time, the star changes in brightness in a periodic and distinctive way. During the first part of the cycle, its luminosity increases very rapidly, whereas during the remainder of the cycle, the luminosity of the Cepheid decreases slowly. On average, Cepheid variables are about 10,000 times brighter than the Sun.

Remarkably, the distance to a Cepheid can be calculated from its period (the length of its cycle) and its average apparent brightness (its luminosity as observed from Earth). In 1908 Henrietta S. Leavitt of Harvard College Observatory discovered that the period of a Cepheid correlates closely with its brightness. She found that the longer the period, the brighter the star. This relation arises from the fact that a Cepheid's brightness is proportional to its surface area. Large, bright Cepheids pulsate over a long period just as, for example, large bells resonate at a low frequency (or longer period).

By observing a Cepheid's variations in luminosity over time, astronomers can obtain its period and average apparent luminosity, thereby calculating its absolute luminosity (that is, the apparent brightness the star would have if it were a standard distance of 10 parsecs away). Furthermore, they know that the apparent luminosity decreases as the distance it travels increases—because the apparent luminosity falls off in proportion to the square of the distance to an object. Therefore, we can compute the distance to the Cepheid from the ratio of its absolute brightness to its apparent brightness.

During the 1920s, Hubble used Cepheid variables to establish that other galaxies existed far beyond the Milky Way. By measuring apparent brightnesses and periods of faint, starlike images that he discovered on photographs of objects such as the Andromeda Nebula (also known as M31), the Triangulum Nebula (M33) and NGC 6882, he could show that these objects were located more than several hundred thousand light-years from the Sun, well outside the Milky Way. From the 1930s to the 1960s, Hubble, Sandage and others struggled to find Cepheids in nearby galaxies. They succeeded in measuring the distances to about a dozen galaxies. About half these galaxies are useful for the derivation of the Hubble constant.

Were it feasible, we would use Cepheids directly to measure distances associated with the universe's expansion. Unfortunately, so far we cannot detect Cepheids in galaxies sufficiently far away so that we know they are part of a "pure" Hubble expansion of the universe.

Nevertheless, astronomers have developed several other methods for measuring relative distances between galaxies on vast scales, well beyond Cepheid range. Because we must use the Cepheid distance scale to calibrate these techniques, they are considered secondary distance indicators.

Astronomers have made great strides developing techniques to measure such relative distances. These methods include observing and measuring a special category of supernovae: catastrophic explosions signaling the death of certain low-mass stars. Sandage and his collaborators are now determining the Hubble constant by studying such supernovae based on the calibration of Cepheids. Other secondary distance-determining methods include measuring the brightnesses and rotations of velocities of entire spiral galaxies, the fluctuations (or graininess) in the light of elliptical galaxies and the analysis and measurement of the expansion properties of another category of younger, more massive supernovae. The key to measuring the Hubble constant using these techniques is to determine the distance to selected galaxies using Cepheids; their distances can, in turn, be used to calibrate the relative extragalactic distance scale by applying secondary methods.

ESTABLISHING A DISTANCE SCALE

One technique for measuring great distances, the Tully-Fisher relation, relies on a correlation between a galaxy's brightness and its rotation rate. High-luminosity galaxies typically have more mass than low-luminosity galaxies, and so bright galaxies rotate slower than dim galaxies. Several groups have tested the Tully-Fisher method and shown that the relation does not appear to depend on environment; it remains the same in the dense and outer parts of rich clusters and for relatively isolated galaxies. The Tully-Fisher relation can be used to estimate distances as far away as 300 million light-years. A disadvantage is that astronomers lack a detailed theoretical understanding of the Tully-Fisher relation.

Another distance indicator that has great potential is a particular kind of supernova known as Type Ia. Type Ia supernovae, astronomers believe, occur in double-star systems in which one of the stars is a very dense object known as a white dwarf. When a companion star transfers its mass to a white dwarf, it triggers an explosion. Because supernovae release tremendous amounts of radiation, astronomers should be able to see supernovae as far away as five billion light-years—that is, a distance spanning a radius of half the visible universe.

Type Ia supernovae make good distance indicators because, at the peak of their brightness, they all produce roughly the same amount of light. Using this information, astronomers can infer their distance.

If supernovae are also observed in galaxies for which Cepheid distances can be measured, then the brightnesses of supernovae can be used to infer distances. In practice, however, the brightnesses of supernovae are not all the same; there is a range of brightnesses that must be taken into account. A difficulty is that supernovae are very rare events, so the chance of seeing one nearby is very small. Unfortunately, a current limitation of this method is that about half of all supernovae observed in galaxies close enough to have Cepheid distances were observed decades ago, and these measurements are of low quality.

For decades, astronomers have recognized that the solution to the impasse on the extragalactic distance scale would require observations made at very high spatial resolution. The Hubble telescope can now resolve Cepheids at distances 10 times farther (and therefore in a volume 1,000 times larger) than we can do from the ground. A primary motivation for building an orbiting optical telescope was to enable the discovery of Cepheids in remoter galaxies and to measure accurately the Hubble constant.

More than a decade ago several colleagues and I were awarded time on the Hubble telescope to undertake this project. This program involves 26 astronomers, led by me, Jeremy R. Mould of Mount Stromlo and Siding Springs Observatory and Robert C. Kennicutt of Steward Observatory. Our effort involves measuring Cepheid distances to about 20 galaxies, enough to calibrate a wide range of secondary distance methods. We aim to compare and contrast results from many techniques and to assess the true uncertainties in the measurement of the Hubble constant.

Though still incomplete, new Cepheid distances to a dozen galaxies have been measured as part of this project. Preliminary results yield a value of the Hubble constant of about 70 km/s/Mpc with an uncertainty of about 15 percent. This value is based on a number of methods, including the Tully-Fisher relation, Type Ia supernovae, Type II supernovae, surface-brightness fluctuations, and Cepheid measurements to galaxies in the nearby Virgo and Fornax clusters.

Sandage and his collaborators have reported a value of 59 km/s/Mpc, based on Type Ia supernovae. Other groups (including our own) have found a value in the middle 60 range, based on the same Type Ia supernovae. Nevertheless, these current disagreements are much smaller than the earlier discrepancies of a factor of two, which have existed until now. This progress is encouraging.

HOW OLD IS THE UNIVERSE?

The value of the Hubble constant has many implications for the age, evolution and fate of the universe. A low value for the Hubble constant implies an old age for the universe, whereas a high value suggests a young age. For example, a value of 100 km/s/Mpc indicates the universe is about 6.5 to 8.5 billion years old (depending on the amount of matter in the universe and the corresponding deceleration caused by that matter). A value of 50 km/s/Mpc suggests, however, an age of 13 to 16.5 billion years.

And what of the ultimate fate of the universe? If the average density of matter in the universe is low, as current observations indicate, the standard cosmological model predicts that the universe will expand forever.

Nevertheless, theory and observations suggest that the universe contains more mass than what can be attributed to luminous matter. A very active area of cosmological research is the search for this additional "dark" matter in the universe. To answer the question about the fate of the universe unambiguously, cosmologists require not only a knowledge of the Hubble constant and the average mass density of the universe but also an independent measure of the age of the universe. These three quantities are needed to specify uniquely the geometry and the evolution of the universe.

If the Hubble constant turns out to be high, it would have profound implications for our understanding of the evolution of galaxies and the universe. A Hubble constant of 70 km/s/Mpc yields an age estimate of nine to 12 billion years (allowing for uncertainty in the value of the average density of the universe). A high-density universe corresponds to an age of about nine billion years. A low-density universe corresponds to an age of about 12 billion years for this same value of the Hubble constant.

These estimates are all shorter than what theoretical models suggest for the age of old stellar systems known as globular clusters. Globular clusters are believed to be among the first objects to form in our galaxy, and their age is estimated to be between 13 and 17 billion years. Obviously, the ages of the globular clusters cannot be older than the age of the universe itself.

Age estimates for globular clusters are often cited as a reason to prefer a low value for the Hubble constant and therefore an older age of the universe. Some astronomers argue, however, that the theoretical models of globular clusters on which these estimates depend may not be complete and may be based on inaccurate assumptions. For instance, the models rely on knowing precise ratios of certain elements in globular clusters, particularly oxygen and iron. Moreover, accurate ages require accurate measures of luminosities of globular cluster stars, which in turn require accurate measurements of the distances to the globular clusters.

Spiral galaxy appears in the Fornax cluster of galaxies, or NGC1365, visible from Earth's Southern Hemisphere. Using some of the 50 or so Cepheid variable stars in this galaxy's spiral arms as distance markers, NASA's Key Project team estimated the distance from Earth to the Fornax cluster to be about 60 million light-years.

A high value for the Hubble constant raises another potentially serious problem: it disagrees with standard theories of how galaxies are formed and distributed in space. For example, the theories predict how much time is required for large-scale clustering, which has been observed in the distribution of galaxies, to occur. If the Hubble constant is large (that is, the universe is young), the models cannot reproduce the observed distribution of galaxies.

Although the history of science suggests that ours will not be the last generation to wrestle with these questions, the next decade promises much excitement. There are many reasons to be optimistic that the current disagreement over values of the cosmological parameters governing the evolution of the universe will soon be resolved.

The Self-Reproducing Inflationary Universe

Andrei Linde

If my colleagues and I are right, we may soon be saying good-bye to the idea that our universe was a single fireball created in the big bang. We are exploring a new theory based on a notion that the universe went through a stage of inflation. During that time, the theory holds, the cosmos became exponentially large within an infinitesimal fraction of a second. At the end of this period, the universe continued its evolution according to the big bang model. As workers refined this inflationary scenario, they uncovered some surprising consequences. One of them constitutes a fundamental change in how the cosmos is seen. Recent versions of inflationary theory assert that instead of being an expanding ball of fire the universe is a huge, growing fractal. It consists of many inflating balls that produce new balls, which in turn produce more balls, ad infinitum.

Cosmologists did not arbitrarily invent this rather peculiar vision of the universe. Several workers, first in Russia and later in the United States, proposed the inflationary hypothesis that is the basis of its foundation. We did so to solve some of the complications left by the old big bang idea. In its standard form, the big bang theory maintains that the universe was born about 15 billion years ago from a cosmological singularity—a state in which the temperature and density are infinitely high. Of course, one cannot really speak in physical terms about these quantities as being infinite. One usually assumes that the current laws of physics did not apply then. They took hold only after the density of the universe dropped below the so-called Planck density, which equals about 10^{94} grams per cubic centimeter.

As the universe expanded, it gradually cooled. Remnants of the primordial cosmic fire still surround us in the form of the microwave background radiation. This radiation indicates that the temperature of the universe has dropped to 2.7 kelvins. The 1965 discovery of this background radiation by Arno A. Penzias and Robert W. Wilson of Bell Laboratories proved to be the crucial evidence in establishing the big bang theory as the preeminent theory of cosmology. The big bang theory also explained the abundances of hydrogen, helium and other elements in the universe.

As investigators developed the theory, they uncovered complicated problems. For example, the standard big bang theory, coupled with the modern theory of elementary particles, predicts the existence of many superheavy particles carrying magnetic charge—that is, objects that have only one magnetic pole. These magnetic monopoles would have a typical mass 10^{16} times that of the proton, or about 0.00001 milligram. According to the standard big bang theory, monopoles should have emerged very early in the evolution of the universe and should now be as abundant as protons. In that case, the mean density of matter in the universe would be about 15 orders of magnitude greater than its present value, which is about 10^{-29} gram per cubic centimeter.

QUESTIONING STANDARD THEORY

This and other puzzles forced physicists to look more attentively at the basic assumptions underlying the standard cosmological theory. And we found many to be highly suspicious. I will review six of the most difficult. The first, and main, problem is the very existence of the big bang. One may wonder, What came before? If space-time did not exist then, how could everything appear from nothing? What arose first: The universe or the laws determining its evolution? Explaining this initial singularity—where and when it all began—still remains the most intractable problem of modern cosmology.

A second trouble spot is the flatness of space. General relativity suggests that space may be very curved, with a typical radius on the order of the Planck length, or 10^{-33} centimeter. We see, however, that our universe is just about flat on a scale of 10^{28} centimeters, the radius of the observable part of the universe. This result of our observation differs from theoretical expectations by more than 60 orders of magnitude.

A similar discrepancy between theory and observations concerns the size of the universe, a third problem. Cosmological examinations show that our part of the universe contains at least 10^{88} elementary particles. But why is the universe so big? If one takes a universe of a typical initial size given by the Planck length and a typical initial density equal to the Planck density, then, using the standard big bang theory, one can calculate how many elementary particles such a universe might encompass. The answer is rather unexpected: the entire universe should only be large enough to accommodate just one elementary particle—or at most 10 of them. Obviously, something is wrong with this theory.

The fourth problem deals with the timing of the expansion. In its standard form, the big bang theory assumes that all parts of the universe began expanding simultaneously. But how could all the different parts of the universe synchronize the beginning of their expansion?

Fifth, there is the question about the distribution of matter in the universe. On the very large scale, matter has spread out with remarkable uniformity. Across more than 10 billion light-years, its distribution departs from perfect homogeneity by less than one part in 10,000. For a long time, nobody had any idea why the universe was so homogeneous. But those who do not have ideas sometimes have principles. One of the cornerstones of the standard cosmology was the "cosmological principle," which asserts that the universe must be homogeneous. This assumption, however, does not help much, because the universe incorporates important deviations from homogeneity, namely, stars, galaxies and other agglomerations of matter. Hence, we must explain why the universe is so uniform on large scales and at the same time suggest some mechanism that produces galaxies.

Finally, there is what I call the uniqueness problem. Albert Einstein captured its essence when he said, "What really interests me is whether God had any choice in the creation of the world." Indeed, slight changes in the physical constants of nature could have made the universe unfold in a completely different manner. For example, many popular theories of elementary particles assume that space-time originally had considerably more than four dimensions (three spatial and one temporal). In order to square theoretical calculations with the physical world in which we live, these models state that the extra dimensions have been "compactified," or shrunk to a small size and tucked away. But one may wonder why compactification stopped with four dimensions, not two or five.

Moreover, the manner in which the other dimensions become rolled up is significant, for it determines the values of the constants of nature and the masses of particles. In some theories, compactification can occur in billions of different ways.

All these problems (and others I have not mentioned) are extremely perplexing. That is why it is encouraging that many of these puzzles can be resolved in the context of the theory of the self-reproducing, inflationary universe.

The basic features of the inflationary scenario are rooted in the physics of elementary particles. So I would like to take you on a brief excursion into this realm—in particular, to the unified theory of weak and electromagnetic interactions. Both these forces exert themselves through particles. Photons mediate the electromagnetic force; the W and Z particles are responsible for the weak force. But whereas photons are massless, the W and Z particles are extremely heavy. To unify the weak and electromagnetic interactions despite the obvious differences between photons and the W and Z particles, physicists introduced what are called scalar fields.

Although scalar fields are not the stuff of everyday life, a familiar analogue exists. That is the electrostatic potential—the voltage in a circuit is an example. Electrical fields appear only if this potential is uneven, as it is between the poles of a battery or if the potential changes in time. If the entire universe had the same electrostatic potential—say, 110 volts—then nobody would notice it; the potential would seem to be just another vacuum state. Similarly, a constant scalar field looks like a vacuum: we do not see it even if we are surrounded by it.

SCALAR FIELDS

Scalar fields play a crucial role in cosmology as well as in particle physics. They provide the mechanism that generates the rapid inflation of the universe. Indeed, according to general relativity, the universe expands at a rate (approximately) proportional to the square root of its density. If the universe were filled by ordinary matter, then the density would rapidly decrease as the universe expanded. Thus, the expansion of the universe would rapidly slow down as density decreased. But because of the equivalence of mass and energy established by Einstein, the potential energy of the scalar field also contributes to the expansion. In certain cases, this energy decreases much more slowly than does the density of ordinary matter.

The persistence of this energy may lead to a stage of extremely rapid expansion, or inflation, of the universe. This possibility emerges even if one considers the very simplest version of the theory of a scalar field. In this version the potential energy reaches a minimum at the point where the scalar field vanishes. In this case, the larger the scalar field, the greater the potential energy. According to Einstein's theory of gravity, the energy of the scalar field must have caused the universe to expand very rapidly. The expansion slowed down when the scalar field reached the minimum of its potential energy.

One way to imagine the situation is to picture a ball rolling down the side of a large bowl. The bottom of the bowl represents the energy minimum. The position of the ball corresponds to the value of the scalar field. Of course, the equations describing the motion of the scalar field in an expanding universe are somewhat more complicated than the equations for the ball in an empty bowl. They contain an extra term corresponding to friction, or viscosity. This friction is akin to having molasses in the bowl. The viscosity of this liquid depends on the energy of the field: the higher the ball in the bowl is, the thicker the liquid will be. Therefore, if the field initially was very large, the energy dropped extremely slowly.

The sluggishness of the energy drop in the scalar field has a crucial implication in the expansion rate. The decline was so gradual that the potential energy of the scalar field remained almost constant as the universe expanded. This behavior contrasts sharply with that of ordinary matter, whose density rapidly decreases in an expanding universe. Thanks to the large energy of the scalar field, the universe continued to expand at a speed much greater than that predicted by preinflation cosmological theories. The size of the universe in this regime grew exponentially.

This stage of self-sustained, exponentially rapid inflation did not last long. Its duration could have been as short as 10^{-35} second. Once the energy of the field declined, the viscosity nearly disappeared, and inflation ended. Like the ball as it reaches the bottom of the bowl, the scalar field began to oscillate near the minimum of its potential energy. As the scalar field oscillated, it lost energy, giving it up in the form of elementary particles. These particles interacted with one another and eventually settled down to some equilibrium temperature. From this time on, the standard big bang theory can describe the evolution of the universe.

The main difference between inflationary theory and the old cosmology becomes clear when one calculates the size of the universe at the end of inflation. Even if the universe at the beginning of inflation was as small as 10^{-33} centimeter, after 10^{-35} second of inflation this domain acquires an unbelievable size. According to some inflationary models, this size in centimeters can equal $10^{10^{12}}$—that is, a 1 followed by a trillion zeros. These numbers depend on the models used, but in most versions, this size is many orders of magnitude greater than the size of the observable universe, or 10^{28} centimeters.

This tremendous spurt immediately solves most of the problems of the old cosmological theory. Our universe appears smooth and uniform because all inhomogeneities were stretched $10^{10^{12}}$ times. The density of primordial monopoles and other undesirable "defects" becomes exponentially diluted. The universe has become so large that we can now see just a tiny fraction of it. That is why, just like a small area on a surface of a huge inflated balloon, our part looks flat. That is why we do not need to insist that all parts of the universe began expanding simultaneously. One domain of a smallest possible size of 10^{-33} centimeter is more than enough to produce everything we see now.

AN INFLATIONARY UNIVERSE

In 1982 I introduced the so-called new inflationary universe scenario, which Andreas Albrecht and Paul J. Steinhardt of the University of Pennsylvania also later discovered. This scenario shrugged off the main problems of Guth's model. But it was still rather complicated and not very realistic.

Only a year later did I realize that inflation is a naturally emerging feature in many theories of elementary particles, including the simplest model of the scalar field discussed earlier. There is no need for quantum gravity effects, phase transitions, supercooling or even the standard assumption that the universe originally was hot. One just considers all possible kinds and values of scalar fields in the early universe and then checks to see if any of them leads to inflation. Those places where inflation does not occur remain small. Those domains where inflation takes place become exponentially large and dominate the total volume of the universe. Because the scalar fields can take arbitrary values in the early universe, I called this scenario chaotic inflation.

In many ways, chaotic inflation is so simple that it is hard to understand why the idea was not discovered sooner. I think the reason was purely psychological. The glorious successes of the big bang theory hypnotized cosmologists. We assumed that the entire universe was created at the same moment, that initially it was hot and that the scalar field from the beginning resided close to the minimum of its potential energy. Once we began relaxing these assumptions, we immediately found that inflation is not an exotic phenomenon invoked by theorists for solving their problems. It is a general regime that occurs in a wide class of theories of elementary particles.

That a rapid stretching of the universe can simultaneously resolve many difficult cosmological problems may seem too good to be true. Indeed, if all inhomogeneities were stretched away, how did galaxies form? The answer is that while removing previously existing inhomogeneities, inflation at the same time made new ones.

These inhomogeneities arise from quantum effects. According to quantum mechanics, empty space is not entirely empty. The vacuum is filled with small quantum fluctuations. These fluctuations can be regarded as waves, or undulations in physical fields. The waves have all possible wavelengths and move in all directions. We cannot detect these waves, because they live only briefly and are microscopic.

In the inflationary universe the vacuum structure becomes even more complicated. Inflation rapidly stretches the waves. Once their wavelengths become sufficiently large, the undulations begin to "feel" the curvature of the universe. At this moment, they stop moving because of the viscosity of the scalar field (recall that the equations describing the field contain a friction term).

The first fluctuations to freeze are those that have large wavelengths. As the universe continues to expand, new fluctuations become stretched and freeze on top of other frozen waves. At this stage one cannot call these waves quantum fluctuations anymore. Most of them have extremely large wavelengths. Because these waves do not move and do not disappear, they enhance the value of the scalar field in some areas and depress it in others, thus creating inhomogeneities. These disturbances in the scalar field cause the density perturbations in the universe that are crucial for the subsequent formation of galaxies.

TESTING INFLATIONARY THEORY

In addition to explaining many features of our world, inflationary theory makes several important and testable predictions. First, density perturbations produced during inflation affect the distribution of matter in the universe. They may also accompany gravitational waves. Both density perturbations and gravitational waves make their imprint on the microwave background radiation. They render the temperature of this radiation slightly different in various places in the sky. This nonuniformity was found in 1992 by the Cosmic Background Explorer (COBE) satellite, a finding later confirmed by several other experiments.

Inflation also predicts that the universe should be nearly flat. Flatness of the universe can be experimentally verified because the density of a flat universe is related in a sim-

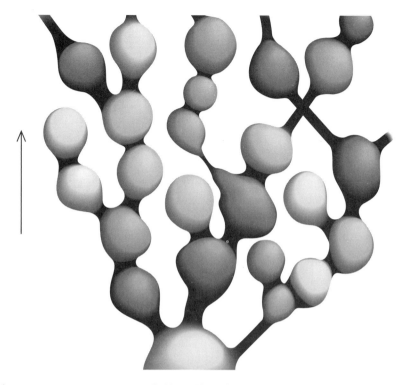

Self-reproducing cosmos appears as an extended branching of inflationary bubbles. Changes in tone represent "mutations" in the laws of physics from parent universes. The properties of space in each bubble do not depend on the time when the bubble formed. In this sense, the universe as a whole may be stationary, even though the interior of each bubble can be described by the big bang theory.

ple way to the speed of its expansion. So far observational data are consistent with this prediction. A few years ago it seemed that if someone were to show that the universe is open rather than flat, then inflationary theory would fall apart. Recently, however, several models of an open inflationary universe have been found. The only consistent description of a large homogeneous open universe that we currently know is based on inflationary theory. Thus, even if the universe is open, inflation is still the best theory to describe it. One may argue that the only way to disprove the theory of inflation is to propose a better theory.

One should remember that inflationary models are based on the theory of elementary particles, and this theory is not completely established. Some versions (most notably, superstring theory) do not automatically lead to inflation. Pulling inflation out of the superstring model may require radically new ideas.

Here we come to the most interesting part of our story, to the theory of an eternally existing, self-reproducing inflationary universe. This theory is rather general, but it looks especially promising and leads to the most dramatic consequences in the context of the chaotic inflation scenario.

As I already mentioned, one can visualize quantum fluctuations of the scalar field in an inflationary universe as waves. They first moved in all possible directions and then froze on top of one another. Each frozen wave slightly increased the scalar field in some parts of the universe and decreased it in others.

Now consider those places of the universe where these newly frozen waves persistently increased the scalar field. Such regions are extremely rare, but still they do exist. And they can be extremely important. Those rare domains of the universe where the field jumps high enough begin exponentially expanding with ever-increasing speed. The higher the scalar field jumps, the faster the universe expands. Very soon those rare domains will acquire a much greater volume than other domains.

From this theory it follows that if the universe contains at least one inflationary domain of a sufficiently large size, it begins unceasingly producing new inflationary domains. Inflation in each particular point may end quickly, but many other places will continue to expand. The total volume of all these domains will grow without end. In essence, one inflationary universe sprouts other inflationary bubbles, which in turn produce other inflationary bubbles.

This process, which I have called eternal inflation, keeps going as a chain reaction, producing a fractallike pattern of universes. In this scenario the universe as a whole is immortal. Each particular part of the universe may stem from a singularity somewhere in the past, and it may end up in a singularity somewhere in the future. There is, however, no end for the evolution of the entire universe.

The situation with the very beginning is less certain. There is a chance that all parts of the universe were created simultaneously in an initial big bang singularity. The necessity of this assumption, however, is no longer obvious.

Furthermore, the total number of inflationary bubbles on our "cosmic tree" grows exponentially in time. Therefore, most bubbles (including our own part of the universe) grow indefinitely far away from the trunk of this tree. Although this scenario makes the existence of the initial big bang almost irrelevant, for all practical purposes, one can consider the moment of formation of each inflationary bubble as a new "big bang."

A NEW COSMOLOGY

Could matters become even more curious? The answer is yes. Until now, we have considered the simplest inflationary model with only one scalar field, which has only one minimum of its potential energy. Meanwhile realistic models of elementary particles propound many kinds of scalar fields. For example, in the unified theories of weak, strong and electromagnetic interactions, at least two other scalar fields exist. The potential energy of these scalar fields may have several different minima. This condition means that the same theory may have different "vacuum states," corresponding to different types of symmetry breaking between fundamental interactions and, as a result, to different laws of low-energy physics. (Interactions of particles at extremely large energies do not depend on symmetry breaking.)

Such complexities in the scalar field mean that after inflation the universe may become divided into exponentially large domains that have different laws of low-energy physics. Note that this division occurs even if the entire universe originally began in the same state, corresponding to one particular minimum of potential energy. Indeed, large quantum fluctuations can cause scalar fields to jump out of their minima. That is, they jiggle some of the balls out of their bowls and into other ones. Each bowl corresponds to alternative laws of particle interactions. In some inflationary models, quantum fluctuations are so strong that even the number of dimensions of space and time can change.

If this model is correct, then physics alone cannot provide a complete explanation for all properties of our allotment of the universe. Does this mean that understanding all the properties of our region of the universe will require, besides a knowledge of physics, a deep investigation of our own nature, perhaps even including the nature of our consciousness? This conclusion would certainly be one of the most unexpected that one could draw from the recent developments in inflationary cosmology.

The evolution of inflationary theory has given rise to a completely new cosmological paradigm, which differs considerably from the old big bang theory and even from the first versions of the inflationary scenario. In it the universe appears to be both chaotic and homogeneous, expanding and stationary. Our cosmic home grows, fluctuates and eternally reproduces itself in all possible forms, as if adjusting itself for all possible types of life.

Some parts of the new theory, we hope, will stay with us for years to come. Many others will have to be considerably modified to fit with new observational data and with the ever-changing theory of elementary particles.

About the Authors

CLAUDE J. ALLÈGRE and STEPHEN H. SCHNEIDER ("The Evolution of the Earth") study various aspects of the Earth's geologic history and its climate. Allègre is a professor at the University of Paris and directs the Department of Geochemistry at the Institut de Physique du Globe de Paris. He is a foreign member of the National Academy of Sciences. Schneider is a professor in the Department of Biological Sciences at Stanford University and is a senior fellow at the university's Institute of International Studies. He is also a senior scientist at the National Center for Atmospheric Research in Boulder, Colo.

MAX P. BERNSTEIN, SCOTT A. SANDFORD and LOUIS J. ALLAMANDOLA ("Life's Far-Flung Raw Materials") work in the Astrochemistry Laboratory at the National Aeronautics and Space Administration Ames Research Center. Bernstein is a contractor to NASA Ames and a member of the Search for Extraterrestrial Intelligence Institute in Mountain View, Calif. He simulates the organic chemistry of comets and interstellar ice grains and ponders their connection to the origins of life. Sandford and Allamandola are both civil servants at NASA Ames. Sandford performed seminal work on interplanetary dust particles, is an associate editor of the journal Meteoritics and Planetary Science and is a co-investigator in NASA's Stardust mission. Allamandola, the founder and director of the Ames Astrochemistry Laboratory, has 20 years of experience in pioneering studies of interstellar and solar system ices and is an originator of the polycyclic aromatic hydrocarbon hypothesis. Bernstein, Sandford and Allamandola all live in Moffett Field, California.

RICHARD P. BINZEL, M. ANTONIETTA BARUCCI and MARCELLO FULCHIGNONI ("The Origins of the Asteroids") have asteroids bearing their names: 2873, 3485 and 3486, respectively. Binzel, an associate professor of earth, atmospheric and planetary science at the Massachusetts Institute of Technology, received a National Science Foundation Young Investigator Award in 1990. He described Pluto in the June 1990 issue of Scientific American. Barucci, an astronomer at the Parts Observatory, received her doctorate from the University of Rome. She is involved with the asteroid encounters of the Cassini mission to Saturn. Fulchignoni began his research in space science during the Apollo program, analyzing lunar samples. A professesor of physics of the solar system at the University of Rome, he is currently editor in chief of the international jounral Annales Geophysicae. Binzel lives in Cambridge, and Fulchignoni lives in Rome, Italy. Barucci lives in Paris, France.

ALAN P. BOSS ("Collapse and Formation of Stars") began modeling the formation of stellar and planetary systems as a physics graduate at the University of California, Santa Barbara, where he received a doctorate in 1979. After two years at the National Aeronautics and Space Administration Ames Research Center, he joined the Department of Terrestrial Magnetism at the Carnegie Institution of Washington (where, despite its name, terrestrial management has not been studied for decades). Boss chairs a committee that advises NASA about searches for planets outside the solar system. He lives in the Washington, D. C., area.

MARK A. BULLOCK and DAVID H. GRINSPOON ("Global Climate Change on Venus") are planetary scientists at the University of Colorado at Boulder. Bullock began his career studying the destruction of organic compounds on Mars and now analyzes the destruction of clement conditions on Venus. At night he takes his young sons, Sean and Brian, outside and shows them the points of light he studies. Grinspoon, in addition to studying the evolution of planetary atmospheres and of life, is a member of the Solar System Exploration Subcommittee, which advises NASA on space policy. He has played electric guitar and percussion in a variety of world-beat and trip-hop bands and lived in Zimbabwe for two months learning chimurenga music. Both Bullock and Grinspoon live in Boulder, Colorado.

JOHN K. CANNIZZO and RONALD H. KAITCHUCK ("Accretion Disks in Interacting Binary Stars") have been drawn together by a shared interest in accretion disks and in interacting binary stars. Cannizzo received his Ph.D. in theoretical astrophysics in 1984 from the University of Texas at Austin. He is a Humboldt Fellow at the Max Planck Institute for Astrophysics in Garching, Germany. He has attempted to model accretion disks in cataclysmic variables and around supermassive black holes; he is also an avid tennis player. Kaitchuck earned a Ph.D. in astronomy in 1981 from Indiana University. He is a faculty member in the Department of Physics and Astronomy at Ball State University. Kaitchuck specializes in time-resolved spectroscopy of binary stars. Cannizzo lives in Hamilton, Ontario, and Kaitchuck lives in Muncie, Indiana.

FRANCIS H. C. CRICK ("The Structure of Hereditary Material") is a British biologist who was originally trained as a physicist. After designing mines for the British Admiralty during World War II, he decided to go into molecular biology. He obtained a Medical Research Council studentship, entered the Strangeways Laboratory in Cambridge, worked on the viscosity of the cytoplasm of chick fibroblasts and "read everything I could lay my hands on." He then joined a molecular biology unit sponsored by the Medical Research Council, where he was able to concentrate on molecular structure. In 1962 Crick and his colleagues James D. Watson and Maurice Wilkins were awarded the Nobel Prize for Physiology or Medicine for the work described here. Since 1977 Crick has been Kleckhefer Distinguished Professor at the Salk Institute for Biological Studies in San Diego. Crick lives in San Diego, California.

ALBERT EINSTEIN (1879–1955) ("On the Generalized Theory of Gravitation") formulated the general theory of relativity and the special theory of relativity and profoundly influenced modern physics in many other ways. He was awarded the Nobel Prize for Physics in 1921.

WENDY L. FREEDMAN ("The Expansion Rate and Size of the Universe") is a staff member at the Carnegie Institution's observatories in Pasadena, Calif. Born in Toronto, she received a Ph.D. in astronomy and astrophysics from the University of Toronto in 1984 and in 1987 became the first woman to join Carnegie's scientific staff. In 1994 she received the Marc Aaronson Prize for her contributions to the study of extragalactic distance and stellar populations of galaxies. A coleader of the Hubble Space Telescope Key Project to measure the Hubble constant, she is also a member of the National Research Council's Committee on Astronomy and Astrophysics, the executive board of the Center for Particle Astrophysics and the National Aeronautics and Space Administration's scientific oversight committee planning the Next Generation Space Telescope. Beyond astronomy, her main interest is her family: husband Barry, daughter Rachael and son Daniel. Her essay updates a version that appeared in Scientific American in November 1992. She lives in Pasadena.

EVERETT K. GIBSON, JR. ,DAVID S. McKAY, KATHIE THOMAS-KEPTRA, and CHRISTOPHER S. ROMANEK ("The Case for Relic Life on Mars") were members of the team that first reported evidence of past biological activity within the ALH84001 meteorite. Gibson, McKay and Thomas-Keprta work at the National Aeronautics and Space Administration Johnson Space Center in Houston, Texas; Romanek, a former National Research Council postdoctoral fellow at the Johnson center, is with the Department of Geology and the Savannah River Ecology Laboratory at the University of Georgia. Gibson, a giochemist and meteorite specialist, and McKay, a geologist and expert on planetary regoliths, are senior scientists in the Johnson Center's Earth Sciences and Solar System Exploration Division. Thomas-Keptra, a senior scientist at Lockheed Martin, is a biologist who applies electron microscopy to the study of meteorites, interplanetary dust particles and lunar samples. Romanek's specialty is low-temperature geochemistry and stable-isotope mass spectrometry. Gibson, McKay, and Thomas-Keprta live in Houseton, Texas, and Romanek lives in Aiken, South Carolina.

STEPHEN JAY GOULD ("The Evolution of Life on Earth") teaches biology, geology and the history of science at Harvard University, where he has been on the faculty since 1967. He received an A.B. from Antioch College and a Ph.D. in paleontology from Columbia University. Well known for his popular scientific writings, in particular his monthly column in Natural History magazine, he is the author of over 20 books. He lives in Cambridge, Massachusetts.

ALAN H. GUTH and PAUL J. STEINHARDT ("The Inflationary Universe") are physicists who share an interest in the early history of the universe and particularly the first 10-45 second of that history. Guth attended the Massachusetts Institute of Technology as an undergraduate and a graduate student; his Ph.D. in physics was awarded by M.I.T. in 1972. He writes: "I held postdoctoral positions at Princeton University, Columbia University, Cornell University and the Stanford Linear Accelerator Center (SLAC). During most of that period of years. I worked on rather abstract mathematical problems in elementary particle theory and knew no more about developments in cosmology than the average layman does. While I was at Cornell, however, Henry Tye, a fellow postdoctoral worker, persuaded me (with great difficulty) to join him in studying the production of magnetic monopoles in the early universe, and that was how my career changed direction. I continued the work in the following academic year at SLAC. Shortly thereafter I returned to M.I.T." Guth is the Victor F. Weisskopf Professor of Physics at MIT. He has been elected to the National Academy of Arts and Sciences, has been awarded the Lilienfeld Prize of the American Physical Society, the Eddington Medal of the Royal Astronomical Society, and the Robinson Prize in Cosmology of the University of Newcastle upon Tyne. His book *The Inflationary Universe: The Quest for a New Theory of Cosmic Origins* was published in 1997 by Addison-Wesley and was recently released in paperback by Perseus Books. Steinhardt was graduated from the California Institute of Technology with a B.S. in 1974. His M.A. (1975) and Ph.D. (1978) in physics are from Harvard University. From 1979 to 1981 he was a junior fellow in the Society of Fellows at Harvard. In 1981 he moved to the University of Pennsylvania, where he is professor of physics. Guth lives in Brookline, Massachusetts, and Steinhardt lives in Philadelphia, Pennsylvania.

J. PATRICK HENRY, ULRICH G. BRIEL and HANS BÖHRINGER ("The Evolution of Galaxy Clusters") are x-ray astronomers who study clusters of galaxies. The first two met in the late 1970s while working at the Smithsonian Astrophysical Observatory on one of the instruments on the Einstein X-ray Observatory satellite. Henry is now an astronomy professor at the University of Hawaii. He says he enjoys sitting on his lanai and thinking about large-scale structure while watching the sailboats off Diamond Head. Briel and Böhringer are staff members of the Max Planck Institute for Extraterrestrial Physics in Garching, Germany. Briel is an observer who tested and calibrated the ROSAT instrument that made the temperature maps discussed in this essay. Böhringer is a theorist who studies galaxy clusters, cosmology and the interstellar medium. Henry lives in Honolulu, Hawaii, and Briel and Böhringer live in Garching.

JOHN HORGAN was a senior writer at *Scientific American*. He is the author of *The End of Science: Facing the Limits of Knowledge in the Twilight of theScientific Age* and *The Undiscovered Mind: How the Human Brain Defies Replication, Medication and Explanation.*

PETER JEDICKE teaches computer programming at Fanshawe College of Applied Arts and Technology in London, Ontario. He has been an amateur astronomer for over twenty years. In 1987, he won the Simon Newcomb Award of the Royal Astronomical Society of Canada for his essay, "Neutrinos, the Sun and Canada." His writing has appeared in *Spaceflight*, *Astronomy* and *Sky & Telescope* magazines, as well as in newspapers and numerous club newsletters. Peter has presented the weather on a local television news program and hosted two community cable television programs featuring local scientists and astronomy. Peter has also contributed frequently to radio programs in London and across Canada, and given lectures about astronomy to clubs and star parties in many parts of Canada and the United States. Peter was born in Wiesbaden, Germany, and grew up in Niagra Falls, Ontario. He graduated from the University of Western Ontario with and M. A. in philosophy in 1997 and a B. S. in physics in 1976. He and his wife, Diane, live in London.

JEFFREY S. KARGEL and ROBERT G. STROM ("Global Climate Changes on Mars") have worked together on various projects in planetary science for more than a decade. Kargel met Strom soon after beginning graduate studies at the University of Arizona, where he received a doctorate in planetary sciences in 1990. Kargel remained at the University of Arizona's Lunar and Planetary Science Laboratory for two years doing postdoctoral research on the icy moons of the outer solar system and then joined the U.S. Geological Survey's astrogeological group in Flagstaff, Ariz. Strom began his career working as a petroleum geologist, but he became involved in lunar exploration efforts during the 1960s and joined the faculty of the University of Arizona, where he continues to teach and conduct research. He has participated on National Aeronautics and Space Administration science teams assembled for the Apollo program, for the Mariner missions to Venus and Mercury and for the Voyager missions to the outer solar system. Kargel lives in Flagstaff, and Strom lives in Tuscon, Arizona.

ROBERT P. KIRSHNER ("The Earth's Elements") pursues an eventful career in astronomy in addition to one in bicycle racing. After receiving a Ph.D. in 1975 from the California Institute of Technology, he went to Kitt Peak National Observatory as a postdoctoral fellow. In 1976 Kirshner became an assistant professor at the University of Michigan and in 1985 moved to the Harvard-Smithsonian Center for Astrophysics. He now chairs the Astronomy Department at Harvard University. Kirshner's work concentrates on supernovae and extragalactic astronomy. In 1992 he was elected a Fellow of the American Academy of Arts and Sciences. He lives in Cambridge, Massachusetts.

ANDREW H. KNOLL ("End of the Proterozoic Eon") is professor of organismic and evolutionary biology and earth and planetary sciences at Harvard University. His interest in the evolution of life, which was sparked during his undergraduate days at Lehigh University, was fanned to a flame under the tutelage of the late Elso Barghoorn, who was Knoll's adviser and predecessor at Harvard. An avid field geologist, he has traversed the globe, spending nights in such remote areas as the high Arctic, the Australian Outback and the Namib desert of Namibia. Knoll is recipient of the Charles Schuchert Award of the Paleontological Society of America and the Charles Doolittle Walcott Medal of the National Academy of Sciences. In 1991 he was elected to the National Academy of Sciences. He lives in Cambridge, Massachusetts.

RENÉE C. KRAAN-KORTEWEG and OFER LAHAV ("Galaxies Behind the Milky Way") joined forces in 1990, after they met in Durham, England, at a conference on cosmology; independently, they both had discovered a previously unknown cluster behind the Milky Way in the constellation Puppis. Kraan-Korteweg is a professor in the Department of Astronomy of the University of Guanajuato in Mexico. Lahav is a faculty member of the Institute of Astronomy at the University of Cambridge and a Fellow of St. Catharine's College. Kraan-Korteweg explores the zone of avoidance by direct observation, whereas Lahav utilizes theoretical and computational techniques. Kraan-Korteweg lives in Meudon, France, and Lahav lives in Cambridge, England.

DAVID H. LEVY, EUGENE M. SHOEMAKER (1928–1997) and CAROLYN S. SHOE-MAKER ("Comet Shoemaker-Levy 9 Meets Jupiter") have been searching the skies together since 1989. Levy writes and lectures and is also an avid amateur astronomer. He has written 15 books—somewhat fewer than the number of comets he has discovered. Eugene and Carolyn Shoemaker worked as a husband-and-wife team since 1982 until Eugene's death. Eugene served as a geologist for the U.S. Geological Survey (USGS) from 1948 to 1993, where he organized the Branch of Astrogeology. He became scientist emeritus with the USGS and held a staff position at Lowell

Observatory, where Carolyn is now a staff member. She also works as a visiting scientist at the USGS and is a research professor of astronomy at Northern Arizona University. Levy lives in Tucson, Arizona, Carolyn Shoemaker lives in Flagstaff, Arizona.

ANDREI LINDE ("The Self-Reproducing Inflamatory Universe") is one of the originators of inflationary theory. After graduating from Moscow University, he received his Ph.D. at the P. N. Lebedev Physics Institute in Moscow, where he began probing the connections between particle physics and cosmology. He became a professor of physics at Stanford University in 1990. He lives in Stanford, Calif. with his wife, Renata Kallosh (also a professor of physics at Stanford), and his sons, Dmitri and Alex. A detailed description of inflationary theory is given in his book *Particle Physics and Inflationary Cosmology* (1990).

TONY M. LISS and PAUL L. TIPTON ("The Discovery of the Top Quark") helped to build key elements of the Collider Detector at Fermilab (CDF) and have both served as conveners of the search group for the top quark. For his Ph.D. at the University of California, Berkeley, Liss participated in a search for monopoles. In 1988 he joined the faculty at the University of Illinois at Urbana-Champaign and in 1990 was awarded an Alfred P. Sloan Fellowship. Tipton received his Ph.D. from the University of Rochester in 1987 studying bottom quarks and is now on the faculty there. He is a recipient of the U.S. Department of Energy's Outstanding Junior Investigator Award and the National Science Foundation's Young Investigator Award. The authors would like to thank Lynne Orr and Scott Willenbrock for helpful discussions as well as all their colleagues at CDF and D0. Liss lives in Urbana and Tipton lives in Rochester, New York.

JANE X. LUU and DAVID C. JEWITT ("The Kuiper Belt") came to study astronomy in different ways. For Jewitt, astronomy was a passion he developed as a youngster in England. Luu's childhood years were filled with more practical concerns: as a refugee from Vietnam, she had to learn to speak English and adjust to life in southern California. She became enamored of astronomy almost by accident, during a summer spent at the Jet Propulsion Laboratory in Pasadena. Luu and Jewitt began their collaborative work in 1986 at the Massachusetts Institute of Technology. Jewitt moved to the University of Hawaii in 1988. It was during Luu's postdoctoral fellowship at the Harvard-Smithsonian Center for Astrophysics that Luu and Jewitt discovered the first Kuiper Belt object. In 1994 Luu joined the faculty of Harvard University.

MICHAEL V. MAGEE ("What's New in the Milky Way") received his B. S. in Speech and Hearing Sciences from the University of Arizona in 1983. He has worked for the Flandrau Science Center & Planetarium in Tucson, Arizona, since 1981 having produced or coproduced over 40 Planetarium theater shows involving multimedia presentations about astronomy and related sciences. He has participated in a research project on photometry of asteroids with the Planetary Science Institute in Tucson. He is co-author of several articles on Photometric Geodesy of Main-Belt Asteroids published in Icarus, the journal of planetary science. He currently serves as Planetarium Operations Manager for Flandrau. He lives in Tucson.

RENU MALHOTRA ("Migrating Planets") did her undergraduate studies at the Indian Institute of Technology in Delhi and received a Ph.D. in physics from Cornell University in 1988. After completing postdoctoral research at the California Institute of Technology, she moved to her current position as a staff scientist at the Lunar and Planetary Institute in Houston. In her research, she has followed her passionate interest in the dynamics and evolution of the solar system and other planetary systems.

GEOFFREY W. MARCY and R. PAUL BUTLER ("Giant Planets Orbiting Faraway Stars") together have found six of the eight planets around sunlike stars reported to date. Marcy is a Distinguished University Professor at San Francisco State University and an adjunct professor at the

University of California, Berkeley. Butler is a staff astronomer at the Anglo-Australian Observatory. Marcy lives in Berkeley, and Butler lives in Epping, Australia.

ROBERT M. NELSON ("Mercury the Forgotten Planet") has been a research scientist at the Jet Propulsion Laboratory in Pasadena, Calif., since 1979. He received his Ph.D. in planetary astronomy from the University of Pittsburgh in 1977. Nelson was coinvestigator for the Voyager spacecraft's photopolarimeter and is on the science team for the Visual and Infrared Mapping Spectrometer of the Cassini Saturn Orbiter mission. He was also the principal investigator on the Hermes '94 and '96 proposals for a Mercury orbiter and is the flight scientist for the experimental New Millennium Deep Space One mission. The author expresses his gratitude to the Hermes team members for their enlightening contributions. He lives in Pasadena, California.

DONALD E. OSTERBROCK, JOEL A. GWINN and RONALD S. BRASHEAR ("Edwin Hubble and the Expanding Universe") share a keen interest in Hubble's place in the history of astronomy. Osterbrock is a former director of Lick Observatory at Mount Hamilton, Calif., where he now researches and teaches astronomy. He has written many books and is working on a biography of George Willis Ritchey, a noted American telescope maker and designer. Gwinn has spent more than 30 years at the University of Louisville, where he is a professor of physics; he has concentrated on optics, atomic spectroscopy and the history of astronomy, including Hubble's early life. Brashear earned his master's degree in 1984 from Louisville, where he met Gwinn. Brashear is curator of the history of science at the Henry E. Huntington Library in San Marino, Calif. Osterbrock lives in Princeton, New Jersey., Gwinn lives in Louisville, Kentucky and Brashear lives in San Marino, California.

P. JAMES E. PEEBLES, DAVID N. SCHRAMM, EDWIN L. TURNER and RICHARD G. KRON ("The Evolution of the Universe") have individually earned top honors for their work on the evolution of the universe. Peebles is the Einstein Professor of Science at Princeton University, where in 1958 he began an illustrious career in gravitational physics. Turner is chair of astrophysical sciences at Princeton and director of the 3.5-meter ARC telescope in New Mexico. Since 1978 Kron has served on the faculty of the Department of Astronomy and Astrophysics at the University of Chicago, and he is also a member of the experimental astrophysics group at Fermi National Accelerator Laboratory. He enjoys observing distant galaxies almost as much as directing Yerkes Observatory near Lake Geneva, Wisconsin. Schramm, who was Louis Block Distinguished Service Professor in the Physical Sciences and vice president for research at the University of Chicago, died in a tragic airplane accident while this book was being prepared for publication. Peebles and Turner live in Princeton, New Jersey and Kron lives in Batavia, Illinois.

J. ROGER, P. ANGEL and NEVILLE J. WOOLF ("Searching for Life on Other Planets") have collaborated for the past 15 years on methods for making better telescopes. They are based at Steward Observatory at the University of Arizona. Angel is a fellow of the Royal Society and directs the Steward Observatory Mirror Laboratory. Woolf has pioneered techniques to minimize the distortion of images caused by the atmosphere. Angel and Woolf consider the quest for distant planets to be the ultimate test for telescope builders; they are meeting this challenge by pushing the limits of outer-space observation technology, such as adaptive optics and space telescopes. They both live in Tucson, Arizona.

VERA RUBIN ("Dark Matter in the Universe") is a staff member at the Department of Terrestrial Magnetism of the Carnegie Institution of Washington, where she has been since 1965. That same year, she became the first woman permitted to observe at Palomar Observatory. The author of more than 200 papers on the structure of the Milky Way, motions within galaxies and large-scale motions in the universe, she received Carnegie Mellon University's Dickson Prize for Science in 1994 and the Royal Astronomical Society's Gold Medal in 1996. President Bill Clinton awarded her the National Medal of Science in 1993 and appointed her to the President's Committee on the National Medal of Science in 1995.

CARL SAGAN (1934–1996) and FRANK DRAKE ("The Search for Extraterrestrial Intelligence") are astronomers. Sagan was professor of astronomy at Cornell University and director of the Laboratory for Planetary Studies there. Drake is professor of astronomy and astrophysics at the University of California, Santa Cruz, where he also serves as acting associate chancellor for university advancement. Drake lives in Ithaca, New York.

ALEXANDER SCHEELINE has spent over two decades doing research in atomic spectroscopy in laboratory plasmas, beginning as a chemistry graduate student at The University of Wisconsin-Madison, where he received his doctorate in 1978. After a year of post-doctoral research at the National Institute for Standards and Technology, he was Assistant Professor of Chemistry at the University of Illinois at Urbana-Champaign. He has served as a program officer at the National Science Foundation. His research has recently turned to studies of nonlinear chemical and biochemical systems, but he continues to do flame and plasma spectroscopy, particularly with respect to environmental remediation. Advances in astronomical instrumentation have been key to improvements he and his collaborators have made to chemical measurements. He resides in Champaign, IL.

ERWIN SCHRÖDINGER (1887–1961) ("What Is Matter") was one of the founders of modern physics. For developing the theory of wave mechanics he shared the Nobel Prize in 1933 with the British physicist P. A. M. Dirac. Schrödinger, born in Vienna, came from the distinguished Austrian school of physics that produced Ernst Mach and Ludwig Boltzmann. He succeeded Max Planck in the chair of theoretical physics at the University of Berlin in 1927. Upon Hitler's rise to power he went to Dublin to join the Institute for Advanced Study, where he remained until 1956. In his later work he sought to combine the field theories of physics into a unified structure. He was also interested in more general unifications of science, and perhaps his most famous book was *What Is Life?*

JAMES VERNON SCOTTI ("On Comets") is a Senior Research Specialist at the Lunar and Planetary Laboratory in Tuscon, Arizona, where he has been conducting research since 1982. He received his B. S. in Astronomy from the University of Arizona in 1983, and is currently at work on his Ph. D. at Queens University in Belfast. He is a member of the Division for Planetary Sciences, the International Association for Astronomical Arts and the Association of Lunar and Planetary Observers. He is also the Assistant Recorder for the Comets Section and an assoiate member of the American Astronomical Society. He has researched the size distribution of comets and of the origin and dynamical evolution of comets and their likely source populations and was the Principal Investigator on a successful proposal to use the Infrared Space Observatory (ISO) to observe the thermal cross-section of a selection of short-period comets.

LEONARD SUSSKIND ("Black Holes and the Informational Paradox")is one of the early inventors of string theory. He holds a Ph.D. from Cornell University and has been a professor at Stanford University since 1978. He has made many contributions to elementary particle physics, quantum field theory, cosmology and, most recently, to the theory of black holes. His current studies in gravitation have led him to suggest that information can be compressed into one lower dimension, a concept he calls the holographic universe. He lives in Stanford, California.

SIDNEY VAN DEN BERGH ("How the Milky Way was Formed") is an astronomer and an adjunct professor at the University of Victoria in British Columbia, Canada. He was born in the Netherlands where he attended Leiden University from 1947 to 1948 before emigrating to the United States in 1948. He received his A. B. at Princeton University in 1950; his Masters of Science from Ohio State University in 1952. He was an Assistant Professor at Perkins Observatory at Ohio State

University at Columbus from 1956–1958; a research associate at Mount Wilson Observatory, Palomar Observatory in Pasadena, California from 1968–1969; a professor of astronomy at the David Dunlap Observatory at the University of Toronto in Ontario, Canada, from 1958–1977; the director of the Dominion Astrophysics Observatory in Victoria, British Columbia, from 1977–1986; the principal research officer for the National Research Council of Canada from 1977–1998; and he has been an adjunct professor at the University of Victoria since 1977. He resides in British Columbia.

SYLVAIN VEILLEUX, GERALD CECIL and JONATHAN BLAND-HAWTHORN ("The Lives of Quasars" and "Colossal Galactic Explosions") met while working at observatories in Hawaii and were drawn to collaborate by a shared interest in peculiar galaxies. Veilleux, now an assistant professor of astronomy at the University of Maryland, received his Ph.D. from the University of California, Santa Cruz. Cecil, an associate professor of astronomy and physics at the University of North Carolina at Chapel Hill and project scientist of the SOAR four-meter telescope in Chile, received his doctorate from the University of Hawaii. Bland-Hawthorn received his Ph.D. in astronomy and astrophysics from the University of Sussex and the Royal Greenwich Observatory. He is now a research astronomer at the Anglo-Australian Observatory in Sydney, Australia.

Further Reading and Website Listing

CHAPTER TWO: THE BIRTH OF THE UNIVERSE

Berendzen, Richard, Richard Hart and Daniel Seeley. 1984. *Man Discovers the Galaxies.* Columbia University Press.

Burns, Jack O. Stormy "Weather in Galaxy Clusters." *Science* 280 (April 1998): 400-404.

Fabian, A. C., ed. 1992. *Clusters and Superclusters of Galaxies.* Kluwer Academic Publishers.

Lemonick, Michael D. 1993. *The Light at the Edge of the Universe: Dark Matter and the Structure of the Universe.* Villard Books.

Osterbrock Donald E. 1990. "The Observational Approach to Cosmology: U.S. Observatories Pre-World War II." In *Modern Cosmology in Retrospect,* eds. R. Bertotti, R. Balbinot, S. Bergia and A. Messina. Cambridge University Press.

Osterbrock, Donald E., Ronald S. Brashear and Joel A. Gwinn. 1990. "Self-Made Cosmologist: The Education of Edwin Hubble." In *Evolution of the Universe of Galaxies: Edwin Hubble Centennial Symposium,* ed. Richard G. Kron. Astronomical Society of the Pacific.

Overbye, Dennis. 1991. *Lonely Hearts of the Cosmos: The Scientific Quest for the Secret of the Universe.* HarperCollins.

Peebles, P. J. E. 1993. *Principles of Physical Cosmology.* Princeton University Press.

Riordan, Michael, and David N. Schramm. 1991. *The Shadows of Creation: Dark matter and the Structure of the universe.* W. H. Freeman and Company.

Sarazin, Craig L. 1988. *X-ray Emission from Clusters of Galaxies.* Cambridge University Press.

Smith, Robert. 1982. *The Expanding Universe: Astronomy 's "Great Debate," 1900-1931.* Cambridge University Press.

Winn, Joshua N. "An X-rated View of the Sky." *Mercury* 27, No. 1 (January/February 1998): 12-16.

CHAPTER THREE: GALAXY FORMATION

Balkowski, Chantal, and R. C. Kraan-Korteweg, eds. "Unveiling Large-Scale Structures behind the Milky Way." In *Astronomical Society of the Pacific Conference Series* 67 (January 1994).

Bland, Jonathan, and R. Brent Tully. "Large-Scale Bipolar Wind in M82." *Nature* 334 (July 7, 1988): 43-45.

Cecil, G., Jonathan Bland and R. Brent Tully. "Imaging Spectrophotometry of Ionized Gas in NGC 1068, Part 1: Kinematics of the Narrow-Line Region." *Astrophysical Journal* 355 (May 20, 1990): 70-87.

Dekel, Avishai. "Dynamics of Cosmic Flows." *Annual Review of Astronomy and Astrophysics* 32 (1994): 3 371-418.

Filippenko, Alexei V., ed. 1992. "Relationships Between Active Galactic Nuclei and Starburst Galaxies." In *Proceedings of the Taipei Astrophysics Workshop*, 1991. Astronomical Society of the Pacific.

Ibata, R. A., G. Gilmore and M. J. Irwin. "A Dwarf Satellite Galaxy in Sagittarius. *Nature* 370 (July 21, 1994): 194-196.

Kraan-Koretweg, R. C., et al. "A Nearby Massive Cluster behind the Milky Way." *Nature* 379 (February 8, 1996): 519-521.

Peebles, P. J. E. 1993. *Principles of Physical Cosmology*. Princeton University Press.

Tully, R. Brent, J. Morse and P. Shopbell. "Dissecting Cosmic Expulsions." *Sky & Telescope* 90, No. 1 (July 1995): 18-21 .

Veilleux, S., G. Cecil, J. Bland-Hawthorn, R. Brent Tully, Alexei V. Filippenko and W. L. W. Sargent. "The Nuclear Superbubble of NGC 3079." *Astrophysical Journal*, part 1, 433 (September 20, 1994): 48-64.

CHAPTER FOUR: THE MILKY WAY

Barbuy, B., and A. Renzini, eds. 1992. *The Stellar Populations of Galaxies*. Kluwer Academic Publishers.

Cannizzo, John K., Allen W. Shaftcer and J. Craig Wheeler. "On the Outburst Recurrence Time for the Accretion Disk Limit Cycle Mechanism in Dwarf Novae." *Astrophysical Journal* 333, No. 1 (October 1, 1988): 227-235.

Frank, J., A. R. King and D. J. Raine. 1985. *Accretion Power in Astrophysics*. Cambridge University Press.

Gilmore, Gerard, Ivan R. King and Pieter C. van der Kruit. *The Milky Way as a Galaxy*. University Science Books.

Hawking, Stephen W. 1996. *The Illustrated "A Brief History of Time."* Bantam Books.

Janes, Kenneth, ed. 1991. *The Formation and Evolution of Star Clusters*. Astronomical Society of the Pacific.

Mansperger, Cathy S., and Ronald H. Kaitchuck. "Spectroscopy of the Dwarf Nova TW Virginis Caught on the Rise to Outburst." *Astrophysical Journal* 358 (July 20, 1990): 268-273.

Mihalas, Dimitri, and James Binney. 1981. *Galactic Astronomy: Structure and Kinematics*. W. H. Freeman and Company.

Mineshige, Shin, and Janet H. Wood. "Viscous Evolution of Accretion Disks in the Quiescence of Dwarf Novae." *Monthly Notices of the Royal Astronomical Society* 241, No. 2 (November 15, 1989): 259-280.

Mukrjee, Madhusree. "Trends in Theoretical Physics: Explaining Everything." *Scientific American* 274, No. 1 (January 1996): 88-94.

Thorne, Kip S. 1994. *Black Holes and Time Warps: Einstein 's Outrageous Legacy*. W. W. Norton.

CHAPTER FIVE: THE GENESIS OF THE SOLAR SYSTEM

Beatty, J. Kelly, and Andrew Chaiken, eds. 1990. *The New Solar System*. Cambridge University Press.

Binzel, R. P., T. Gehrels and M. Shapley Matthews, eds. 1989. *Asteroids II*. University of Arizona Press.

"Comet Shoemaker Levy 9." Special section in *Science* 267 (March 3, 1995): 1277-1323.

Cunningham, Clifford J. 1988. *Introduction to Asteroids*. Willmann-Bell, Inc.

Duncan, Martin, Thomas Quinn and Scott Tremaine. "The Origin of Short-Period Comets." *Astrophysical Journal* 328 (May 15, 1988): L69-L73.

"Impact!: Comet Shoemaker-Levy 9 Collides with Jupiter." Special issue of *Sky & Telescope* 88, No. 4 (October 1994).

Jewitt, D. C., and J. X. Luu. "The Solar System Beyond Neptune." *Astronomical Journal* 109, No. 4 (April 1995): 1867-1876.

Levy, David H. 1994. *The Quest for Comets: An Explosive Trail of Beauty and Danger*. Plenum Press.

ibid. 1995. *Impact Jupiter: The Crash of Shoemaker-Levy 9*. Plenum Press.

ibid. 1998. *Comets: Creators and Destroyers*. Touchstone.

Luu, J. X. 1993. "The Kuiper Belt Objects." In *Asteroids, Comets, Meteors, 1993*, eds. A. Milani, M. Di Martino and A. Cellino. Kluwer Academic Publishers.

Malhotra, Renu. "The Origin of Pluto's Orbit: Implications for the Solar System Beyond Neptune." *Astronomical Journal* 110 (July 1995): 420-429.

Malhotra, Renu, et al. (in press). "Dynamics of the Kuiper Belt." *In Protostars and Planets IV*, eds. V. Mannings et al. University of Arizona Press.

Marcy, Geoffrey W., and R. Paul Butler. "Detection of Extrasolar Giant Planets." *Annual Review of Astronomy and Astrophysics* 36 (1998): 57-98.

Peterson, Ivars. 1993. *Newton's Clock: Chaos in the Solar System*. W. H. Freeman and Company.

Whipple, F. L. 1984. *The Mystery of Comets*. Smithsonian Institution Press.

Yeomans, D. K. 1991. *Comets*. John Wiley & Sons.

CHAPTER SIX: THE PLANETARY TOUR

Allegre, Claude J. 1992. *From Stone to Star: A View of Modern Geology*. Harvard University Press.

Baker, V. R., R. G. Strom, V. C. Gulick, J. S. Kargel, G. Komastsu and V. S. Kale. "Ancient Oceans, Ice Sheets and the Hydrological Cycle on Mars." *Nature* 352 (August 15,1991): 589-594.

Bartusiak, Marcia. 1993. *Through a Universe Darkly: A Cosmic Tale of Ancient Ethers, Dark Matter, and the Fate of the Universe*. HarperCollins.

Beatty, J. Kelly, Carolyn Collins Petersen, and Andrew Chaikin, eds. 1998. *The New Solar System*, 4th ed. Cambridge University Press.

Bougher, Stephen W., Donald M. Hunten and Roger J. Phillips, eds. 1997. *Venus II: Geology, Geophysics, Atmosphere and Solar Wind Environment*. University of Arizona Press.

Broecker, Wallace. 1990. *How to Build a Habitable Planet*. Lamont-Doherty Geological Observatory Press.

Bullock, Mark A., and David H. Grinspoon. "The Stability of Climate on Venus." *Journal of Geophysical Research* 101, No. E3 (March 1996): 7521-7530.

Davies, M. E., D. E. Gault, S. E. Dwornik and R. G. Strom, eds. 1978. *Atlas of Mercury*. NASA Scientific and Technical Information Office.

Ferris, Timothy. 1988. *Coming of Age in the Milky Way*. William Morrow and Company.

Grinspoon, David H. 1997. *Venus Revealed: A New Look below the Clouds of Our Mysterious Twin Planet*. Perseus Books.

Kargel, J. S., and R. G. Strom. "Ancient Glaciation on Mars." *Geology* 20, No. 1 (January 1992): 3-7. "The Ice Ages on Mars." *Astronomy* 20, No. 12 (December 1992):4-5.

Kasting, James F. "Earth's Early Atmosphere." *Science* 259 (February 12, 1993): 920-926.

Kirshner, Robert P. 1992. "Supernovae and Stellar Catastrophe." In *Understanding Catastrophe*, ed. J. Bourriau. Cambridge University Press.

Murdin, Paul. 1990. *End in Fire: The Supernova in the Large Magellanic Cloud*. Cambridge University Press.

Parker, T. J., D. S. Gorsline, R. S. Saunders, D. C. Pieri and D. M. Schneeberger. "Coastal Geomorphology of the Martian Northern Plains." *Journal of Geophysical Research* E (Planets) 98, No. 6 (June 25, 1993): 11061-11078.

Schneider, Stephen H., and Penelope J. Boston. 1991. *Scientists on GAIA*. MIT Press.

Vilas, F., C. R. Chapman and M. S. Matthews, eds. 1988. *Mercury*. University of Arizona Press.

CHAPTER SEVEN: LIFE ON EARTH . . . AND ELSEWHERE

Angel, J. Roger P. 1990. "Use of 16-Meter Telescope to Detect Earthlike Planets." In *The Next Generation Space Telescope*, eds. P. Bely and C. J. Burrows. Space Telescope Science Institute.

Angel, J. Roger P., A. Y. S. Cheng and N. J. Woolf. "A Space Telescope for Infrared Spec troscopy of Earthlike Planets." *Nature* 322 (July 24, 1986): 341-343.

Bakes, Emma L. O. 1997. *The Astrochemical Evolution of the Interstellar Medium*. Twin Press Astronomy Publishers.

Barbree, Jay, and Martin Caidin with Susan Wright. 1997. *Destination Mars: In Art, Myth and Science*. Penguin Studio.

Bernstein, Max P. "UV Irradiation of Polycyclic Aromatic Hydrocarbons in Ices: Production of Alcohols, Quinones, and Ethers." *Science* 283 (February 19, 1999): 1135-1138.

Carr, Michael H. 1996. *Water on Mars*. Oxford University Press.

Cronin, John. Pasteur, "Light and Life." *Physics World* 11, No. 10 (October 1998): 23-24.

Frederickson, James K., and Tullis C. Onstott. "Microbes Deep inside the Earth. *Scientific American* 275, No. 4 (October 1996): 68-73.

Gould, Stephen J. 1989. *Wonderful Life: The Burgess Shale and the Nature of History*. W. W. Norton.

Gould, Stephen J., ed. 1993. *The Book of Life*. W. W. Norton.

Kieffer, Hugh H., Bruce M. Jakosky, Conway W. Snyder and Mildred S. Matthews, eds. 1992. *Mars*. University of Arizona Press.

Knoll, A. H. 1989. "Paleomicrobiological Information in Proterozoic Rocks." In *Microbial Mars: Physiological Ecology of Benthic Microbial Communities*, eds. Yehuda Cohen and Eugene Rosenberg. American Society for Microbiology.

Knoll, A. H., J. M. Hayes, A. J. Kaufman, K. Swett and I. B. Lambert. "Secular Variation in Carbon Isotope Ratios from Upper Proterozoic Successions of Svalbard and East Greenland." *Nature* 321, No. 6073 (June 26, 1986): 832-838.

Life in the Universe. 1995. Special issue of *Scientific American* (October 1994). W. H. Freeman and Company.

Lipps, J. H., and P. W. Signor. (in press). *Origin and Early Evolution of the Metozoa*. Plenum Press.

McKay, David S., et al. "Search for Past Life on Mars: Possible Relic Biogenic Activity in Martian Meteorite ALH84001." *Science* 273 (August 16, 1996): 924-930.

MeSween, Harry Y., Jr. "What We Have Learned about Mars from SNC Meteorites." *Meteoritics* 29, No. 6 (November 1994): 757-779.

Nieman, H. W., and C. Wells Nieman. "What Shall We Say to Mars?" *Scientific American* 122, No. 12 (March 20, 1920): 298.

Oliver, B. M., et al. 1972. *Project Cyclops*. NASA Publication CR-114445.

Runnegar, Bruce. "The Cambrian Explosion: Animals or Fossils?" *Journal of the Geological Society of Australia* 29 (1982): 395-411.

Sagan, Carl, ed. 1973. *Communication with Extraterrestrial Intelligence*. MIT Press.

Stanley, Steven M. 1987. *Extinction: A Scientific American Book*. W. H. Freeman and Company.

Thomas, Paul J., Christopher F. Chyba and Christopher P. McKay, eds. 1997. *Comets and the Origin and Evolution of Life*. Springer.

Whittington, Henry B. 1985. *The Burgess Shale*. Yale University Press.

CHAPTER EIGHT: THE MICROVERSE

Abachi, S., et al. "Observation of the Top Quark." *Physical Review Letters* 74, No. 14 (April 3, 1995): 2632-2637.

Abe, F., et al. "Observation of Top Quark Production in Collisions with the Collider Detector at Fermilab." *Physical Review Letters* 74, No. 14 (April 3, 1995): 2626-2631.

Hartquist, T. W., and D. A. Williams. 1995. *The Chemically Controlled Cosmos: Astronomical Molecules from the Big Bang to Exploding Stars*. Cambridge University Press.

Herzberg, Gerhard. 1950. *Molecular Spectra and Molecular Structure I: The Spectra of Diatomic Molecules*, 2d ed. Van Nostrand.

Quigg, Chris. Topology. *Physics Today* 50, No. 5 (May 1997): 20-26.

Thorne, Anne P. 1988. *Spectrophysics*, 2d ed. Chapman and Hall.

Weinberg, Steven. 1988. *The First Three Minutes: A Modern View of the Origin of the Universe*, updated ed. Basic Books.

ibid. 1992. *Dreams of a Final Theory*. Pantheon Books.

CHAPTER NINE: SPECULATION ON ENDINGS AND BEGINNINGS

Albrecht, Andreas, and Paul J. Steinhardt. "Cosmology for Grand Unified Theories with Radiatively Induced Symmetry Breaking." *Physical Review Letters* 48, No. 17 (April 1982): 1220-1223.

Guth, Alan H. "Inflationary Universe: A Possible Solution to the Horizon and Flatness Problems." *Physical Review* 23, No. 2 (January 15, 1981): 347-356.

Linde, A. D. "A New Inflationary Universe Scenario: A Possible Solution of the Horizon, Homogeneity, Isotropy and Primordial Monopole Problems." *Physical Review Letters* 108B, No. 6 (February 4, 1982): 389-393.

Weinberg, Steven. 1988. "The First Three Minutes: A Modern View of the Origin of the Universe," updated ed. Basic Books.

USEFUL WEBSITES

Scientific American
http://www.sciam.com/

NASA
http://www.nasa.gov/

Space Telescope Science Institute (Hubble space telescope)
http://www.stsci.edu/top.html

NASA Chandra X-ray Observatory
http://chandra.nasa.gov/chandra.html

Jet Propulsion Laboratory Picture Archive
http://www.jpl.nasa.gov/pictures/

AsNational Air and Space Museum
http://www.nasm.edu/

Clementine Deep Space Program Science Experiment
http://www.nrl.navy.mil/clementine/c

European Space Agency
http://www.esa.it/

Astronomy Picture of the Day Archive
http://antwrp.gsfc.nasa.gov/apod/archivepix.html

NASA Office of Space Science
http://www.hq.nasa.gov/office/oss/osshome.htm

Goddard Space Flight Center
http://www.gsfc.nasa.gov/GSFC_homepage.html

Marshal Space Flight Center
http://www.msfc.nasa.gov/news

National Optical Astronomy Observatories
http://www.noao.edu/

Association of Universities for Research in Astronomy, Inc.
http://www.aura-astronomy.org/

PHOTO/ILLUSTRATION CREDITS

Chapter One

Page 11: R. W. Porter; Page 15: UPI/Corbis-Bettmann

Chapter Two

Page 39: Johnny Johnson; Page 45: Slim Films; Page 47: J. Patrick Henry, Ulrich G. Briel and Hans Bohringer; Page 51: Patrick Shopbell; Page 53: Tom Sebring, SOAR Telescope Project, and Frank Bash and Frank Ray, McDonald Observatory; Page 59: UPI/Corbis-Bettmann; Page 63: Patricia J. Wynne; Page 64: Patricia J. Wynne

Chapter Three

Page 76: Renee C. Khaan-Korteweg, Patrick A. Woudt and Patricia A. Henning; Page 78: Rrodrigo A. Ibata, Rosemary F.G. Wyse and Richard W. Sword; Page 87: Alfred T. Kamajian

Chapter Four

Page 127: Ian Worpole

Chapter Five

Page 153: Peter Samek; Page 155: Peter Samek; Page 161: Don Dixon and Laurie Grace; Page 173: Space Telescope Science Institute, NASA, and David Levy

Chapter Six

Page 197: Slim Films; Page 199 top: Alfred T. Kamajian, P. H. Schultz and D.E. Gault; Page 199 center: NASA; Page 204: NASA/Jet Propulsion Laboratory; Page 205: NASA/Jet Propulsion Laboratory; Page 126: Ian Worpole; Page 217: Ian Worpole; Page 229: NASA; Page 233: NASA; Page 246: NASA; Page 247: NASA; Page 253: NASA

Chapter Seven

Page 259: UPI/Corbis-Bettmann; Page 275: David Starwood; Page 287: Michael Goodman; Page 295: Michael Goodman; Page 296: Jared Schneidman Design; Pages 304 - 305: Courtesy NASA; Page 313: Jared Schneidman Design; Page 315: Jared Schneidman Design

Chapter Eight

Page 333: Michael Goodman; Page 343 top: Corbis-Bettmann/UPI; Page 343 bottom: Boris Starosta

Chapter Nine

Page 363: Ian Worpole; Page 365: Ian Worpole; Page 383: Wendy L. Freedman, Hubble Space Telescope Key Project Team and NASA; Page 391: Jared Schneidman Design

Color Section

Page 1 top: Space Telescope Science Institute; Page 1 bottom left and bottom right: J. Patrick Henry, Ulrich G. Briel and Hans Bohringer; Pages 2 and 3: Don Dixon; Page 4: Tsafrin S. Kolatt, Avishai Dekel and Ofer Lahav; Page 5 top: CORBIS/STSci/NASA/Ressmeyer; Page 5 bottom: Hale Observatories; Page 6: R. Gilmozzi, Space Telescope Science Institute/European Space 0, Shawn Ewald, JPL, and NASA; Page 7 top: Alfred Kamajian; Page 7 bottom: E. J. Schreier, Space Telescope Science Institute and NASA; Page 8: George Retseck; Page 9 top: Space Telescope Science Institute and NASA; Page 9 bottom: Space Telescope Science Institute, NASA, and David Levy; Page 10 top: NASA/JPL; Page 10 middle left: NASA/JPL; NASA; Page 10 middle right and bottom left and right: Maribeth Price, South Dakota School of Mines and Technology; Page 11 top: Jack Harris, Visual Logic and Dimitry Schidlovsky; Page 11 bottom: NASA/Jet Propulsion Laboratory, Cynthia Phillips and Moses Milazzo University of Arizona; Page 12 top and bottom: Alfred McEwen, NASA, and Slim Films; Page 13 top: Chandra X-ray Observatory, NASA; Page 13 bottom: Roeland P. van der Marel (STScI), Frank C. van den Bosch (University of Washington), and NASA; Page 14 top and bottom: Slim Films; Page 15 top and bottom: NASA Johnson Space Center; Page 16: Hubble Heritage Team (AURA/StScI/NASA)